Praise for *Virtual Reality*

"Covers everything from the history of computers to the history of the human race. . . . [A] hands-on tour of the ingenious and sometimes wacky ways people have tried to humanize our relationship to computers."

—*The New York Times*

"Traces the rollicking history of alternate-reality fabrication from Plato and Dionysus to the highest tech labs in Japan, America, and Europe. This great encyclopedia is a precious compendium of essential facts and opinions that we shall all be referring to in the years to come."

—*Timothy Leary*

"An exhaustive, thorough look at the evolution of this technological talisman for the electronic fast lane. . . . Rheingold has written a book about inventors and their inventions, [and] about the excitement that occurs among those on the cutting edge of high-tech."

—*San Francisco Chronicle*

"An important book. It's rare for there to be so definitive a coverage of a new technology so early in its growth. Required reading for all the players and would-be players in the games as well as a goodly quantity of technology watchers."

—Stewart Brand, author of *The Media Lab,*
      creator of *Whole Earth Catalog*

"There are those who think that virtual reality may be the most important development since man first chipped flint, and there are those who don't know what it is yet. Anybody in doubt should read this book."

—Douglas Adams, author of *The Hitchhiker's Guide to the Galaxy*

BOOKS BY HOWARD RHEINGOLD

Talking Tech: A Conversational Guide to Science and Technology
(with Howard Levine)

New Technology Coloring Book (with Rita Aero and Scott Bartlett)

Higher Creativity (with Willis Harman)

Tools for Thought

The Cognitive Connection: Thought and Language in Man and Machine
(with Howard Levine)

*They Have a Word For It: A Lighthearted Lexicon of Untranslatable
Words and Phrases*

Excursions to the Far Side of the Mind

Exploring the World of Lucid Dreaming (with Stephen LaBerge)

VIRTUAL REALITY

HOWARD RHEINGOLD

A TOUCHSTONE BOOK
Published by Simon & Schuster
New York   London   Toronto   Sydney   Tokyo   Singapore

TOUCHSTONE
Simon & Schuster Building
Rockefeller Center
1230 Avenue of the Americas
New York, New York 10020

Copyright © 1991 by Howard Rheingold

First Touchstone Edition 1992

TOUCHSTONE and colophon are registered trademarks
of Simon & Schuster Inc.

Designed by Irving Perkins Associates
Manufactured in the United States of America

3 5 7 9 10 8 6 4

Library of Congress Cataloging-in-Publication Data is available

ISBN 0-671-69363-8
ISBN 0-671-77897-8 (pbk)

ACKNOWLEDGMENTS

My most heartfelt gratitude is due to my wife, Judy Rheingold, and my daughter, Mamie Rheingold, who put up with a husband and father who was either on the road or lost in thought for much of 1989 and 1990.

A special acknowledgment is due Bob Asahina, my editor, for grace under pressure.

To the staff of the *Whole Earth Review,* who thought they were getting a full-time editor in the summer of 1990: thanks for your patience.

Many others were helpful on my quest, especially Izumi Aizu, ACROE (Claude Cadoz, Annie Luciani, Jean-loup Florens), Sarah Bayliss, John Brockman, Frederick Brooks, Mary Clemmey, Steve Ditlea, Scott Fisher, Patrice Gelband, Tom Furness, Eric Gullichsen, Katsura Hattori, Bob Jacobson, Kevin Kelly, Myron Krueger, Jaron Lanier, Brenda Laurel, Margaret Minsky, Mike Naimark, Mike Miller, Warren Robinett, Robert Stone, Randal Walser.

For Mamie Rheingold.
Flower of my heart.
My hope for the future.

CONTENTS

10   Contents

Part Four
VIRTUAL REALITY AND THE FUTURE
343

Part One

A MICROSCOPE FOR THE MIND

Chapter One

"GRASPING REALITY THROUGH ILLUSION"

The primary research instrument of the sciences of complexity is the computer. It is altering the architectonic of the sciences and the picture we have of material reality. Ever since the rise of modern science three centuries ago, the instruments of investigation such as telescopes and microscopes were analytic and promoted the reductionist view of science. Physics, because it dealt with the smallest and most reduced entities, was the most fundamental science. From the laws of physics one could deduce the laws of chemistry, then of life, and so on up the ladder. This view of nature is not wrong; but it has been powerfully shaped by available instruments and technology. The computer, with its ability to manage enormous amounts of data and to simulate reality, provides a new window on that view of nature. We may begin to see reality differently simply because the computer produces knowledge differently from the traditional analytic instruments. It provides a different angle on reality.

HEINZ PAGELS,
The Dreams of Reason, 1988

We live in a physical world whose properties we have come to know well through long familiarity. We sense an involvement with this physical world which gives us the ability to predict . . . where objects will fall, how well-known shapes look from other angles, and how much force is required to push objects against friction. We lack corresponding familiarity with the forces on charged particles, forces in non-uniform fields, the effects of nonprojective geometric transformations, and high-inertia, low-friction motion. A display connected to a digital computer gives us a chance to gain familiarity with concepts not realizable in the physical world. It is a looking glass into a mathematical wonderland.

If the task of the display is to serve as a looking glass into the mathematical wonderland constructed in a computer memory, it should serve as many senses as possible. So far as I know, no one seriously proposes computer displays of smell or taste. Excellent audio displays exist, but unfortunately we have little ability to have the computer produce meaningful sounds. I want to describe for you a kinesthetic display.

The force required to move a joystick could be computer controlled, just as the actuation force on the controls of a Link Trainer are changed to give the feel of a real airplane. With such a display, a computer model of particles in an electric field could combine manual control of the position of a moving charge, replete with the sensation of forces on the charge, with a visual presentation of the charge's position . . . By use of such an input/output device, we can add a force display to our sight and sound capability.

IVAN SUTHERLAND,
"The Ultimate Display," 1965

At the university of North Carolina, I had a conversion experience akin to the experience that had bonded many of the personal computer pioneers of the 1960s and 1970s—a compelling vision of the future. But this time, the creative impulse had a tinge of awe.

I was standing in a carpeted room, gripping a handle, but I was also staring into microscopic space and directly maneuvering two molecules with my hands. Perhaps someone in an earlier century experienced something similar looking through Leeuwenhoek's microscope or Galileo's telescope. It felt like a microscope for the mind, not just the eye.

I didn't know the rules of "molecular docking"—a tool for helping chemists find molecules shaped like the keys to specific proteins—the way a chemist knows them, but I could *feel* them, through my hand and the force-reflective feedback mechanism built into the ARM, the Argonne remote manipulator. The metal grip felt like the handlebar of a gargantuan, well-lubricated Harley. I looked up at a million dollars' worth of articulated joints, encoiled by umbilicals of electrical cables. The entire apparatus was suspended from the ceiling a few feet in front of me. I wore lightweight goggles connected by a wire to a computer.

I looked toward the computer screen while using my shoulder, elbow, wrist, and fingers to explore the ARM's six degrees of freedom. A few pushes, pulls, nudges, and rotations gave me a feel for the ARM's dimensions of movement. I could move the grip and the rest of the ARM around in a sphere of at least arm's length diameter. The computer display and the glasses I wore were synchronized so that each eye was blanked every 1/60 of a second. Because slightly different views of the same image were scanned on the screen every 1/60 of a second, each eye saw only the images appropriate to a left- or right-eye view of a three-dimensional scene. "Stereoscopy"—presenting different views to each eye in order to give the illusion of three dimensions—is an old trick; the twist here was that the source of the images of the molecules was not pairs of still photographs or drawings, but electronic representations of computer-simulated molecules. The ARM granted me the power to reach out and manipulate these illusory 3D objects, and granted the objects the power to resist my attempts to move them.

The simulated molecular forces were translated into physical forces that resisted my attempts to maneuver the molecule; the force-reflection felt like the molecules themselves were pushing back at me through the ARM, matching my own arm's motions with analogous force. It was like playing with lumpy magnets that pushed back very strongly when I put similar poles near one another, and began attracting one another when I put opposite poles in proximity. There are many ways to fit the molecules together, but only a few juxtapositions that bring them close enough to bond. The task of searching through the universe of possibilities for a fit has a high degree of difficulty and a high payoff—one of those possibilities in a haystack might encode the shape of a cure for a carcinoma.

I had traveled to North Carolina because I had heard about people who are already creating medicines by using their eyes, ears, and muscles as well as their minds to invent new chemical compounds. I tried my hand at puzzling out a molecular docking problem with the ARM, the same way chemists at this lab do. It's part of something called "virtual reality," also known as "VR." Although I lacked the knowledge that distinguishes an expert in a field such as chemistry, I did not lack for skill at sensing my own way into the problem. Because the apparatus had made aspects of molecular docking directly perceptible to my eyes, ears, and hand, I was able to use all my experience in the world of gravity and manipulable objects, my gut-feel of the world, to advance a hard problem farther than most chemists could have done without the aid of computer modeling. For a truly skilled chemist, it must feel like an intellectual power-tool. I was beginning to understand why so many of the researchers in VR talk about the field's potential with such fervor. But the molecular level isn't the only place this new technology is creating a new kind of window on reality.

I had entered virtual reality for the first time in December, 1988, through a portal in NASA's Ames Research Center in Mountain View, California. Garments with wires played a part in it. So did a computer they called a "reality engine." A headpiece that looked and felt like an aluminum SCUBA mask covered my face, and a three-dimensional binocular television filled my field of view with electronic mirages, no matter which direction I swiveled my head. My body wasn't in the computer world I could see around me, but one of my hands had accompanied my point of view onto the vast electronic plain that seemed to surround me, replacing the crowded laboratory I had left behind, where my body groped and probed. A ghostly cube of light floated in front of me. I reached for it, and picked it up. A sensor-webbed glove synched my physical gestures in the room where my body was located to the movements of a cartoonlike glove that floated in the computer-created world. "Cyberspace," I had heard it called. It's a place, all right. What kind of place it is, is a big question.

The word *cyberspace* was coined by novelist William Gibson in his 1984 book *Neuromancer:* "Cyberspace. A consensual hallucination experienced daily by billions of legitimate operators, in every nation, by children being taught mathematical concepts. . . . A graphic representation of data abstracted from the banks of every computer in the human system. Unthinkable complexity. Lines of light ranged in the nonspace of the mind, clusters and constellations of data. Like city lights, receding." There are cowboys in Gibson's cyberspace, who make their way through the vastness of data by directly "jacking in" their nervous systems to "the Matrix," Gibson's name for the global communications and computing infrastructure that engendered this new realm. I first heard the terms "virtual reality," "VR," and "reality engine" from a computer scientist named Jaron Lanier, who will play a prominent role in later chapters.

Imagine a wraparound television with three-dimensional programs, including three-dimensional sound, and solid objects that you can pick up and manipulate, even feel with your fingers and hands. Imagine immersing yourself in an artificial world and actively exploring it, rather than peering in at it from a fixed perspective through a flat screen in a movie theater, on a television set, or on a computer display. Imagine that you are the creator as well as the consumer of your artificial experience, with the power to use a gesture or word to remold the world you see and hear and feel. That part is not fiction. The head-mounted displays (HMDs) and three-dimensional computer graphics, input/output devices, computer models that constitute a VR system make it possible, today, to immerse yourself in an artificial world and to reach in and reshape it.

If you had to choose one old-fashioned word to describe the general category of what this new thing might be, "simulator" would be my candidate. VR technology resembles, and is partially derived from, the flight simulators that the Air Force and commercial airlines use to train pilots. In conventional flight simulators, pilots learn something about flying an aircraft without leaving the ground, by practicing with a replica of airplane controls; the "windshield" of a flight simulator is a computer graphics display screen upon which changing scenery is presented according to the course the pilot steers. The entire simulated cockpit is mounted on a motion platform that moves in accord with the motions of the simulated airplane. Virtual reality is also a simulator, but instead of looking at a flat, two-dimensional screen and operating a joystick, the person who experiences VR is surrounded by a three-dimensional computer-generated representation, and is able to move around in the virtual world and see it from different angles, to reach into it, grab it, and reshape it.

Right now, it is necessary to put a high-tech helmet on your head or don a pair of electronic-shutter glasses, the way I did, to see that world, and slide a special glove on your hand or grasp a mechanical input device in order to manipulate the objects you see there. Lenses and two miniature display screens in the NASA helmet, linked with a device that tracked my head position, created the illusion that the screen surrounded me on every side. The reality engine updated the way I saw the world when I moved my gaze. I could look behind computer-generated objects, pick them up and examine them, walk around and see things from a different angle. The complex visual model of the virtual world that changed every time I moved was mediated by a simulation program in a powerful computer, to which the helmet and glove were linked by cables. In the future, less intrusive technologies will be used to create the same experience, and the computers will be both more powerful and less expensive, which means the virtualities will be more realistic and more people will be able to afford to visit them.

Although I stayed in cyberspace for just a few minutes, that first brief flight through a computer-created universe launched me on my own odyssey to the outposts of a new scientific frontier. This book is a report from those outposts and an advance glimpse of a possible new world in which reality itself might become a manufactured and metered commodity. Although it sounds like science fiction, and the word "cyberspace" in fact originated in a science-fiction novel, virtual reality is already a science, a technology, and a business, supported by significant funding from the computer, communications, design, and entertainment industries worldwide.

My quick peek through that dimensional doorway in Silicon Valley propelled me all the way around the planet on a treasure-hunt for bigger, better, faster, more realistic artificial worlds. I ended up traveling via 747s and rented cars, bullet trains, turnpikes, subways, taxicabs, limousines, and shuttle buses from one reality lab to another, from Chapel Hill, North Carolina, to the suburbs of Kyoto, Japan, from central Texas to the south of France. This "reality-industrial complex" is too young to be visible to most people, but it has already spread around the world and across disciplines; although few people know about it yet, it appears to have picked up a momentum of its own.

- At the University of North Carolina, I strolled through a building that existed in cyberspace before it was constructed in physical reality.
- In Kansai Science City outside Kyoto, I sat in a prototype of a "responsive environment" that watched where I looked and how I moved, and I talked to the Japanese researchers who were using VR to build the communication systems of the twenty-first century.
- At NASA, I commanded repair robots in virtual "outer space."
- In Cambridge, Massachusetts, I ran my fingertips over "virtual sandpaper" by means of a texture-sensing joystick and watched scientists create animated creatures who will live in tomorrow's semisentient virtual worlds.
- In Vancouver, British Columbia, I played with my first "teleoperator"—a robot instead of a computer simulator at the other end of a telepresence system. Imagine a pair of goggles and glove that take you—through the eyes and hands of a robot—under the ocean or into your own bloodstream.
- In Tsukuba, one of the original "science cities" in Japan, I had the eerie experience of watching myself through the eyes of a tele-robot—an electronically augmented out-of-the-body experience.
- In Honolulu, I inspected a teleoperated machine-gun on wheels at a high-security US Marine Corps research facility.
- In Santa Monica, I tried out a 40-year-old reality simulator—a "Sensorama" arcade game—in its present resting place on the patio of its inventor.
- An hour's drive outside Hartford, in a corner of a natural history museum, I found the man who invented the term "artificial reality" decades ago and pioneered the techniques that are being perfected by research and development laboratories worldwide today.
- In Santa Barbara, I watched a new science precipitate like a crystal in a supersaturated solution, when computer graphics specialists, experts on robotics and control theory, cognitive scientists, and computer programmers from around the world got together for the first time to explore their common interest. By the end of the conference, they had decided to take a major step toward creating a new science where their specialities now intersect, by founding an MIT Press journal devoted to virtual worlds studies.
- In London, I found a roboticist in a political puppet factory, an inventor who claims to be creating a glove capable of transmitting the sensation of touch over a distance.
- In Silicon Valley, I visited an unusual company where the goggles and gloves I had used in many of the other sites were manufactured, and there I danced with a woman who had taken the form of a twelve-foot-tall three-dimensional purple lobster.

- In Grenoble, a four-hour train ride from Paris, I visited a computer science laboratory within a computer science laboratory, and put my hands on a machine they had built. Although it was made of metal and microchips, manipulating their device felt and sounded like drawing a bow against a taut violin string.
- I journeyed back in time to the prehistoric underground paintings at Lascaux, explored Plato's cave, and peered into a technological future that deserves our attention now—because when today's infant VR technology matures in a few years, it promises (and threatens) to change what it means to be human.

One way to see VR is as a magical window onto other worlds, from molecules to minds. Another way to see VR is to recognize that in the closing decades of the twentieth century, reality is disappearing behind a screen. Is the mass marketing of artificial reality experiences going to result in the kind of world we would want our grandchildren to live in? What are the most powerful, most troubling, least predictable potentials of VR? If we could discern a clear view of the potentials and pitfalls of VR, how would we go about optimizing one and avoiding the other? The genie is out of the bottle, and there is no way to reverse the momentum of VR research; but these are young jinn, and still partially trainable. We can't stop VR, even if that is what we discover is the best thing to do. But we might be able to guide it, if we start thinking about it now.

The most lurid implications of VR have already been trumpeted in the mass media, via reports of what it just might make possible—such as "teledildonics" (simulated sex at a distance) or "electronic LSD" (simulations so powerfully addictive that they replace reality). And many of the reports in the popular press give the impression that the technology is "the latest thing to come out of California." But laboratories at places like the University of North Carolina have been conducting solid scientific research and developing potentially lifesaving applications for more than twenty years: anticancer medicines that already have been created with the VR molecular docking system that I tried out, and VR-based radiology treatment planning that is alleviating the suffering of real patients right now.

My own odyssey through the realms of virtual reality research and development actually began years ago, with my interest in the way computers seem to be evolving from powerful calculators to media that extend human intellectual capabilities—"mind amplifiers," I called them. While I was conducting research for a previous book, several years ago, I met a number of people who used personal computers to build tools capable of enhancing human thought, communication, and imagination: The "infonauts" I profiled in *Tools for*

Thought saw how to use computers as fantasy amplifiers and intellectual levers instead of just number crunchers or data-processing engines.

Today's personal computers are only beginning to approach the threshold of mind amplification as it was originally conceived, decades ago. But the personal computer as we know it is not the only possible vision for computer-based mind amplifiers. Even in the early days of the PC revolution of the 1970s and 1980s, a few of the new industry's pioneers wanted to push the development of personal computer technology toward something much more interesting, at least in theory, than the little boxes people now have on their desks—something they called "personal simulator" technology.

In the early 1980s, when I first talked with certain young computer wizards about their dreams of immersing themselves in simulated realities, it didn't occur to me that they would foment a technological and cultural revolution in the 1990s. Years later, when I heard that the same wizards had combined efforts to build a reality machine for NASA, I had to try it. I knew that one of them had been dreaming of 3D goggles since his MIT days, that another had been perfecting a glovelike computer input device, and that another, a legendary programmer of educational software and adventure games, had been hired to put together the world-modeling software for the NASA system. They were all in their twenties and thirties when we first met— the new post-Apple generation of computer designers, those who grew up using the personal computers created by their elders. Their latest venture sounded wild but intriguing.

After my first demo at NASA in 1988, I spent some time at the library. One of my first big surprises while finding my way around cyberspace was my discovery that the intellectual hotspot of the VR research world is neither Mountain View, Tokyo, Cambridge, nor even Salt Lake City. It didn't take more than a quick review of the VR literature to notice that the most important work in VR as a scientific visualization aid, as a "medical imaging" technology, and as an architectural tool, has all come from the same place. Not only these practical applications, but equally important work in the enabling technologies of head-mounted displays, reality engine architectures, 3D computer graphics, position sensors, have been coming from the same laboratory for more than twenty years. Chapel Hill, the home of the Department of Computer Science at the University of North Carolina (UNC), turned out to be the first of many places I never thought I'd visit before I embarked on my quest for VR lore.

A dedicated group has been conducting virtual worlds research there since the late 1960s, and growing in numbers, prestige, and funding for the past two decades. Both the pharmaceutical chemistry and medical imaging projects are conducted at one of the world's most advanced VR research centers, certainly the one that has been at the job for the longest time. It also is a laboratory where the creation of scientific and medical tools, not the marketing of consumer electronics gadgets, is the primary focus of the research. I ended up there because, having peeked at the rather crude state of the art of today's VR prototypes, I wanted to see what a determined VR team with powerful computing resources could conjure in the way of a convincing synthetic reality when they put their cybernetic pedals to the virtual metal.

Frederick Brooks was already a figure of quasi-mythical stature in the computer software world when he left IBM for UNC. I had read Brooks's book about software engineering years ago. Although I didn't know anything about the nuts and bolts of organizing large programming projects, I was attracted to *The Mythical Man-Month* because so many programmers seemed to regard it in an almost metaphysical way. Brooks had directed the IBM team that created what was then the most ambitious programming task in history, the operating system software for the IBM 360 series of computers—the mainframe revolution that computerized the business and scientific world. The "mythical man-month" is a techno-bureaucratic measure of productivity that represented a way of thinking Brooks has been trying to discredit for years—the idea that the hard, necessarily trial-and-error intellectual labor of producing good software can be circumvented by institutional or technical means. He demonstrated that if you are in charge of a software project that is 3 months behind, and you decide to double the number of "man-months" devoted to it by hiring twice as many programmers, you are actually guaranteeing that the project will be at least 6 months behind. It struck me as significant that the elder statesman of VR, a field in which the realization of the potential of the technology is dependent on the creation of a great deal of software that does not exist yet, is also a man who is known for his pessimism about accomplishing great things in short periods of time when complex software is involved.

Brooks is a modest man who finds popular interest in the sensational aspects of VR to be a distraction from the task of building thinking tools for scientists, doctors, and architects. He would be the first to point out that his colleagues deserve a great deal of credit for the work that has emerged from Chapel Hill since the late 1960s. When I talked to his colleagues, I discovered that Brooks is correct, and I found that his colleagues are quick to point to Brooks's leadership as a central reason for the strength of the research effort at UNC. It's a strong and diverse group: Henry Fuchs, a legendary figure in the computer graphics world, had been a student of Alan Kay's at the University of

Utah when VR pioneer Ivan Sutherland's "head-mounted display" (HMD) was still working; Fuchs and others had created a new VR architecture at UNC by fabricating their own microchips and creating a network of 250,000 processors. Stephen Pizer is at the leading edge of 3D interactive medical imaging; together with Dr. Julian Rosenman and other medical specialists, he has been pioneering VR imaging techniques—the "X-ray glasses" that diagnosticians have secretly wished for since the days of Hippocrates, and which UNC research might well deliver in the next ten years. Other principal investigators on the UNC faculty are experts in hardware, software, sensing devices, feedback devices, computer architecture, graphics programming—all the building blocks of VR systems. When I found out that Warren Robinett, yet another infonaut I had encountered a Silicon Valley century (seven years) ago, and a key member of the NASA VR team, had joined the UNC group, I stopped procrastinating and made plans to visit Chapel Hill.

The North Carolina license plates on the rented car I picked up at the Raleigh-Durham airport bore the state motto: "First in Flight." It's hard not to think of the VR technology of the early 1990s as the Kitty Hawk stage of cyberspace. My test flights of today's cybervehicles were not as thrilling as the mental model they inspired me to imagine, visions of what the 747 version will be like a few years down the road. The progress of VR technology is probably going to be at least twice as fast as the progress of aviation, however. We're in an era where "jet-powered" sounds vaguely archaic. I remember seeing advertisements for the Altair "personal computer" in science magazines in 1974; the input device for the Altair was an array of switches, the output device was a row of lights. I couldn't figure out what I might want to do with an Altair, but it looked vaguely interesting. A few years later, two homebrew computer hobbyists who did know what they wanted to do started Apple Computer. Ten years after the Altair came the first Macintosh. Today's VR is not quite at the Altair stage. The Apple, Inc., of the VR industry probably doesn't exist yet. But the research that Brooks, Fuchs, Pizer, Robinett, and others are doing in North Carolina today is creating the knowledge base upon which industries could be built in the future. And just as Stanford was a key element in the success of Silicon Valley, UNC and the nearby Research Triangle might be big winners if a true VR industrial boom does materialize. I call that scenario "the rise of the reality-industrial complex."

The leaves had just begun to turn when I drove from the airport through the humid, forested, fairly flat countryside to Chapel Hill. Shirtsleeve weather. Honeysuckle evenings. I spent most of a week there, and found Chapel Hill to be a pleasant, bucolic college town with good bookstores and good coffee, two important measures of an intellectual nexus. When I parked and made my way to Sitterson Hall, I underwent an adventure in *virtual-vu*: At SIGGRAPH 1989, the Association for Computing Machinery's annual computer graphics convention, earlier that summer, Margaret Minsky showed a videotape of the VR "architectural walkthrough" at UNC. A consummately lifelike 3D rendering of a building interior had been converted to a virtual world that you could explore by wearing an HMD and walking on a treadmill connected to a computer. Now, walking into the real hallway for the first time was like walking into the virtual world in the tape I had seen that summer. Entering the dizzying universe of VR, I was looking for the real room where they would put an HMD on me and let me walk through the 3D model of the building I already was in.

I found Warren Robinett's office—his physical office. I hadn't seen him in years. He's still soft-spoken, unshakably confident that he knows what he is doing, still enthralled by the sheer pleasure of crafting a universe out of programming code. For as long as I've known him, Robinett has wanted to build a world people can crawl around in, have adventures, and learn things. *Rocky's Boots*, his computer game that taught Boole's algebraic logic to ten-year-olds, first attracted my attention in 1983, when I was writing an article about educational software. I had first heard his name in connection with one of the classic legends of the video game industry. Robinett was a game programmer at Atari during their glory years. He was also an aficionado of a text-based game called *Adventure* that computer researchers played on mainframe computers. *Adventure* is a virtual world in a conceptual way: The game is played in an imaginary underground world of caves and chambers, and players proceed through that world, learn its cartography, accumulate weapons and wealth, overcome obstacles, slay dragons—all accomplished by typing commands such as "go north" and "pick up sword." The physical space is not displayed but described; the only visual scenes are in the player's mind's eye. The computer keeps track of where you are.

Robinett believed *Adventure* would be a great theme for a graphical video game, in which the game could be transformed from the conceptual world of text to the pictorial world of computer graphics. There was a technical problem, however. The home video game machines Atari was selling at that time had extremely small memory capacity. The programming code that made a game work had to be extremely terse—"compact," as programmers would say it. The original version of Robinett's proposed product occupied several hundred kilobytes (K) of RAM memory on a mainframe at Stanford's artificial intelligence lab. The Atari game cartridge could hold no more than 4K. Warren's boss was convinced that it would be impossible to convert the mainframe version of *Adventure* to a program compact enough for an Atari product, and killed Robinett's request. So Robinett created the game anyway, against his boss's orders, and it ended up as one of Atari's best-sellers.

Robinett's Missouri accent somehow seemed a bit broader and deeper in North Carolina than it had in Silicon Valley. He recalled that he had jumped into the video game programming boom of the late 1970s for the same reason teenage boys were pouring fortunes in quarters through the coin slots of Pac Man and Space Invaders. He believes that cyberspace has the same kind of allure, for the same kind of reasons, as well-designed video games.

"Eleven-year-old boys took to video games immediately, instinctively," Robinett explained. "I think their instincts were right. There's something about interactive graphics on a screen that really grabs you. I think that the process of learning requires concrete examples, and wonderfully responsive and detailed graphical models can be made on a computer. And it has been only recently, after video games and educational software were established industries, that the scientific world has accepted the idea that computer graphics can help them understand their piles of numbers. This is now called 'scientific visualization.' It took scientists a decade to see what was obvious to every kid the first time he touched a video game—the power of interactive computer graphics."

The UNC faculty saw the value in having a veteran of the video game boom and the NASA/Ames project on a VR research team. Brooks is from the old school of mainframes and computer science. Henry Fuchs had also gone the university research route. But Warren Robinett came from the culture of seat-of-the-pants programming wizards who liked to accomplish software feats that others deemed impossible. A few of the best made extraordinarily good money for a few years. Robinett worked for Atari in the late 1970s, made some money, quit his job, backpacked around Europe, and did a lot of thinking in the early 1980s about what he was going to do next with the power of interactive graphical simulations.

Robinett was focused on bringing into existence really neat software that *demonstrated* what the heck he was excited about, instead of just theorizing about the wonderful potential of computers. It was a beautiful hack to create something because he wanted to play with it, and to design it to have innate educational value, to enhance people's lives, and to make him money at the same time. I reminded him that when I first interviewed him, seven years ago, one of the first things he told me was: "What I'm really doing is making *interactive graphical simulations,* and you should write that down because every one of those three words is important." I remember that he explained how a truly powerful video game has to be a simulation of some kind of world, has to represent its key concepts graphically, and there has to be a way for the player to interact with the graphics on the screen and the simulated world.

Rocky's Boots, for example, was a graphical adventure game, but in this case the adventure was not a matter of finding your way through a cave, chasing snakes, and snagging treasure. Warren devised a way to use a cursor to touch different graphic objects and drag them into position to lock together, thus constructing simple "machines" that accomplished a task. Young children found that they could learn to be proficient at building these video game "machines"—just as they learned to be proficient at shooting down space invaders or eating power dots. The symbols in *Rocky's Boots,* which Robinett developed together with educators Ann Piestrup, Teri Perl, and Leslie Grimm at an educational software developer known as *The Learning Company,* are not just arbitrary symbols. In fact, they are the same symbols, accomplishing the same symbolic functions, as the symbols of algebraic or "Boolean" logic. The "machines" that young players constructed are actually fully functional logic circuits. This combination of mathematics and logic is usually not taught until high school or college, and is one of the theoretical foundations of computer design and computer programming.

After *Adventure* and *Rocky's Boots,* programming the software for a virtual world seemed to Robinett a natural next step. In 1986, Scott Fisher's vision of a VR research platform drew Robinett to NASA. Two years later, when the NASA project grew to a point where he wasn't in control of the software engineering, when it became a matter of going to meetings and advocating and selling a point of view on how the project ought to be done—instead of just doing it, as had been his wont—Robinett quit NASA and went sailing in the South Pacific. When he returned, he traveled around the USA and decided to see what was happening at UNC. After meeting Frederick Brooks, Henry Fuchs, and Stephen Pizer, he decided to move to Chapel Hill and join the VR laboratory. Part of his enthusiasm stemmed from his mistrust of bosses or bureaucracies. Frederick Brooks, of all people in the VR world, knows how to make a programmer happy.

"When you talk to Fred Brooks," Robinett told me, "be sure to ask him about IA—intelligence amplification. That's why I came here. It dovetails perfectly with my take on computer graphics—I want to use computers to expand human perception. We have certain built-in senses such as vision, hearing, and smell, but there are many phenomena which are completely imperceptible to us. Some examples are X-rays, radioactivity, electricity, and the inside of opaque objects. You could say that we lack the senses to perceive these things. But using electronic sensors and computer displays, we can make these imperceptible phenomena visible. Or audible or touchable. I call the apparatus for doing this a sensory transducer.

"One thing we're working on here at UNC is hooking up an ultrasound scanner to the head-mounted display, so that instead of looking at an ultrasound image on a monitor screen, a doctor could put on the headset and see directly into the living tissue. This would be, in effect, a synthetic sense similar to Superman's X-ray vision. I realized after the fact that what I was doing with *Rocky's Boots* was making the electricity *visible* in simulated circuits, so my earlier work was also using computers to expand human perceptions."

Robinett brought me to meet Ming Ouh-Young, the senior graduate student for the project, in the corner of the computer graphics laboratory where the ARM was set up. The Computer Science Department includes several floors of classrooms, offices, lecture halls, hardware and software workshops. The computer graphics laboratory itself includes a large common space in which students and researchers sit at state-of-the-art graphics workstations arrayed on desktops and stare at their screens with that million-mile gaze you find in places like this. As usual, a few of them were playing *SpaceWar,* that historic and inevitable pastime of computer graphics programmers. An anteroom in one corner of the open space housed the ARM and other virtual world displays.

First Robinett and Ouh-Young gave me a warm-up exercise. Before they turned on the graphics, I had to grope around in the virtual reality world blindly, with my hand, wrist, arm, and shoulder.

"The ARM is connected to a virtual fishing pole," Robinett instructed.

"If you lift the fish too suddenly, you'll lose it. If you don't resist, it will reel away from you," Ouh-Young added.

The ARM twitched into action. I did feel a sensation spookily similar to that feeling of something *alive* at the other end of the line that you get from a spirited trout. After I moved the virtual fish to a virtual holding tank and released it to rejoin the virtual stream, I was ready to arm-wrestle molecular forces.

Floating in space in front of the screen at eye level was a computer-generated visual model of the receptor site within the human protein dihydrofolate reductase. The protein looked like a lumpy, airy, sculpture made of clouds of blue and red points aggregated into the shape of beach balls and tennis balls melded together. It seemed to be about two feet across, looked to be a little more than an arm's length away from me, and appeared to float in midair. The colored sphere-clouds were clumped, folded and twisted into a geometrically complex pocket—the docking site. Think of it as similar to the kind of three-dimensional puzzle found inside a lock; the proper key opens a door by "solving" the lock.

Between the protein model and my hand floated a smaller molecule model in the form of thin yellow lines, like glowing, rigid wires, representing the angles between the atoms of a synthetic molecule, methotrexate—the "key" candidate. These were not just computer graphic representations of molecules, but multidimensional simulations of those molecules, including all the measures that chemists use to describe molecular behavior—visible bonds could stretch only so far, and when the two molecules were brought together, neighborhing atoms attracted or repelled one another according to the physical rules governing the electromagnetic forces at the molecular level. The cloud-solids and stick figures were just two of many possible representations of the modeled entities. There were invisible but literally palpable representations, as well. When I tried to push the models together by exerting force on the ARM, the force I felt in reaction was an analog of the way the physical molecules would behave if pushed together in physical space.

The grip enabled me to use arm and hand movements to manipulate the models in relation to one another. The more I used the grip to rotate the models that floated in midair, the more three-dimensional they seemed. My job was to use the ARM grip to maneuver the 3D puzzle pieces together close enough, in precise spatial relationship, to bond together—a successful dock. If a chemical can be found that makes such a fit on the protein shape found on a tumor cell or a pathogenic bacterium, then it is a potentially medically useful compound.

Although I'm not a chemist, I was able to find quickly the rough position in which the molecular forces between the two molecules are in the kind of alignment necessary for docking. I was playing a molecular game with specific rules; the rules were translated into visible changes in the graphic display, such as color or shape changes, audibly translated into "pings" when two virtual molecules bumped against one another, and "haptically" translated into a model of electromagnetic forces that I literally grasped with my hands. Haptic perception involves that melange of senses we lump together under the category of "touch," but haptic is not strictly tactile in the same way one's fingertips convey tactile information about the outside world; rather, haptic tasks like landing a fish or docking a molecule also use the body's internal sense of *proprioception* that informs us about the position of our own limbs in relation to one another and to the space around us.

Human proprioception includes a system of internal sensors at joints and in muscles to detect changes in pressure and position. A higher-level processing system detects significant patterns among the messages from the body's proprioceptors (e.g., *this* pattern of messages from this particular set of sensors means that your body is going to topple forward if you don't do something about it; *that* pattern of messages means that you are pushing something heavy and polished across a low-friction surface). Proprioception's third information system consists of the effectors for transmitting commands from the sensing and sense-making systems to the muscles—the microadjustments that keep us upright and guide our movements. Part of daily life that is so ordinary we hardly notice it is the fast, silent, information-processing and fine muscular coordination skill that enables you to move your hand in exactly the right direction when you decide to reach for a glass of water. A ballet dancer is a virtuoso of proprioception. Haptics involves both proprioceptive and tactile senses, in concert with other senses.

The ARM and molecular manipulation system were a way of using the highly refined capabilities of the human haptic system as a handle on a virtual world with rules of its own. That is where intellectual amplification can happen, where the display can help raise the understanding of a novice and nudge an expert toward insight. The reason for going to the trouble of translating computer models to haptic displays is to take advantage of the fact that human haptic systems are known to be extremely good at decoding the rules of their environment from moment to moment. In a human-computer haptic system, the human is the part of the system that finds the significant patterns; the haptic part translates invisible forces to human-sensible form. Finding the right 3D pattern-related problem to solve with a computer-aided haptic system is the most important issue; it sounds terribly abstract until you get to anticancer medicines, medical imaging, aircraft design tools. Indeed, much of the scientific and engineering value of haptic VR systems derives from the real-world problems that require a system with both haptic and computational capabilities to accomplish tasks we very much want to do. Frederick Brooks refers to these applications that stimulate the progress of sciences and technologies as "driving problems."

The ARM exerted enough force to tire my arm when I actively wrestled with it for many minutes. I tried to twist, rotate, jam, and twiddle the thing into place by looking at the 3D jigsaw puzzle and manipulating it with my hand. Knowing next to nothing about the chemistry symbolized by the colored clouds floating in virtual space and the tinkertoy bonds that I could feel in my arm bones, I was able to find my way into a place where the ARM resisted at a minimum amount between all its degrees of freedom. It was more like playing a video game or a trombone than any problem-solving I remembered from high-school chemistry. It feels like there is a little pocket of relaxation in the middle of the force-puzzle-cloud, and if you can find your way into it, the embarrassingly telltale pinging sounds stop and your arm has to work a lot less vigorously.

When I nudged the molecule into a relatively satisfactory haptic-visual zone, little white vectors shot out from the corners of the methotrexate skeleton. Ming Ouh-Young directed my attention to a series of metal knobs on the arm. I was gripping the molecule in place with my right arm. With my left hand, I could tweak the drug molecule until the white lines disappeared. Whenever my manipulations caused a boundary to cross a threshold value for molecular "bump forces," I heard an audible "ping." The combination of the sound and the ARM's physical resistance that made it feel as if I had actually bumped into something solid "out there" might have been far more meaningful if I knew anything about chemistry.

After the ARM demo, I traded the clear glasses of the molecular docking display for an HMD, let go of the ARM, and stepped up onto a treadmill. Thus, I had the opportunity to walk through Sitterson Hall in cyberspace, as I had foreseen when I walked into the (physical) building.

The virtual version of Sitterson Hall was another one of the applications that drew me to Chapel Hill. The idea behind using architectural walkthroughs as one of the "driving problems" for VR development is that the main thing architects do is envision models of three-dimensional structures. This is a complex cognitive and perceptual task, exactly the kind of problem that humans still solve better than computers. Architects build up mental models of the projects they want to construct, but those models only exist in the individual architect's mind's eye and that explains why so much sketching and drafting go into the job of communicating those mental models to clients. When they need to communicate their specifications to contractors, the architects' models become formalized—but they are still depicted in only two dimensions on a flat sheet of paper or display screen. Because they spend a lot of time thinking about three dimensions, architects tend to have better skills at envisioning three-dimensional spaces from two-dimensional renderings. But there are many aspects of any design—the effects of different kinds of lighting on complex spaces, or acoustics, or seismic stress, for example—that are beyond the abilities of even the most skilled mental modeler. A three-dimensional model large enough to walk through could boost the architect's ability to conceive of three-dimensional spaces and could make it easier for clients to understand something of what the architects see in their imagination.

Scaling is one great power of virtual objects. You can fit a building into a computer, shrink it down small enough to look down into it from a God's-eye view, or expand it enough to stroll through it. With a sufficiently powerful computer, you could shrink the building you are in and put it in your virtual pocket in cyberspace, pull it out, and expand it to building-size on voice command. If you want to sacrifice detail, you can model an entire city. When Sitterson Hall was in the planning stages, the VR researchers who were going to work in the multimillion-dollar building after it was completed converted the architect's plans to a full-scale 3D model—that existed only in cyberspace. When the people who were going to spend their days in the building "walked" through the model, many of them felt that one particular partition in the lobby created a cramped feeling in a busy hallway. The architects didn't agree until the future occupants of the building used the 3D model to give the planners a walkthrough. The partition was moved. The building was built.

Months after I first saw a videotape of the virtual building, I walked into the physical building, found the room where I put the cybergoggles over my eyes, then entered the virtual building and walked through the virtual lobby, past the offending partition in the originally proposed position. When I walked down the virtual hall, it did feel more cramped than the configuration they ended up with. I was able to stroll the corridors of an entire building while physically never leaving one small room because I was pacing along a treadmill, holding on to a pair of handlebars. When I wanted to steer to the right and discover what was down the hallway I could see stretching out in that direction, I turned the handlebars and kept walking straight. It took a minute to get used to. Finding a way to actually pace about in the physical world while exploring a large virtual world is one of the engineering challenges in this application area. At UNC, the treadmill offered one solution. Six months later, in Japan, I met a man who used a stationary harness, a universal joint, and roller skates to solve the same problem.

With the 1989 entry into the VR industry of Autodesk and VPL, one large and one small software vendor in California, the notion of architectural walkthroughs left the realm of research and entered the world of commercial development. If the collective gasp I witnessed at a convention of computer-aided designers in Anaheim, or the interest I found in Kawasaki, at the top levels of Fujitsu's research and development laboratory, is any indication of the future, the VR design-tool revolution is likely to pick up speed over the next five years, and that could have a major effect on other businesses over the next ten

years. The next phase of development will be driven by those commercial enterprises that take advantage of VR capabilities for computer-aided design, itself a thriving industry that drives productivity
in other industries. The economic effects of adding an intellectual
amplifier to the earliest part of the planning process can ripple through
any industry that deals with three-dimensional objects. (The world of
design and the use of VR architectural walkthroughs are the subject
of a later chapter.)

When I passed through the computer graphics laboratory again, after
the HMD demos, I saw a fleeting image on a screen on the other side
of the room that looked like something from a slasher movie, like
somebody had peeled the flesh off a quadrant of a human skull. I
moved in for a closer look. Marc Levoy was demonstrating the medical
imaging technique known as "volumetric rendering." A closer examination of the color image on the display screen revealed the different
layers of soft and hard tissues in the skull, portrayed in different
degrees of transparency. Levoy tapped a command on the keyboard,
and layers of cartilage or bony casings appeared or disappeared, grew
opaque or transparent. The focus of effort here was not displaying
full 3D, but in solving the "visualization" problem inherent in trying
to look through one kind of material at another kind of material: what
colors, shadow textures, levels of transparency work best at rendering
anatomical structures visible?
    Everything from obstetrics to orthopedics has been affected by the
tremendous progress in medical imaging technologies—driven by the
introduction of the computer as a means of manipulating digitized
image information. X-rays can enable physicians to see through soft
tissue, but they are limited to a single "slice" through the patient. CAT
scans employ computer control to assemble a "fan" of consecutive
image slices; the 3D reconstruction of the patient's anatomy from that
fan takes place in the minds of diagnosticians. Ultrasound can be done
in real time, but it is also a two-dimensional look at a three-dimensional
structure. Thus medical imaging is a special case of scientific visualization, and seems to be benefitting from the same computer graphics
breakthroughs that are driving visualization in the physical sciences.
Levoy's work was part of a coordinated effort by a number of specialists
from different disciplines to build VR medical imaging tools. I walked
down the hall again—noticing that I couldn't help noticing how realistic the lighting in real life can be—and spent the rest of my first
afternoon in Chapel Hill talking with Stephen Pizer, the leader of a
team that includes computer scientists, radiologists, radiation oncol-
ogists, surgeons, and perceptual psychologists.
    "We want to know how to make better medical images," Pizer told

me in his office that afternoon. "That includes not only the presentation of medical images in 3D, but their use in medical practice for radiation treatment planning, surgery planning, and diagnosis."

It is fairly easy to see why a surgeon would like to be able to previsualize specific patients in three dimensions before operating. I wasn't as clear about their concentration on radiation treatment planning, so I asked him.

"The goal of radiation treatment planning is to find a plan that delivers a lot of radiation to a tumor and minimizes radiation exposure in other regions," Pizer explained. Radiation beams pass through the patient's healthy tissue on their way to the target, but if beams are crossfired at a tumor from a number of different angles, the tumor is irradiated far more than surrounding tissue. Each patient, each tumor, is different, and so is each treatment plan. Like molecular docking and architectural walkthrough, radiation treatment planning is inherently a 3D problem.

"One problem in planning external-beam radiation," Pizer continued, "is to find all the possible ways to shoot a beam at a tumor without passing through the most highly radiosensitive tissue. You want to be able to visualize the beam in relation to the patient's healthy tissue and to the tumor, and you want to be able to visualize anatomy relative to some visual representation of the dosage."

James Chung, one of the graduate students working with Pizer's team, later showed me a prototype system for radiation treatment planning. The key idea is that the planner can use an HMD or other 3D display to sight along the proposed radiation beam. The planner can then sight the beam through other tissues in the best manner, given the best kind of selectively transparent, three-dimensional rendering techniques.

Pizer stressed the importance of finding a way not only for diagnosticians and treatment planners to visualize what is inside a patient, but to be able to look at it from all kinds of angles: "The process of moving our head and eyes, of looking around, is an important part of vision," he pointed out. "The depth cues associated with natural motion are very strong, and creating a virtual world in which you move and change your view of things as you would in the ordinary world is important not simply to spur the imagination and suspend disbelief, but to give you adequate perception. We have a perceptual psychologist on our team, to help us work on the problems of determining which aspects of an object are important for an expert such as a diagnostician or surgeon to perceive. We're working with surgeons to find out what kind of visual and other information they need to place incisions, drill, insert probes for biopsies, and so on."

At some time in the future it will be possible to model and rehearse tricky operations, using the actual diagnostic imaging of each patient, mimicking not only the placement of organs but the way the scalpel feels when cutting through tissue, as well. Such simulation, already in its early research stages at Stanford as well as UNC, is years, perhaps decades, away. While the more advanced work on surgical simulation is still in the planning stages, Pizer pointed out that "the members of the radiation treatment planning team are in fact using 3D treatment planning to treat patients." It's too early for definitive medical results, but early reports from radiation treatment planners are enthusiastic.

Medical imaging consists of two kinds of systems. The "transducer" detects information about otherwise hidden structures beneath the skin. X-rays shoot high-energy radiation beams through the body, exposing photographic film on the other side. Ultrasound scanners use high-frequency sound waves as a kind of internal "sonar," bouncing the inaudible frequencies off tissues inside the body to provide information that is translated into a video image. Because imaging techniques that use X-radiation are not healthy for growing fetuses, ultrasound imaging is used widely in obstetrics. Although the quality of images is often fuzzy, the most exciting part of ultrasound to expectant parents (and their doctors) is the capability of watching the fetus move. If ultrasound transducers could somehow transmit sharply focused three-dimensional structural information, perhaps VR could enhance the other system involved in 3D imagery—the visual representation of the information provided by the sensing technology.

In regard to future developments, Pizer noted, "One of our colleagues is developing an ultrasound sensor that will produce 3D data at the rate of a few tens of 3D images per second. Think of the ultrasound image of a human heart beating in real time as a kind of virtual world. From our point of view, the challenge is to detect three-dimensional data and process it into a viewable image fast enough to be useful, which isn't easy when the transducer is providing information at the rate of two billion bytes per second. The visualization problem of representing the various tissues as they are moving inside the heart is a hard one. The problem of dealing with the computational load caused by all that data coming in that quickly is a hard one. The time lags and perceptual irregularities in the HMD are hard problems. None of the problems are insoluble in principle, but I think they mean we aren't going to get X-ray spectacles right away," Pizer cautioned.

"The X-ray spectacles are more Henry Fuchs's vision of where this technology ought to go than it is mine, although I think it's a nice one," said Pizer. "Once you are putting 3D virtual worlds in front of the surgeon or diagnostician, why not put them where they belong—namely, in the patient, superimposed on where the organs are located? One could imagine a situation where surgeons can see their surgical instruments, can see the real tissue of the patients as they operate, and can simultaneously see an augmented image that allows them to see behind all the blood and opaque surfaces." How long will it take for this vision to mature from successful pilot projects through working prototypes to a medical instrument? "I would be happy with having a solution in no more than ten years, and I can't see having it in less than five years," he answered.

The next morning, at the restored Colonial building that now housed one of the UNC cafeterias, I met with Henry Fuchs, a red-headed, mustachioed, energetic fellow who doesn't waste time and doesn't mince words. He told me right away that he thought all the interest in VR by the popular media was a bad idea. Not that the promise of VR technology isn't truly as spectacular as the predictions of X-ray spectacles and scientific visualization aids seem to be. It's just that there are many problems to be solved before any of these possibilities proves to be practical. The breathless speculations in the mass media, Fuchs seemed to believe, were arousing people's expectations for imminent breakthroughs in a technology that would take years, perhaps decades, to mature.

Fuchs ought to know. He's been working on the hard problems of building workable VR systems since the early 1970s. He started his virtual worlds quest in the same laboratory where VR pioneer Ivan Sutherland and others had built the first working HMDs in the early 1970s.

"The lag problem, for example," Fuchs argued, "comes across in popular articles as an artifact of today's systems that will probably be solved by some chip one day soon. It isn't that easy."

Fuchs was referring to the way the visual world seems to lag just the tiniest bit when you move your head or walk down a virtual hallway. The delay built into the position-sensing technology, combined with the delay built into processing signals from the VPL DataGlove and head tracker, combined with the delay built into the computations necessary to maintain the world-model, produces lag. If you want your world to be more detailed and realistic, you add more time to the lag, and if you want it to move, you add more time to the lag. At UNC, the strategy has been to temporarily abandon the fine detail in favor of movement, and try to get natural motion whenever the operator's point of view is moving, then crank up the detail when the operator's point of view stabilizes.

"It was obvious to me and to everybody else who was working with Sutherland's HMD that tracking was the biggest problem. You really have to spend a few years trying to beat the lag to get a feeling for how big the difference feels between 100 and 200 milliseconds," he added. A special project at UNC is developing a new kind of position tracker that will shave some delay: an array of lights on the ceiling are tracked through electronic-optical sensors. And another one of Fuchs's projects, "Pixel-Planes," is attacking the computation problem. The point he kept making was that although VR itself is a kind of break-through from the realm of the flat screen, the development of truly useful 3D systems was going to proceed in an undramatic manner that will move at a pace more incremental than revolutionary. If you've been through four generations of microprocessor architecture over ten years in quest of an adequate engine, as Henry Fuchs has, I suppose you are likely to take a dim view of the people who pop up in the popular press as the "promoters" of VR, and who portray it as a "brand-new technology from Silicon Valley" destined to change the world before next television season.

Even with these cautions in mind, Fuchs agreed that the idea of "X-ray spectacles" was not only feasible but constituted a goal worth working toward: "I believe that in twenty or thirty years, maybe fifty years from now, the dominant method of medical imaging might well be an HMD with sufficient resolution in which you could see the data super-imposed on the patient: you could see where the ribs are, see the position of a fetus, and move the transducer around like a flashlight in order to track some complex organ, like an umbilical cord," he told me.

Fuchs, John Poulton, and others have been working on the "boiler-room" level of VR systems, as well—the question of marshaling the vast, fast computation power required by all the fanciful and useful VR application systems people are dreaming up. If you are concerned with 3D computer-generated graphics, you have to be concerned about the raw horsepower you can generate, computer-wise. And while everybody who works with computer technology has come to accept the fact that today's desktop technology will be replaced by something ten or a hundred times more powerful in a few years, there always seems to be more to reality than you can fit in your computer. That's where radically new ways to perform the computing aspects of VR begin to enter the picture. What Fuchs, Poulton, and their colleagues have spent the past ten years doing is creating custom processing chips that link into a network that functions as a single computer. UNC's monster computer-community (or, more formally, their "massively parallel computing architecture") is known as "Pixel-planes." When I visited UNC last summer, Pixel-planes 4 consisted of 250,000 processors. That means, in effect, that every pixel, every particle of the world-model, has its own computer.

Even the fastest computers accomplish their tasks one at a time. The first computers could perform hundreds of operations per second. Today, computing speed is measured in MIPS (millions of instructions per second), and strange new rates, such as "gigaflops," are proposed for the fastest of tomorrow's computer technology. But VR systems eat MIPS for breakfast. And as all who have ventured into the realm of real-time three-dimensional simulations have discovered, VR requires a hybrid of general purpose and special processors to create even a crude approximation of reality. One general pathway to extremely high rates of processing speed is the computer architecture known as "massive parallelism," in which many computers instead of one are devoted to the same task, which is logically parceled out in some manner. The fact that the cost of computer chips keeps dropping as their power increases is making it possible to link hundreds, even thousands, of computer chips in special systems. Pixel-planes has been the name of several generations of VR engines that have been custom-built at UNC from the chip level on up to the parallel architecture. Pixel-planes 4 made it possible for me to stroll down a virtual hallway in which I could discern the texture of the ceiling and the light diffusion through an open office door. Pixel-planes 5, which is in the final testing stages at the time of this writing, is expected to be twenty times faster than Pixel-planes 4.

I had met Frederick Brooks in the hallway the first day I was at UNC. We exchanged pleasantries. The next night, we met again at a party for the VR team that Warren Robinett gave. At the party, we talked about our mutual acquaintance with Douglas Engelbart, a man whose quest for intellectual augmentation was an intriguing parallel to Brooks's work on "intelligence amplification" (IA). Before we sat down for a serious talk about VR, I spent hours rereading the exquisitely polished scientific reports Brooks has published about his work over the past decades. Scientific prose can be deadly, but Brooks has the touch of a literary man, combining logic, rhetoric, and a touch of poetic imagery: "We graphicists choreograph colored dots on a glass bottle so as to fool eye and mind into seeing desktops, spacecraft, molecules, and worlds that are not and never can be," he wrote in his dramatically titled "Grasping Reality Through Illusion: Interactive Graphics Serving Science." Speaking of the role of computer scientists as toolsmiths, he wrote: "If we perceive our role aright, we then see more clearly the proper criterion for success: a toolmaker succeeds as, and only as, the *users* of his tool succeed with his aid. However shining the blade, however jeweled the hilt, however perfect the heft, a sword is tested only by cutting. That swordsmith is successful whose clients die of old age."

Frederick Brooks is a soft-spoken gentleman whose writing and verbal presentations take on a formal yet down-home elegance; he has a storyteller's cadence, and he habitually blends logic, rhetoric, and a touch of poetry. He sounds like the senator from the great state of VR at times. He definitely anchors the room to solid earth when you get him together with latter-day VR luminaries such as Jaron Lanier, and it strains my imagination to think of the possibility of encountering Frederick Brooks and Timothy Leary in the same panel discussion. Although some might consider virtual worlds research to be unorthodox and esoteric, Brooks's credentials belie any such thoughts. He knows what he is doing, he knows why he is doing it, and he has known it for some time. I took Warren Robinett's cue, and asked Brooks to tell me the significance of "IA." I found him to be as eloquent, sitting in his office in Sitterson Hall, leaning back in his chair, hands behind his neck, as his formal voice sounds in his scientific writing. He smiled. He knew I had been prompted, and he didn't mind.

"I believe the use of computer systems for intelligence amplification is much more powerful today, and will be at any given point in the future, than the use of computers for artificial intelligence (AI)," Brooks declared. "In the AI community, the objective is to replace the human mind by the machine and its program and its data base. In the IA community, the objective is to build systems that amplify the human mind by providing it with computer-based auxiliaries that do the things that the mind has trouble doing."

Brooks sees three areas in which human minds are more powerful than any computer algorithms yet designed. "The first of these is *pattern recognition*, whether visual or aural," he said. "Computer scientists don't even have good ways of approximating the pattern-recognition power a one-week-old baby uses to recognize its mother's face from an angle and with a lighting it has never seen before." In which case, Brooks believes, it is possible to multiply that power by using the computer to show humans patterns in ways they are not normally able to perceive, and let the human side of the system decide which ones are meaningful.

The second major area of human computational superiority is the realm of what Brooks calls *evaluations:* "Every time you go to the supermarket, you're performing the kind of evaluations that the computer algorithms we have today can only roughly approximate." The third area of human mental superiority is in "the overall sense of *context* that enables us to recall, at the appropriate moment, something that was read in an obscure journal twenty years previously, in reference to a completely different subject, that we suddenly see to be meaningful."

According to Brooks, the three areas in which computers are more skilled than human minds are "evaluations of computations, storing massive amounts of data, and remembering things without forgetting."

I asked Brooks how he thought a human-computer cooperative system ought to be built, and he replied: "I think in an ideal system for tackling a very hard problem, the machine does the calculation and remembering and searching of data bases—and by calculation I mean the evaluation of some very complicated functions—while the human being does the strategy, evaluation, pattern recognition, planning, and fetching information in context." When you start to define the interface for such a system, you bring yourself to the threshold of VR. Brooks presents it as an ineluctable logic, the same logic that computer graphics pioneer Ivan Sutherland followed in 1965 when he made the first head-mounted display and mapped out the agenda for all those in the future who would seek ways to put the user inside a computer-created world, instead of peering in at it through a narrow window.

"The question has always been how do we couple the human and machine parts of the system?" Brooks continued. "I remember an ad in a magazine, years ago, that showed a human head viewed from the top. The caption said, 'the hardest part of communication is the last four inches.' That's the story in a nutshell: how do we get information from the machine into the head, and how do we get information from the head into the machine? And that leads one inevitably to computer graphics, because the eye is a broad-bandwidth information channel and it's already designed to absorb all kinds of specialized information processing and real-time pattern recognition. In building intelligence-amplifying systems, from earliest times, people have focused on computer graphics as the means to go from the machines to the mind.

"As early as 1965, Ivan Sutherland, in his great speech at the IFIP [International Federation of Information Processing] Congress, set forth an explicit program for the development of computer graphics, defined the notion of the virtual world, and said we want to go in through not only the human visual sense, we want to go in through the ears as well, we want to go in through the sense of feel as well. We want to use all the channels for communicating with the human being that the mind already knows how to interpret. Sutherland also said you create the mathematical model of the virtual world in the computer and then you want it to look, feel, and sound as much as possible like a real world to the human mind that is coupled with it.

"In the other direction, the same thing is true—the way human beings express themselves with physical objects, pushing and pulling and manipulating with our hands. And so, in virtual worlds research here from earliest times, we've given emphasis to rich manual interfaces with lots of joysticks and sliders and knobs and manipulator arms and cueballs and things in which people directly manipulated objects in three dimensions without doing anything differently from what they would be doing with a *real* object as opposed to a *virtual* one.

"That means speech recognition is important, being able to do things with the hands and feet is important, being able to do things with head motion and eye motion is important. These are the ways we couple to the real world, and these are the right ways to couple to the virtual world."

Like a few others who had the foresight to tackle intelligence amplification before computing hardware was powerful enough, Brooks started working with whatever he could find, which wasn't too shabby at the time. In 1969 he obtained a special graphics engine from IBM, the commercial version of the high-powered systems IBM built for General Motors' design automation project. It was puny hardware compared with the computers you can find on an undergraduate's desktop today, but it was something to start with. More than that, Brooks had a student who was up to the task of answering Sutherland's challenge.

"I had a very able, experienced senior graduate student," Brooks recalled, "so I went to the provost of UNC and said, 'I'm ready to do an intelligence-amplifying system, and I have the two things I need, a graphics engine and a smart graduate student. Who on this faculty most deserves to have his intelligence amplified?' " Brooks smiled. It was a great story, he had told it before, but it didn't matter. He continued. At some parts of the story, Brooks underlined his words with his tone of voice, sometimes in combination with the shape of his eyebrows.

"I explained what I was really looking for, which was a collaborator who had a problem that had a high geometric content, because if you're going to work in three-space you might as well start with the easy problems, in which they're working in three-space directly, as opposed to working in abstract spaces. I wanted a problem that was too hard to do by machine algorithm alone and required human insight, a problem that was too hard to do by human insight alone and required a lot of calculation. I believe very strongly that the way to advance computer science and computer technology is by using them to build prototype tools against the demands of *real* applications that require you to solve *all* the problems and not just the ones that are easy and nice. That's what I mean by the 'driving problem' philosophy. The technology advances best if you *carefully* choose a good driving problem, with good collaborators who will keep you honest and keep your feet on the ground. And then, stretch yourself. We knew that the virtual world problems required real-time interactive three-dimensional computer graphics, which we couldn't do very well in

1969, and it looked to us that these problems would be as demanding as anything that one could do with the computer systems of the time.

"The provost said, 'Well, nobody's ever asked me that question before. Let me go think about that.' So he came back the next day with a surprisingly long list of possible candidates, including astronomers who were studying galactic structure, geologists who were looking at the oil-bearing cavities under the ground, molecular chemists, architects who were designing low-cost housing and wanted real-time interactive cost estimating because if they move a brick it's multiplied by 100 because they're making a hundred units. Highway safety people were concerned with driving simulators. Geographers were worried about city planning people who were worried about what happens to the rainwater as you progressively pave over Greensboro. It was a very interesting set of people."

Brooks and his students knew one of the protein chemists well, from working with him on some scientific programming. Jan Hermans in the department of biochemistry was game to collaborate, so Brooks and his early crew picked molecular structure of the life molecules in nucleic acids as their first driving problem. Full haptic simulation would have to be approached in stages. In the late 1960s, the two-dimensional pilot system of J. J. Batter, one of Brooks's graduate students, had involved a small knob that could move over a two-inch-square area, and would offer resistance via servomotors, to model a very simple force-field. In 1971, another one of Brooks's graduate students, W. V. Wright, worked with biochemist Hermans to design a system for displaying and studying proteins—the GRIP-71 system.

"Now, we're not wedded to protein stucture per se; we're interested in human/machine problem-solving systems," Brook interjected, when he got to this part of his story. "So every five years or so we ask ourselves, 'Have we worked this driving problem for all it's worth, should we move our attention to another, instead?' And the answer keeps coming back, 'This one is very rich.' So here we are, twenty years later, still building tools to help biochemists, with protein and nucleic acid structures. Building tools for biochemists has driven us to a lot of new computer science."

The selection of molecular docking as a driving problem converged with another rapidly developing technology—the use of computer modeling as a tool for investigating chemical structures, particularly the complex geometric structures of the key biological molecules. The first tool for using 2D interactive computer graphics as a molecular docking aid was built and demonstrated in 1966 by C. Levinthal at MIT. About a dozen teams worldwide continue to work with interactive computer graphics and molecular docking. But their dedication to Sutherland's vision led the UNC team inevitably to the ARM and the idea of haptic visualization.

"The force display work was one of the ideas Ivan Sutherland put forward in his 1965 speech, and I always thought it was too good an idea to leave lying around," Brooks recounted. "If watching imaginary objects move and alter as one manipulates them endows them with a kind of real existence, and if this process yields more power for understanding and designing imaginary objects, can we build yet more powerful tools by using more senses?" Brooks had written in 1977.

Brooks and his students started work on combining force-reflection feedback with interactive computer graphics in 1972. Having decided to follow Batter's and Wright's experiments with a 6D (three-force/three-torque) system, Brooks's team "providentially encountered" Raymond Goertz, who had designed the remote manipulator arm at Argonne National Laboratories—a master-slave system that enabled humans behind lead shields to manipulate radioactive substances with remote machines that mimicked their hand movements. Goertz arranged for UNC to obtain a pair of orphaned Argonne remote manipulators (ARMs). The UNC team substituted the computer and its world model—in this case, protein structures—for the remote manipulator's mechanical slave and used the input device as a haptic transducer. The problem was computing power. They had defined a task that was simply beyond the state of the art.

"The most complicated physical world we could maintain had seven child's blocks on a table. That's all we could do in real time. Another student, P. J. Kilpatrick, did a thesis on that, and meanwhile we realized that what we always wanted to do with this was molecule docking, but that would need another factor of 100 in computing power, so we mothballed it. And then in 1986 we got it out again and said, we've got a factor of 100 in computer power, let's go! And so, now, Ming has brought that on as his project, and that's doing very nicely. That's a good example, though, of the power of the driving problem."

Ming Ouh-Young's research, as I had witnessed, was concentrated on measuring the usefulness of VR docking aids in performing real docking problems. The results, published a year after my visit, confirmed what biochemists working with Brooks's teams for two decades have been enthusiastically reporting—haptic VR works approximately twice as well as the next best method, by objective measures. Although Brooks cautions that even with improvements in the state of the system, it doesn't seem likely that tenfold augmentation is achievable; he also reminds fellow researchers that "the purpose of computing is insight, not numbers." If you can cut the time in half that it takes to solve a complicated problem, you might also be able to solve problems that were too complicated to solve before. The GROPE-III system (UNC's name for the latest version of their molecular docking apparatus) has proved its usefulness to those biochemists who have used it in a computer research context. The next goal at UNC is the development of the VIEW system, a 3D visualization workbench for biochemists and other scientists to switch between many different kinds of visualization.

Over the years, Brooks and the growing number of computer science faculty who had joined UNC's VR-centric research program added other driving problems. Medical imaging has turned into another rich field that looks as if it will continue to drive VR systems for years to come. In regard to the architectural walkthrough, Brooks noted: "We decided that this is a good driving problem compared with both the molecular one and the medical imaging one; it has the tremendous advantage that sooner or later the building gets built and you can compare the real and the virtual worlds and see how far away you are from realism, whereas the molecules are too small to see. In the molecular work, the objective is to help people discover the unknown. To understand the structure. In the building work, the objective is to help the architect and the client to see what the consequences of a specification are while it is still a specification."

I asked Brooks how his experience as the leader of a large software design team had informed his outlook in conducting VR research, and he used the architectural walkthrough as an example of the kind of tools he wanted to build: "My experience in designing computers and software is that the hardest part of any design problem is figuring out what it was you wanted to design, getting goals and specifications exactly right. I'm firmly convinced that this *cannot* be done on a one-shot basis in advance. You *have* to iterate. What people do with virtual buildings, if they're willing to take the time, is they take the floor plans and they live through a day and trace all the patterns and see where they run into trouble, where they store things, and where they have traffic problems, and so forth. And they run through the major annual events and whatnot. The architects say they don't have any trouble visualizing three-dimensional buildings from floor plans and elevations, but the *clients* surely do. So if the client and the architect working together can debug the building at what is called the design development stage, while everything is still on paper and you haven't even done the working drawings, you can save a lot of trouble downstream. I don't think that has nearly the potential to make the difference to society that the biochemistry work does, but it's a good driving problem for us because of this testability aspect."

Moving beyond molecular docking, architectural walkthroughs, and medical imaging, the fourth and newest driving problem is related to the desire I heard voiced by Michael McGreevy at NASA, who wants to use VR to fly through planetary imaging data. In the case of James Coggins's work at UNC, the planet is the one we inhabit. Landsat and other satellites furnish a rich stream of data, 24 hours a day, around the world. The discipline of "remote sensing" is devoted to the extraction of meaningful meteorological, geological, ecological, energy-use information from that flood of image data. Again, the idea of literally flying your point of view through a 3D image of the earth as seen by a satellite is to stimulate *insight*. An insight about where a hurricane will hit or where a pool of petroleum can be found can be measured in lives as well as dollars. And the macroscale patterns of life-related processes on planet Earth are bound to be at the center of concern for industrial civilizations for decades to come. Any key insights at the level of problems that big and complex might be worth far more than the equipment required to stimulate the bright ideas.

Scientific visualization is another one of Frederick Brooks's hobby horses. If intellectual amplification is the goal, the vision the UNC laboratory pursues, scientific visualization is the toolbox Brooks and his colleagues offer to those scientists who are seeking to extend their power. Having waited for decades to gain sufficient computing power, Brooks is now convinced that today's computers "empower us to build sophisticated models of complex natural phenomena and to explore them for new insights into models and phenomena." Noting that the great scientific revolutions of Newton's times were empowered by the new mathematical tools of the calculus and other analytic equations, Brooks believes that the power of modern computers to represent complex mathematical models in real time, in human sensible form, can leverage yet further progress in the power of scientific insight. You simply have to look at the kinds of models computers are good at maintaining, and the kinds of complex natural phenomena that scientists are interested in understanding, to see the broad utility of interactive haptics. An example of a massive mathematical model is species population dynamics. A detailed mathematical model is a protein structure. A three-dimensional model is a magnetohydrodynamic simulation. Plate tectonics is a geographical example of a nonlinear model; shockwave propagation is a discontinuous kind of model; quasars, pulsars, and black holes are discrete models; particles and quarks are indeterminate. Other natural phenomena cited by Brooks as appropriate to sophisticated mathematical modeling include oil-field geology and blood flow in the body. The key to plugging the computer's forte into the human partner's special capability is interactive computer graphics, or as Brooks says: "If mathematics is queen of the sciences, computer graphics is the royal interpreter."

Certainly for nonscientists, the usefulness of computer graphics for communicating complex information has proved itself to be a real intellectual amplifier compared with the days of typewriters. Brooks sees the emergence of personal computers as intellectual tools as part of the infrastructure for the kind of scientific visualization that is just beginning to emerge. "All of scientific visualization has been technologically enabled by the interactive graphics commercial base supported by word processing, spreadsheets, desktop publishing, CAD-CAM, flight simulators, and entertainment. Indeed, almost all of computer graphics has been technologically enabled by the commercial base of television," Brooks declared in a presentation to SIGGRAPH 1990, adding: "We expect the same process to pace the development and cost of haptic display hardware. The manufacturing robotics industry, the video game industry, and the vehicle simulator industry will develop the technologies, which will be adapted for teleoperation and for scientific visualization." In other words, progress in pharmaceutical design and medical imaging will benefit indirectly from the success of the true industrial giants of the future VR industry—who are more likely to be video game companies than scientific toolmakers.

Brooks cites several areas as likely hotspots for development of haptic displays for scientific visualization. Molecular design is foremost among them: "We believe the pharmaceutical industry has a substantial potential for using haptic displays, but that such applications will develop very slowly over the decade. Besides drug-enzyme docking, one can imagine applications in DNA intercalators [computerized efforts to map the human genetic code], in protein design, and in studies of protein folding and packing. Feeling these subtle force fields could matter a lot to a researcher seeking insight and new hypotheses," he predicted in Dallas in August, 1990. He cited the Disney-Lucasfilm Star Tours motion platform immersion experience and Atari's Hard Drivin' force-reflection feedback video game as the early successes of haptic displays in entertainment, which he believes will be the fastest-growing application in the 1990s.

With all his fervor for the future of haptic displays as a tool for scientific exploration, Brooks voices two deep concerns: First, although the educational value of simulation is very high, because is it "learning by doing," many phenomena require firsthand experience in order to know the difference between theory and practice. He is also concerned that as VR simulations grow more realistic, their potential for being dangerously misleading also increases. No model can ever be as complex as the phenomena it models, no map can ever be as detailed as the territory it describes, and more importantly, as semanticist Korzybski noted, "the map is not territory."

In regard to limitations of VR as an educational tool, Brooks told me: "I don't have any trouble imagining the mechanic wearing something like bifocals, and the upper half shows him a "manual" of full-motion 3D pictures on how you test something or how you disassemble something or how you diagnose something, while he's looking at the something through the lower half, and trying it with his own hands. The manual's probably pictures and voice, presented on voice command as he works, and that way anybody might work knowledgeably on a car, a kind they've never seen before. But that's never going to be a completely effective substitute for having worked a lot on cars, to know the difference between theory and practice so deeply that it's not conscious."

Certain kinds of hyperrealistic simulations could be dangerously misleading. One example Brooks cited was the fashion of using fractal-generated mathematical models to simulate the natural irregularities in mountains. The old kind of computer-generated mountains are too regular looking to be natural. But mountains, like coastlines and snowflakes, are known to take the kinds of shapes that can be described by mathematical models known as fractals. You could take a computer model of the Rocky Mountains, for example, and make it look a lot more realistic by adding a fractal element. The fractal would not be a direct measure of the irregularity of the real Rockies, but would be mathematically characteristic of the Rockies' irregularity, as far as the human eye perceives it. But, as Brooks points out, you wouldn't want to lead an army through a landscape, based solely on their practice in a fractal model.

"The potentials for misleading are very great in that kind of instance," Brooks emphasized at the end of our time together, "and the fractal mountains are a good clear visual image of an important distinction between *realism* and *truthfulness*. The danger of more and more realism is that if you don't have corresponding truthfulness, you teach people things that are not so. In business scenarios, or war games, to the extent that your model of the business world or the war world is not real, you can make the mistake of teaching people very effectively how to apply tactics and strategy that won't work in the real world."

Brooks's hopes for a new scientific lever for hard and useful problems, and his warnings about the limitations of simulations, no matter how lifelike they might be, reflect the manifold nature of VR. The citizen of the twenty-second century might find it hard to understand how the human race ever managed to make do without the assistance of VR systems, just as we take the usefulness of antibiotics, modern plumbing, electrical refrigerators, and literacy for granted today.

Better medicines, new thinking tools, more intelligent robots, safer buildings, improved communications systems, marvelously effective educational media, and unprecedented wealth could result from intelligent application of VR. And a number of social effects, less pleasant to late-twentieth-century sensibilities, might also result from the same technologies.

Our most intimate and heretofore most stable personal characteristics—our sense of where we are in space, who we are personally, and how we define "human" attributes—are now open to redefinition. The technology that can replicate the human mind's fanciest trick—weaving sense-mediated signal streams into the fine-grained, three-dimensional, full-color, more-or-less consistent model we call "reality"—is in its infancy today. But technologies evolve more quickly these days than ever before. What will we think of each other, and ourselves, when we begin to live in computer-created worlds for large portions of our waking hours?

The early days of any technological revolution are filled with uncertainties. For a brief period, before industries, infrastructures, and belief systems grow out of and around a communication technology, the course of the technology is unknown. VR represents a unique historical opportunity. In retrospect, we now understand something about the way telephones, television, and computers expanded far beyond the expectations of their inventors and changed the way humans live. We can begin to see how better decisions might have been made twenty and fifty years ago, knowing what we know now about the social impact of new technologies. The ten to twenty years we still have to wait before the full impact of virtual reality technology begins to hit affords a chance to apply foresight—our only tool for getting a grip on runaway technologies.

At the heart of VR is an *experience*—the experience of being in a virtual world or a remote location—and the problems inherent in creating artificial experiences are older than computers. While MIT and the Defense Department might know a thing or two about spurring new computer technologies, the center of the illusion industry is closer to Hollywood, California. If it had not been for the vicissitudes of research funding, Morton Heilig, rather than Ivan Sutherland, might be considered the founder of VR. Although he was dreaming about "experience theaters" and arcade simulators thirty years ago, Heilig was trying to move Hollywood into the same world of 3D illusions that the computer scientists are exploring today. Ironically, the entertainment industry may come to be the largest driving force in the future development of VR, a generation after Heilig tried to point the way.

Part Two

BREAKING THE REALITY BARRIER

Chapter Two

THE EXPERIENCE THEATER AND THE ART OF BINOCULAR ILLUSION

The present invention, generally, relates to simulator apparatus and, more particularly, to apparatus to stimulate the senses of an individual to simulate an actual experience realistically.

There are increasing demands today for ways and means to teach and train individuals without actually subjecting the individuals to the hazards of particular simulations. . . .

Industry, on the other hand, is faced with a similar problem due to present day rapid rate of development of automatic machines. . . .

The above outlined problem has arisen also in educational institutions due to such factors as increasingly complex subject matter being taught, larger groups of students and an inadequate number of teachers. As a result of this situation, there has developed an increased demand for teaching devices which will relieve, if not supplant, the teacher's burden.

Accordingly, it is an object of the present invention to provide an apparatus to simulate a desired experience by developing sensations in a plurality of the senses.

It is also an object of the invention to provide an apparatus for simulating an actual, predetermined experience in the senses of an individual.

A further object of the invention is to provide an apparatus for use by one or more persons to experience a simulated situation.

Another object of the invention is to provide a new and improved apparatus to develop realism in a simulated situation.

Morton Heilig,
"Sensorama Simulator,"
US Patent #3,050,870, 1962

It looked more like a vintage pinball machine than a prehistoric VR prototype, but it still worked, after all those years. One of the original

"Sensorama" games, it was still in remarkable working condition, slowly deteriorating in the cabana next to Morton Heilig's pool in West L.A. By virtue of its longevity, it was a time machine of sorts. I sat down, put my hands and eyes and ears in the right places, and peered through the eyes of a motorcycle passenger at the streets of a city as they appeared decades ago. For thirty seconds, in southern California, the first week of March, 1990, I was transported to the driver's seat of a motorcycle in Brooklyn in the 1950s. I heard the engine start. I felt a growing vibration through the handlebar, and the 3D photo that filled much of my field of view came alive, animating into a yellowed, scratchy, but still effective 3D motion picture. I was on my way through the streets of a city that hasn't looked like this for a generation. It didn't make me bite my tongue or scream aloud, but that wasn't the point of Sensorama. It was meant to be a proof of concept, a place to start, a demo. In terms of VR history, putting my hands and head into Sensorama was a bit like looking up the Wright Brothers and taking their original prototype out for a spin.

The Sensorama is where an alternate probability world could have branched off, a scenario in which the entertainment industry, not the computer industry, succeeded in cracking the reality barrier with predigital technology. In the early 1960s, while the rest of America was settling back into the cathode glow of network television, Heilig in fact did develop, patent, and attempt to market an arcade version of virtual reality. He also published detailed plans for an "Experience Theater" in 1955 and patented a head-mounted stereophonic television display in 1960. If it wasn't for the requirement that his machinery would have to stand up to Times Square arcade treatment, the repeated failures of vision on the part of those who financed the experience industry, his own irrepressible iconoclastic tendencies, and a series of bad business breaks, Heilig might have ushered in the cyberspace age thirty years ago.

Heilig invented "reality for a nickel" so long ago that his patents expired in the 1970s, but he is still game for designing a new, digital-electronic version of what he had envisioned in his youth. His son, a computer science student, has been helping him get up to speed. In the 1950s, when he started drawing up his plans, 3D multisensory cinema packed into an arcade device was just another crazy idea by yet another garage inventor.

Sensorama first came to my attention in 1988, the same afternoon I took my first test-flight in cyberspace. In Scott Fisher's NASA/Ames office, down the corridor from the VR lab where I had ventured into the virtual world earlier that morning, I noticed the reproduction of an old advertisement for Sensorama posted prominently on his bulletin board. It looked like something from a science fiction movie about time travel: A man was sitting in a large wooden booth that semisurrounded him. He was leaning forward at a 45-degree angle, his face swallowed up by a nickelodeonlike viewer. His hands gripped a set of handlebars. The man on the poster was dressed in the style of the early 1960s, but the image depicted a kind of precomputer virtual reality. A little more than a year later, I sat down in the very booth portrayed on that ancient poster and test-drove an aging but mostly functional Sensorama for myself.

I ended up at Morton Heilig's place through a kind of cyberspace serendipity. Early in 1990, I knew that I would be attending a virtual reality conference in Santa Barbara, and my travel agent said it would be cheaper to fly to LA and spend the night. A week before I left, Eric Gullichsen, a VR programmer and entrepreneur I had met at Autodesk, a VR software company, brought me copies of two of Morton Heilig's patents. One look at the descriptions of his inventions, the detailed drawings, and the date, convinced me that Heilig was the true pioneer Scott Fisher believed he was. Although the most recent patent (which listed his address) was close to thirty years old, I tried calling for directory assistance in the last place Heilig lived. It turned out that he was still there. His house was conveniently located between LAX and Beverly Hills, and he was indeed still interested in talking about Sensorama. He even promised to crank it up for me.

Pulling up at the curb of Heilig's comfortable but far from lavish home, I was reminded of the middle-class Phoenix suburb where I grew up—two or three bedrooms, "ranch-style," front lawn, driveway, carport. The swimming pool and cabana out back, where the Sensorama prototype resides, are rather more modest than what one imagines "out by the pool" to be. Heilig, a wiry, salty-tongued, gray-goateed fellow who appeared to be in his sixties, met me at the door. He's a man who enjoys conversation. We sat and chatted in his living room most of the afternoon. He wanted to know everything I could tell him that could bring him up to speed in the "new" world of VR that seemed to be emerging. Above all, the word I would have to use to describe Morton Heilig is "enthusiastic"—he's enthusiastic when he remembers the dreams he had, acerbic when he recalls the reasons they never quite materialized the way he had planned, then enthusiastic again when he talks about the reasons the "experience theater" he started imagining the day he saw Cinerama might still happen with today's technology.

Heilig wasn't a computer programmer or electrical engineer. He had been a cinematographer, photographer, inventor of projection and camera apparatus—a Hollywood, California, visionary rather than a Cambridge, Massachusetts, computerist or a university-based artist-technologist. The more I listened to him, glanced at the publications he had stashed away, going as far back as 1955, and tried out his handmade prototypes, the more clear it became that Hollywood could have been the original driving force behind VR development, rather than the Defense Department and NASA. By now, I've talked to others who had ideas that might have led to cyberspace. Some succeeded in realizing their dreams, some did not. It appears that technological revolutions don't happen simply because progress is inevitable; true technological paradigm-shifts require a visionary or two of Heilig's caliber, access to one or more enabling technologies that would make the visionaries' gadgets possible and, most important, a person with some power to champion the new idea among the conservative decision-makers who can finance its development. The right person has to pursue the right vision at the right time—and the right people have to be convinced to offer financial support.

Heilig's patents and rare published articles don't tell the whole story by a long way. Where and why did Heilig get started with the ideas that led to Sensorama? What had he attempted to do with them? Why didn't they succeed in revolutionizing entertainment and education? What would he do today, given proper financial backing? Yes, there are answers to those questions, involving peculiar twists of fate. Before we sat down to talk about his history and my curiosity, I asked if I could take a look at Sensorama. We walked out through his kitchen door into the bright sunlight of his back yard. The old machine, looking fairly spiffy considering its age and location, was under a tarp in a sheltered corner of a small vine-covered structure, open on three sides, near the swimming pool. Heilig slid back a panel on the side of the machine, looked inside, showed me the film projection mechanism and told me, apologetically, that the system for transporting odors to my nose and whisking them away again needed minor repairs. I would have to settle for the stereophonic audio-visual-tactile version of the Sensorama experience and forego the olfactory dimension.

I put my hands on the handlebars and rested my face against a viewer that looked like a pair of binoculars with a padded faceplate. Right below the eyepiece was a small grill, near my nose, where the odors would have been pumped in and out of smelling range. Other grills to either side of my face emitted unscented breezes at appropriate times. Small speakers were positioned on either side of my ears. The machine started. I heard an automobile engine, apparently with the muffler removed, saw an expanse of sand dunes, felt my seat lurch, and found myself looking from the driver's seat at a stereoscopic view of a dune-buggy ride. The film had begun to turn yellowish-brown.

It looked as if I were sitting in the front seat and holding on to the handlebars, but I had no way of steering any of the vehicles I found myself riding; I was strictly a passenger. I spent a few loud minutes meandering over sand dunes the size of houses, then I found myself on a motorcycle ride through the streets of Brooklyn, as Brooklyn has not appeared for thirty years.

After Brooklyn and a California helicopter ride, I found myself in a convertible with a young blonde woman who smiled at me to the accompaniment of a pop song about "riding with Sabina" that I couldn't swear I had never heard before. I took a bicycle ride with Sabina, followed by a romp on the beach. Then it was time for the grande finale, a belly dancer with a come-hither look. Heilig told me that the smell of cheap perfume was programmed to waft out the nosepiece whenever the belly dancer approached. "The businessmen who came to demonstrations as potential backers always liked the belly dancer," he said, above the din of a decades-old recording of the Middle Eastern band blasting at me stereophonically, as the dancer ching-ching-chinged her finger cymbals next to one ear, then the other. The stereoscopic effect was nothing like the kind of multimillion-dollar displays I had seen at Lucasfilm's Disneyland exhibit a few months earlier, of course, but there was a perceptible sense of depth, the dune buggy did lurch, the motorcyle handlebars did vibrate, and breezes did blow against my temples. The motorcycle driver was reckless, which made me very mildly uncomfortable, much to my delight.

Compared with the hokey imagery that was the usual state of the art in arcade devices of the 1960s, Sensorama was everything its poster claimed. It wasn't the utterly realistic "All-Super-Singing, Synthetic Talking, Coloured, Stereoscopic Feely" that Aldous Huxley had predicted in *Brave New World,* but I couldn't help wonder what it might have evolved into by now if research and development had not halted three decades ago. Something had happened, or more likely, had failed to happen, since the day Heilig took his homemade stereoscopic movie camera on a motorcycle ride in Brooklyn. After my brief demonstration, he shut down Sensorama, unplugged it, replaced the tarp. We returned to the cool darkness of his living room to talk about what might have happened, and how he still hopes to materialize his old dream by means of this bright new cyberspace technology he's been hearing about.

"I got hooked on film-making as a young man, and I started to dream about the possibilities of multisensory experiences soon after that," Heilig recalled. Drafted the day before Hiroshima, he was stationed in postwar Europe. After his hitch was over, he put together his G.I. Bill money with a couple of Fulbright fellowships and studied film-making in Rome, then embarked on a career as a self-employed documentary maker. In the early 1950s, he started to read in the American press about a new film-making system known as "Cinerama," an idea that attracted him immediately, not so much for what it did as what it portended.

"Cinerama was invented by Fred Waller," Heilig explained. Waller wanted to enlarge the field of view of the rather small rectangle presented by even the largest theater screens of the day. A human being sees 155 degrees vertically and 185 degrees horizontally. A movie screen fills only a small portion of the normal human field of view. Waller, in the late 1930s, began experimenting with multiple projectors and multiple screens as a way of presenting a wider field of view for films.

"He finally got a contract with the Air Force," Heilig recounted. "Waller built a motion picture display for the first flight simulators that had five cameras and five projectors, three on the bottom and two on top. His intention was to fill the peripheral areas of the human visual range with as much imagery as possible. He couldn't get large-format cameras or extra-wide-angle lenses, so he did it by having multiple cameras." After the war, Waller scrapped the two upper cameras and tried to get Hollywood interested in a three-camera, three-projector system. Every scene in a Cinerama movie was photographed by three synchronized cameras from slightly different angles, then projected in synchrony onto three screens that curved inward to wrap around the audience members' peripheral visual field. Waller managed to draw producer Mike Todd to an old tennis court on Long Island to get a demonstration. Todd raised $10 million, Waller set up the Cinerama company with Todd, and their first feature, *This is Cinerama,* was a smash hit at the first Cinerama theater on Broadway. The wrap-around screen definitely improved the perceived sense of presence. It certainly had an effect on Heilig, who had a kind of conversion experience when he sat in that theater on Broadway in the early 1950s.

"Cinerama was a revolution because it was a real enlargement of the film experience—something that was badly needed when television started to catch on," Heilig continued. "I read about it in Italy, and it excited me so much I returned to New York and went to Broadway to see it. I realized I was watching something of tremendous importance. I started reading everything I could find about it." There was ferment in Hollywood in the early 1950s because the film industry was threatened by television, and Hollywood finally started looking at what they could do that the television people couldn't do. The 3D movie craze, stereophonic sound, wide screens, and other technical innovations that had been resisted for years became attractive to those who controlled the purse strings for research and development. Distribution, not R & D, had driven Hollywood, but the film medium itself was a technological innovation, and the film industry had been saved once before during a disastrous slump by the advent of a new cinema technology, "talkies."

"I recognized right away why Cinerama and 3D were important," Heilig recalled. "When you watch TV or a movie in a theater, you are sitting in one reality, and at the same time you are looking at another reality through an imaginary transparent wall. However, when you enlarge that window enough, you get a visceral sense of personal involvement. You *feel* the experience, you don't just *see* it. I felt as if I had stepped through that window and was riding the roller coaster myself instead of watching somebody else. I felt vertigo. That, to me, was significant. I thought about where the technology might go in the future, and I was convinced on the spot, sitting in that Cinerama theater on Broadway, that the future of cinema will mean the creation of films that create a total illusion of reality, just as you are sitting across the room from me right now, with no frame between us."

Heilig started to think about what would have to be accomplished to create an artificial experience that could fool people into believing they were actually occupying and experiencing a movie set. "How do I know I'm in a particular environment?" Heilig asked himself in 1954. He sketched a schematic of the brain, sensory channels, the motor network, the rough outlines of the main perceptual inputs that contribute to a sense of reality. He thought that all that was required was to hire engineers to duplicate mechanically or electrically or optically the sensory information that contributes to a sense of reality, find a way to record it, and play it back in specially equipped theaters. Heilig told me that he considered that first sketch a kind of "periodic table of elements" of what he came to call "experience theater." It was an intriguing metaphor: the original periodic table of the elements was a chart of the known chemical elements, created in the nineteenth century, that grouped the basic forms of matter in a manner consistent with their properties. The periodic chart made evident the gaps in knowledge of the elements, and predicted that other elements with certain properities would be found to fill in those holes; the periodic chart served both as a device for organizing what was known about matter and as a research agenda for finding out more.

Heilig compared the state of VR in 1955—and 1990, for that matter—with the state of chemistry in the nineteenth century: "So far, we've duplicated a small fraction of the perceptual elements that convince us we are experiencing reality. There are known elements for stereophonic sound and stereoscopic visual input, but there are large blanks where it comes to our senses of smell and touch. 'Why mess around?' I thought. 'Let's get our act together and do it.' " So he wrote a manifesto, calling for the Hollywood studios or the government to mount a broad-based R & D effort into the sound, peripheral vision, vibration, smell, and wind elements. He was an unknown. Naturally, nobody listened. That didn't stop him from seeing what he saw and putting his plans on paper.

Failing to find a publisher for his manifesto, Heilig, at the age of twenty-six, went to Mexico and started shooting documentary films. Along the way, he met the great Mexican muralist Siqueros, who introduced him to a group of intellectuals, engineers, painters, and architects. Young Heilig started lecturing about his ideas at salons held at Siqueros's house. At the end of his lectures, the group asked him to write his vision down and published it in January, 1955, in a bilingual Mexican journal named *Espacios*. In this article he described in detail, complete with sketches and schematics, his vision of a "cinema of the future" that would be recognized in essence if not in terms of enabling technologies by any of today's cybernauts:

Celluloid film is a very crude and primitive means of recording light and is already being replaced by a combination television camera and magnetic tape recorder. Similarly, sound recording on film or plastic records is being replaced by tape recording . . . a reel of the cinema of the future being a roll of magnetic tape with a separate track for each sense material. With these problems solved it is easy to imagine the cinema of the future. Open your eyes, listen, smell, and feel—sense the world in all its magnificent colors, depth, sounds, odors and textures—this is the cinema of the future!

The screen will not fill only 5% of your visual field as the local movie screen does, or the mere 7.5% of Wide Screen, or 18% of the "miracle mirror" screen of Cinemascope, or the 25% of Cinerama— but 100%. The screen will curve past the spectator's ears on both sides and beyond his sphere of vision above and below. In all the praise about the marvels of "peripheral vision," no one has paused to state that the human eye has a vertical span of 150 degrees as well as a horizontal one of 180 degrees. The vertical field is difficult, but by no means impossible, to provide. . . . This 180 degree by 150 degree oval will be filled with true and not illusory depth. Why? Because as demonstrated above this is another essential element of man's consciousness. Glasses, however, will not be necessary. Electronic and optical means will be devised to create illusory depth without them.

Heilig also noted in the same article that human vision tends to be sharp at the center and less sharply focused at the periphery. I remembered that statement from a 1955 Mexican magazine several weeks after my trip to see Heilig, when I found myself at a research institute outside Kyoto, Japan, watching a new $1 million projection system that tracks the human viewer's gaze and projects a sharp image at the center of the field of view, with a less sharply focused image on the periphery.

It would take years of research and development into the component technologies before the full-scale version could be built, but Heilig's article made enough of a stir in Mexico that the minister of education decided to support the proposed R & D.

"We built a huge semispherical screen and I taught myself optics, designed and built one of the first bug-eye lenses, and took some test shots and projected them onto the screen," Heilig recounted, his grim little smile foreshadowing the rest of his story. "And then something happened that seems to have happened a number of times," he said, the way a man says something he has said before, and still puzzles over every time he repeats it. "The minister of education who had been supporting me and my project, a man who seemed destined to become the President of Mexico, was killed in an aircraft accident. Strangely, years later, when I tried to get the experience theater project going in the States, a man who was the CEO of a big projector company was killed in an airplane accident not long after he decided to back me. Then Mike Todd, the man who made Cinerama possible, was killed in another airplane accident." His Mexican backing dissolved, so he headed for New York, rented a hotel room, and invited investors to see pictures of what he had developed to that point with his large screen and projectors.

"The Cinerama people came to see me," he remembers, somewhat ruefully. "They sent some engineers, and then they came back with the entire organization. I lectured them on all my ideas. I didn't get anything out of it. I've never wanted to grab all the glory. I don't give a shit about that. I just wanted to *do* it. But all I had left were still photos and a spiel. And every time I showed them to an institution, it would come down to one guy who had the power to decide; if he didn't have the imagination to see what I was talking about from the still photographs, my article, my drawings, and what I had to tell him, then I'd never get him to spend the money necessary to develop a theater. I'd have to make a one-man version to convince the one man who would make the decision about the theater version, my ultimate goal. I'd build a booth, an arcade-size version. To me it was obvious that it should be done in 3D and should have sound and wind and even smell. I decided to do it myself, a piece at a time."

In the late 1950s, Heilig took a job teaching a course in film appreciation. He bought a pair of film cameras and a pair of projectors and linked them into a homebrewed stereoptic system. He found a partner, finally, who was willing to sink some money into finishing the prototype. They started a company, named it *Sensorama*, built the first working model, and moved it to Heilig's Greenwich Village apartment. At that time, the Village was at the height of the bohemian ferment of the beatnik era, and Heilig, a goateed film-maker, was in the thick of it. His partner came up with a brochure and sent it to major corporations, including Ford and International Harvester. They pitched it as a showroom display. You could give your prospect the experience of riding in a combine, smell the hay, feel the breeze—or take a ride in a convertible. The big companies didn't bite.

The Sensorama partners used the analogy of restaurants and food vending machines to describe the relationship between the experience theater and the Sensorama. A restaurant and an experience theater both serve hundreds of people at a time, are expensive to develop, and take up a lot of space. A vending machine and a Sensorama both serve one person at a time, are less expensive to develop, and take up a small amount of space. There is room for both restaurants and vending machines in the food marketplace, and there is room for experience theaters and Sensoramas in the experience marketplace.

Carried to its ultimate level of personal presentation, the experience theater could be compressed into a head-mounted display that a person could wear like a pair of exceptionally bulky sunglasses. Heilig's patent #2,955,156, granted October 4, 1960, was for a "Stereoscopic Television Apparatus for Individual Use," and his patent #3,050,870, granted August 28, 1962, was for a "Sensorama Simulator." When I asked about the head-mounted version, Heilig disappeared into a back room and came out with a boxy aluminum prototype, thirty years old, called "the Telesphere mask," which he strapped onto my head. There was no stereographic television projector to animate the mask with imagery, but it was definitely a head-mounted display, built and patented more than five years before Ivan Sutherland assembled the head-mounted computer graphics display at MIT—the one that is most widely known as the historical predecessor to today's VR helmets.

Although they decided to put out their prototype as a coin-operated amusement device, Heilig and his partner were acutely aware of Sensorama's educational and industrial potential. The Sensorama patent predated much of the psychological research into the nature of learning that informs the design of today's educational technologies, but Heilig noted in his Sensorama patent, "A basic concept in teaching is that a person will have a greater efficiency of learning if he can actually experience a situation as compared with merely reading about it or listening to a lecture." Educators weren't interested, however, and even if they had understood what Heilig was proposing, educators were the last people to ask for millions of dollars for R & D. Industry wasn't interested. But the indefatigable Sensorama partners managed to talk an arcade owner into allowing them to place the machine in an arcade on 52nd and Broadway in New York. It broke within hours. They hauled it away and fixed it. Despite more months of work improving its robustness, Sensorama was simply too complex for the arcade treatment.

Heilig's partner lost interest. Heilig continued his career as a documentary film-maker. One night, he got a call from a fellow who had seen the Sensorama machine on Broadway, and was interested in discussing the possibility of mass-manufacturing the miniature experience theaters. The fellow was in the San Francisco area, however, and Heilig still lived in New York.

"I found myself in San Francisco on a film project a few months later," Heilig recalled. "Quite late in the evening before the day we were scheduled to leave California, my wife urged me to call this fellow. Even though it was midnight, he told me to come over. He gave me the directions to a place where he would meet me, about an hour's drive from the city. It was a foggy night. I followed his directions and found myself in a secluded parking lot. There was a bearded guy with a dirty suit, sitting in a beat-up car. He told me to follow him, and I found myself driving up a winding country road in the middle of the foggy night. It was a scene out of Dracula. And then he stopped in front of this huge house, this damn castle with a big gate." The backer turned out to be legitimate, if eccentric, and he put up the $50,000 necessary to build a prototype for a mass-marketable Sensorama. They found a vending machine company and an engineer. Heilig returned to the East Coast. A few months later, invited to be associate producer on a feature film, Heilig moved his wife and babies out to Hollywood. Every weekend, he would fly up to Fresno, where the vending machine company was located. Then Heilig's Hollywood mentor, the film's producer, was fired. The people who took over the project fired Heilig. It was the late 1960s by this time, and he had long since learned how hard it was to become a successful entrepreneur in the entertainment industry.

High and dry in Hollywood, Heilig became a member of the camera operators' union and made a living working on various projects. The new prototype of Sensorama—the one that resides under the vines in his cabana today—once again failed to attract second-round financing. Again, Heilig's backer lost interest. Never one to shrink from declaring that the problems he was facing were the result of the world's misperceptions and not a failure of his own vision, Heilig wrote and self-published another manifesto. This one, entitled "Blueprint for a New Hollywood," proposed that the studios invest the money it cost to produce two feature films in a research and development institute for creating the cinema technologies of tomorrow. It's too bad they didn't do it twenty years ago, as Heilig proposed.

"For the price of a couple of feature films, we could have created virtual reality decades ago," Heilig sighed, when we talked about his 1971 "Blueprint." "For far less than the price of a bomber, we could have distributed marvelous learning environments to our major universities. If I had written a proposal for a theater that would kill people, I guess I might have done better with finding funding." Heilig peppered what would otherwise have been a long, sad story with sardonic comments that signal the survival of his feistiness and iconoclasm through all the ups and down of his quest.

Heilig's vision of a medium for conveying multisensory artificial experiences is on the verge of becoming a reality in the 1990s, but the path to today's VR technologies did not come from cinema. Rather, the development of machines to think with, the extension of computer-based tools to the amplification of human perception and cognition, led to the emergence of a form of experience theater from the midst of a most uncinematic field—computer science. There were others like Heilig who saw the great educational power of direct sensory experience, and who had visions of future technologies that could propel human thinking to a whole new level. Like Heilig, many of the visionaries of the computer revolution labored alone for years, unable to convince their colleagues to pay serious attention to their wild ideas. Unlike Heilig, some of the key visionaries of the computer revolution did manage to find powerful backers for their ideas, in the US Department of Defense.

The effort to find ways to couple human minds more tightly with computing machinery took decades to converge with the older effort to create 3D illusions. It was out of such convergences that VR was born, and it is in such potential convergences that we can read the outlines of future VR developments. The way unrelated scientific questions and very different technologies suddenly find themselves part of something new and different is one of the characteristics of VR research I had to get used to. The closer I looked, the more it seemed this "new" technology was more of an evolving combination of several older technologies that didn't know they were about to collide with some other technology's research path.

The convergent nature of VR technology is one reason why it has the potential to develop very quickly from scientific oddity to way of life. New methods for making stone tools took tens of thousands of years to diffuse to the majority of human populations. It took uncounted centuries to crossbreed corn into its modern form. Steam engines, however, changed the way people lived all over the world in less than a century, electronic communications shortened the period for profound global change to around a decade, and computer revolutions seem to be coming along every four or five years now. If this trend can be extrapolated to the VR revolution of the near future, there is a significant chance that the deep cultural changes suggested here could happen faster than anyone has predicted.

Two of the most powerful driving forces behind the accelerated pace of technological change are the phenomena of *enabling technologies* and scientific-technological *convergence.* Think of these ideas as specific ways of looking at the underpinnings of technological change of any significant kind, conceptual lenses that can reveal the reasons for the surprises and unexpected breakthroughs that have dominated the story of twentieth-century technology. Both phenomena are at work in the rapid emergence of VR research.

An *enabling technology* is one that makes another technology possible. Some powerful technologies spring into existence when they do because some kind of price or performance threshold has been crossed by one or several enabling technologies. Some technologies are enabling elements of many other technologies. The vacuum tube, for example, was the enabling technology for both radio and television, just as the steam engine was the enabling technology for both locomotives and electric dynamos. Dynamos and incandescent bulbs were enabling technologies for the electrification of the world. Virtual reality based on computers and head-mounted displays has been dreamed of for decades, but had to wait for enabling technologies of electronic miniaturization, computer simulation, and computer graphics to mature in the late 1980s.

Convergence is related to enabling technologies, but adds a slightly different flavor of the unpredictable: history shows that apparently unrelated scientific and technological paths may converge unexpectedly to create an entirely new field. Convergence requires an intersection of deeply similar ideas as well as an element of maturation or evolution in power or price of related component technologies. Video technology and computer hardware, both driven by miniaturization of electronic components, crossed a price-performance threshold at the same time computer scientists were dreaming of building personal computers. Although it seems obvious now, only a few visionaries foresaw the possibility of connecting television screens and computers and allowing nonprogrammers to take advantage of computing power to augment their thinking and communicating. These components converged to create something else because they were both necessary components of a new technology people wanted to build—a computer interface that used human perception to facilitate human-machine communication.

Convergence makes for strange bedfellows. The US Army funded the first computer, the Defense Department's Advance Research Projects Agency created the enabling technologies for personal computers, and the American public appetite for manufactured experience drove the development of television. In personal computers, the military and consumer entertainment development paths converged to make a new tool possible. In the future, similar strange alliances between weapon designers, communications vendors, and toy manufacturers are possible drivers of the VR industry.

Sometimes, people working at the esoteric fringes of their fields may not even be aware that their work is converging on the research others are doing in fields that are not obviously related. The digital computer emerged from the unexpected convergence of several factors, including abstruse questions regarding the logical foundations of mathematics, the practical need for computation techniques for managing telephone switching circuits and calculating artillery ballistics equations, and the capability of electronic vacuum tubes to act as switching elements. The electronic digital computer itself is a key enabling technology for cyberspace, as we will see, but is not the only enabling technology: the convergence of computers with optically based viewing devices was important to the birth of VR. As often happens, people whose curiosity drives them to tinker and theorize in one century on one side of the world may find their ideas colliding with somebody else's intellectual expedition, perhaps centuries later, in some other part of the world.

The principles of stereoscopy, first explored more than a century before electronic digital computers were invented, are the foundation for the head-mounted displays that transport today's cybernauts into computer-created simulations. The technology of stereoscopy involves using optical apparatus and pairs of images to create three-dimensional illusions. All you need to know right now about the physiological foundations of stereoscopic 3D can be apprehended directly by means of a simple, brief experiment: Stand up, open your eyes, and look around. Move your head, take a few steps. Watch a moving object.

If you are like most people, you use two eyes, approximately 6.5 centimeters apart, approximately five and a half feet off the floor, to perceive a fused, full-color, three-dimensional image of the world. You can adjust your focus, move your head, move your eyes, walk around. Some of the things you see have shadows. You have a sense of which objects are near and which ones are far away; you know instinctively whether something in your visual field is moving toward you or away from you. If you walk around an object you can see what is behind it. Each of these aspects of human visual perception of space is the basis for one or more of the optical techniques designed to simulate three-dimensional worlds.

Stereoscopic illusions are tricks that people have discovered, using lenses and image pairs, to simulate one or more of the perceptual cues humans use to create three-dimensional scenes. When you look at a three-dimensional object, each of your eyes receives reflected light from that object at precisely displaced angles. The angles at which the photons from the light source bounce back off the objects in a scene and enter your eyes trigger your optical system to feed your brain a stream of signals about the three-dimensionality of the scene your senses view. These critical angles are determined by the distance between your eyes and by the distance to the object. The distance between your eyes never changes, but the distances to objects change, your point of view changes, light sources change, generating cascades of multiple parallel recomputations. All these perceptual streams are what some contemporary psychologists call "affordances"—handles that enable us to create the world as we move through it. It will be a long time, if ever, before silicon jockeys get their hands on a nonhuman CPU as powerfully optimized to world-creation as the one in our cranium.

Close one eye and then open it and close the other and you will see how the scene jumps back and forth. As another VR pioneer, Scott Fisher, suggested in his master's thesis: "An informative exercise is to station oneself in front of a large piece of glass and, holding steady, outline objects seen through the glass by one eye with the other closed. Without moving, reverse the process and trace the view seen by the other eye." We are all autostereoscopes. Our eyes are stereo input devices; our eyeballs and necks are sophisticated, multiple degree-of-freedom gimbals for moving our stereo sensors. We are elements in an informational ecology that creates the useful illusion we call "reality." The photons, the light-reflecting properties of the objects we see, the distance between our eyes, the nature of our visual perception system, our parallel data processors and other brain functions still unknown to science, act in concert to weave the apparently seamless cloth of experience.

Artificial means of presenting stereographic three-dimensional information often use special viewing apparatus or glasses that show the stereo image pairs to the proper eyes. Today's "electronic shutter" lenses work by interleaving "left-eye views" with "right-eye views" every one thirtieth of a second; the shutters selectively block and admit views of the screen in sync with the interleaving, always showing the proper view to each eye. There are ways of creating stereoscopic 3D images without special glasses, however. The use of lenticular lenses, for instance, had been known since the experiments of Herman Ives in 1930. Suppose you could take two images necessary to create a 3D stereo pair and cut them into very thin vertical slices, then interleave them in precise order ("multiplexed" is the technical term for this kind of order-preserving interleaving of signals), and put vertical half-cyclinder shaped lenses in front of them so that when you look directly at the image you would see the left-eye image with your left eye and the right-eye image with your right eye. This carefully crafted illusion of depth is based on only one of the cues humans use to create three-dimensional worlds—*binocular parallax*. Another strong cue comes into play only when we move our head, because of the characteristic differentials in the way the world scene moves (nearby objects are perceived to move more than distant objects, for example)—*motion parallax*. Lenticular screens show viewers the proper stereo pairs when moving their heads, to a certain degree. Adding a head-motion sensor that adjusts the image can improve the effect.

The art and science of stereoscopy is an ancient one. In the second century A.D., the Greek physician Galen described the first theory of left-right eye perspective, the beginning of thinking beyond a planar, two-dimensional representation of what we see. For thousands of years after Galen, only a few isolated specialists looked into the binocular aspects of depth perception. In the Renaissance, European artists took advantage of ways to trick certain aspects of human depth perception to create the illusions of perspective. Later chapters will look more closely at the kind of "pictorial" cues artists use to create the illusion of depth.

The most interesting aspects of anamorphic art and perspective drawings are those that tell us something about the relation between representation and perception. A researcher at Harvard, commenting on theories put forth to explain the realism perceived in painted images, put it this way: "Pictures inform by packaging information in light in essentially the same form that real objects and scenes package it, and the perceiver unwraps the package in essentially the same way."

Stereoscopy was the first technology to package visual information designed to match the binocular aspect of visual "unwrapping." It took a particular invention, a remarkably simple device, once you know how to build one, and the attention of an empress to get modern stereoscopy started. In 1833, Wheatstone's stereoscope was the first major breakthrough, leading to a chain of inventions directly connected to today's head-mounted displays.

Many different kinds of stereoscopic displays have been invented over the past century. They all have in common the use of *stereo pairs*, two slightly different images of the same scene, but the technologies differ in the ways in which the pairs are presented to human eyes. Time-multiplexed systems use various devices to show the different images at short intervals: first the user sees one image; then, a fraction of a second later, the user sees the other image in the pair. In the nineteenth century and into the twentieth century, mechanical methods similar to the machinery of film projection were used to achieve time-multiplexed stereoscopic effects. More recently, electronic methods derived from television technology have been applied to this kind of stereoscopy; the "electronic shutter" glasses I used at UNC and in Japan were time-multiplexed.

Time parallel systems, showing both images at the same time, are either anaglyphic, in which different-colored lenses cause each eye to see each image of the pair in slightly different ways, split screen, in which lenses or mirrors deflect separate images to each eye, or dual CRTs, which use video displays to present different images simultaneously.

Think of your field of vision as consisting of the intersection of two broad cones of perception, one from each of your eyes, each of which sees a slightly different perspective. If you make two drawings (Wheatstone worked before photography was invented, so he can also be thought of as the grandfather of 3D comics), each of them depicting the same scene from slightly different perspectives corresponding to human interocular distance, then present each image to one eye, human vision will fuse the two perspectives into the continuous illusion of a three-dimensional scene. Wheatstone's device exploited this effect ingeniously. The mirror stereoscope consisted of two mirrors at a 45-degree angle that converge two pictures into the viewer's right and left eyes. Imagine yourself looking through the eyepiece of a binocular. The left eyepiece looks at a mirror that is angled out from your nose, thus reflecting a card placed in a holder to your left. The right eye sees the other card. You see one scene, with a perceptible sense of depth. The mirror stereoscope was the predecessor to the very popular postcard stereoscopes that were used through the end of the nineteenth and into the beginning of the twentieth century.

The modern common form of the stereoscope was invented by Sir David Brewster in 1844, who added half-lenses to help each of the two eyes see the precisely different figures on the stereo pairs; Brewster's invention, in which both images are mounted on a single card, survives in different variations, the modern embodiment in 1940 being the ViewMaster. The first stereo photographs were taken with a single camera that was moved laterally two and a half inches for a second exposure. The invention of the stereorealist camera in 1949 enabled interested amateurs to shoot 35 mm film stereophotographs. To this day, Kodak will put photographic pairs on a stereo mounting. There has been a continuity of interest in three-dimensional photography and projection among various optical or photographic subcultures; the SPIE, the Society for Photo-optical Instrumentation Engineering, is an overlapping but mostly very different crowd from SIGGRAPH, the Association for Computing Machinery's Special Interest Group for Graphics, and they are somewhat different from the SID, the Society for Information Display. The enthusiasts of stereoscopic imaging techniques are one of the key subcultures who now find themselves involved in the same interdisciplinary cyberspace as computer graphics, simulation, and human-computer interface research.

In 1851, Queen Victoria expressed enthusiasm over the stereo cards she saw at the exhibition at London's Crystal Palace. Victoria's attention, amplified by the newsprint-based mass media of her day, was powerful enough to make and break products or politicians; her royal enthusiasm pushed stereo cards into mainstream acceptance as a form of home entertainment. In 1891 a man named Duhauron developed the first anaglyphic photograph-based system. The 3D horror movie glasses of the 1950s with green gel over one eye and red gel over the other eye are mass-produced tools for exploiting the anaglyph effect, originally based on work by a man named Dalinrida, who in 1858 was the first to combine the colored filters over the left and right lenses to superimpose images on a screen. The left-eye image, for example, is projected on the screen through a red gel, and because the right eyepiece of the viewer is also covered with red gel, the left image is invisible to the right eye. The right-eye image is projected on the same screen at the same time through a green gel; the left eyepiece admits only the images projected in red.

Then film technology came along and the world's attention was first drawn into the virtual worlds painted on the screen with thousands of photographs, twenty-four per second.

Like today's virtual realities, the films took people by surprise at first: People actually ran screaming from theaters when D. W. Griffith introduced the extreme close-up; we understand what is represented by disembodied "talking heads" on a screen today because it is part of our perceived reality template, our collection of learned perceptual rules that help us make sense of our stream of sensations. But in the first theaters, people had to learn how to perceive film virtuality. Presumably, the same process will take place when mass VR matures.

After silent movies, then talkies, with the rise of the film industry, expectations were high that Hollywood would soon develop a three-dimensional projection system. In 1937, MGM produced a comedy called *The Third Dimension Murder*, shot with the studio's specially designed camera, which was an early two-color process. The first big improvement in the cohesion of stereo images was the invention of polarization processes, which made it possible to view a full-color feature film. The polarity of light is a characteristic that is generally invisible to the human eye, but transparent filters can be created that pass only light of one polarity. Instead of projecting the left and right images through red and green filters, it became possible to make one of the stereo image pairs invisible to each eye by projecting image pairs of different polarities and using appropriate polarized lenses over each eye. The first sheet polarizers were available in the nineteenth century but had very poor quality. In 1928, Edwin Land of Polaroid fame developed the first quality sheet-polarized materials, enabling him to make the Polaroid glasses, using images 120 to 140 degrees out of phase (a measure of the difference in polarities) to create stereo images on the screen. In 1935, Land showed an experimental stereoscopic film. Based on that demonstration, a stereo feature was produced for Chrysler Corporation's exhibit at the New York World's Fair in 1939.

The big boom in 3D films came in 1952 with *Bwana Devil,* the first American feature film in color 3D using what was known as the Naturalvision process. Every 3D film enthusiast Iv'e met seems to wince when *Bwana Devil* is mentioned; it was one of those incredibly bad films that reflect strangely on the culture of their times when viewed thirty or forty years later. Gross receipts for *Bwana Devil,* however, topped $100,000, a lot of money in those days, so studios scrambled to produce a whole wave of 3D films through 1954. *Dial M for Murder, House of Wax, Creature from the Black Lagoon, It Came from Outer Space* were among the best, along with hundreds of truly terrible ones. After 1954, Cinemascope was introduced and film evolution diverged from three-dimensionality to wide fields of view.

The experience theater has not been created, but the old idea seems to be finding new life in the computer world. Although the long and honorable tradition of stereoscopy did not by itself result in a realistic total-immersion synthetic experience, it created an infrastructure for presenting 3D information to a human viewer. When it became possible to use computers to provide that three-dimensional information, and people began to actually enter computer-simulated worlds, the path of Heilig and the 3D freaks of ages past led computer technology deep inside itself. VR as it exists today was born out of computerland, not tinsel town. And the impetus for 3D computer displays was a deliberate, long-term project to create computers that could function as mind amplifiers. The object of mind-amplifier designers was to bring about ever tighter connections between human thought processes and computer information-processing capabilities.

Ultimately, VR is neither strictly a child of computer science nor a form of entertainment, but something that necessarily partakes of both technical legacies, as later chapters will reveal. However, there is little doubt that VR would not have been possible in the 1990s if a small group of visionaries had not conspired to build machines to think with in the 1960s.

Chapter Three

MACHINES TO THINK WITH

The summation of human experience is being expanded at a prodigious rate, and the means we use for threading through the consequent maze to the momentarily important item is the same as was used in the days of square-rigged ships.

But there are signs of a change as new and powerful instrumentalities come into use. Photocells capable of seeing things in a physical sense, advanced photography which can record what is seen or even what is not, thermionic tubes capable of controlling potent forces under the guidance of less power than a mosquito uses to vibrate his wings, cathode ray tubes rendering visible an occurrence so brief that by comparison a microsecond is a long time, relay combinations which will carry out involved sequences of movements more reliably than any human operator and thousand of times as fast—there are plenty of mechanical aids with which to effect a transformation in scientific records. . . . For mature thought there is no mechanical substitute. But creative thought and essentially repetitive thought are very different things. For the latter there are, and may be, powerful mechanical aids.

VANNEVAR BUSH,
"As We May Think," 1945

By "augmenting man's intellect" we mean increasing the capability of a man to approach a complex problem situation, gain comprehension to suit his particular needs, and to derive solutions to problems. Increased capability in this respect is taken to mean . . . that comprehension can be gained more quickly; that better comprehension can be gained; that a useful degree of comprehension can be gained where previously the situation was too complex; that solutions can be produced more quickly; that better solutions can be produced; that solutions can be found where previously the human could find none. And by "complex situations" we include the professional problems of diplomats, executives, social scientists, life scientists, physical scientists, attorneys, designers—whether the problem situation exists for twenty minutes or twenty years. We do not speak of isolated clever tricks that help in particular situations. We refer to a way of life in an integrated domain where hunches, cut-and-try, intangibles, and the human "feel for a situation" usefully coexist with powerful concepts, streamlined terminology and notation, sophisticated methods, and high-powered electronic aids.

DOUGLAS ENGELBART,
"A Conceptual Framework for
Augmenting Man's Intellect," 1963

Although I didn't know it at the time, I took my first steps into cyberspace in 1982. The idea of using a computer as a mind amplifier had long fascinated me, but I didn't get my hands on one until I was hired to help a computer scientist at Xerox Corporation's Palo Alto Research Center write an article for a scientific journal. The device was called an "Alto," and looked like a skinny black and white television screen atop a box half the size of my kitchen stove. A small box, connected to the screen via a wire, rested on the desk next to the keyboard: the "mouse." My computer scientist partner assured me, more or less a computer novice at the time, that it would not only help me make typographical corrections, but could help me think, decide, imagine, in new ways.

At first, it seemed strange. I typed on a keyboard, just as I would type on a regular word processor. But instead of memorizing different alphabetic codes to mark, copy, move, and delete blocks of text, I moved the mouse on the desktop while I watched the cursor move and characters and words and paragraphs highlight on the screen. The "desktop metaphor" that turned files and filing cabinets, electronic mailboxes and digital trashcans, into graphic symbols on the computer screen was a primitive "virtual world" that let me put my thinking tools right here in front of me, where I could see them. By clicking the mouse on the right "icons," I could send my document across the laboratory, to my partner's electronic mailbox, and send an automatic copy to the editor of a scientific journal halfway across the continent. The screen acted as a visual cache for my short-term memory, the way placing documents and file folders on a physical desktop serves as a visual reminder. It was hard to go back to my old-fashioned IBM-PC—at that time, the newest thing in personal computer technology—after my Xerox assignment was finished.

It turned out that the Alto that had been created specifically for helping Xerox programmers perform intellectual work was a historic machine. When I started to look into the origins of this compelling new way of using a computer, I discovered the old quest for "intellectual augmentation" that had led to the creation of mind-amplifying computer technology. I can now see that one of the fundamental principles of the augmentation vision—that it is necessary to use visual communication and gestural input devices in order to closely couple human minds and computer capabilities—remains a fundamental principle of VR, as well. It has to do with the way we create tools for dealing with complexity.

Human culture—the body of languages, methods, and knowledge we preserve and pass on in order to save us from reinventing the wheel every generation—has grown too complex to manage without machines. For better or worse, the maintenance of human civilization for the foreseeable future will require both minds and computers. Right now, humans and our information-processing technologies don't work together especially well. To those who are working on solutions for the problem, the focus of attention has shifted from the way computers work to the way computers are designed to be used by people—the *human-computer interface*. The human-computer interface is where VR development intersects with the evolution of computers.

Human-computer interfaces are tools for helping minds and machines work together more effectively and thus are not primarily matters of bits or bytes, hardware or software, in much the same way that the focus of architecture is not primarily a matter of bricks or girders. A virtual world is a computer that you operate with natural gestures, not by composing computer programs, but by walking around, looking around, and using your hands to manipulate objects—for some tasks, VR is the ultimate computer interface.

Every tool has a human interface: doorknobs are the human interface to a door; steering wheels and speedometers are the human interface to an automobile. An effective prehistoric interface was the hand-shaped grip on a stone axe. As our tools have grown more complex and powerful over the years, the interfaces have evolved as well. Sometimes, the evolution of a tool surges ahead of the evolution of its human interface. Many of the more complex tools of recent history were developed without much thought to the interface; the digital computer is the foremost example of a tool that is useful to the initiated, but less than useful to most people.

The inner workings of most computers are accessible only to experts versed in the communication codes known as programming languages. A radical question occurred to a few people as long as forty years ago: Instead of training people to understand the secret languages of computing machines, why not design computing machines that can communicate with people without the need for secret languages? Personal computers and personal simulators are both potential answers to this question.

If a small number of people had not insisted on pursuing that question against formidable odds, the phrase "personal computer" would sound as outlandish today as it did when it was first proposed. It took at least two technological prophets—J.C.R. Licklider and Douglas Engelbart—who were willing to stake their reputations and careers, and the help of a legion of inspired infonauts who were willing to work outside the bounds of established wisdom, to create personal computers as we know them today. Without personal computers, there would never have been personal simulators. Stereoscopic viewers and arcade devices might have emerged and converged, but it is doubtful that the computation-intensive cyberspace technology we see today would have been built without the momentum of the personal computer revolution.

Devoting a single computer to the use of one person was a revolutionary idea that converged with another revolutionary idea—the notion that the human interfaces for computers should accommodate human needs and abilities, rather than shaping human behavior to fit the demands of computer technology. The way in which a person enters commands or data into a computer is the *input* part of the interface, and the way the computer shows the user the results of computations is the *output* part of the interface; the first input and output devices were designed with the limitations of computers rather than the capabilities of humans in mind. A computer that accepts input only in the form of punched cards and spits out answers only in the form of numbers printed on a roll of paper is an example of a user interface designed to meet the needs of the computing machines of the 1950s. A computer that accepts input by typing on a keyboard is better; a computer that accepts input by pointing to a picture on a television screen is better still.

Computers used to be expensive and massive, so it was widely assumed that methods of using the computers should make the best use of expensive computing resources. When transistors and integrated circuits came along in the 1960s, it became obvious that computers were growing smaller, less expensive, and more powerful. The miniaturization revolution that led from vacuum tubes to transistors to integrated circuits has been one of the most powerful technological driving forces in history; by making the basic switching elements of computers smaller and smaller, it has become possible to reduce the cost of computation while increasing the power of computers by orders of magnitude. This factor is responsible for the fact that the microprocessors in a five-year-old's toy today are millions of times more powerful yet far less expensive than the first electronic digital computer, ENIAC, which crunched its first numbers in 1946.

While the pace of the miniaturization revolution drives the pace of advances in computer technology, the rate at which larger numbers of tiny switches can be inscribed on wafers of silicon also provides a handle on forecasting future computer technologies. The most fundamental component of any digital computer is a switching element— a mechanism that can turn on or off. Computers compute by chunking on and off signals (bits) into more complex symbols. The number of signals a computer can chunk in a given period of time determines the power of that computer to solve problems. Advances in electronic miniaturization inevitably lead to advances in computer hardware a few years later, so it is possible to forecast the gains in computer price/ performance that are likely five or ten years after the introduction of a new generation of electronic devices.

The orthodox computer theorists of the 1950s saw a future in which they would use these advances in electronics to build bigger and more powerful computers, which would still require arcane languages and a priesthood of skilled translators to take advantage of their computing power. Others, outside the mainstream of computer theory, saw something else in the miniaturization trend that was driving the development of computer technology. Sooner or later, probably in the 1970s, it would become economical to devote an entire computer to the use of an individual.

The idea of personal computers doesn't seem so strange today, but in the 1960s and 1970s personal computers did not evolve naturally from the main trends in computer science or the computer industry; rather, they were deliberately created by a community of mavericks who struggled against the tide of orthodox opinion to create the kind of technology they wanted to make. Their goal was not to make typing pools obsolete, to turn accountants into financial forecasters, or to generate new multibillion-dollar industries, although those were some of the effects of their effort. The people who built the first personal computers did it because they wanted mind amplifiers for their own use.

Indeed, the first person who seriously believed that it might one day be possible to use computers for something more widely useful than processing payrolls and crunching scientific calculations couldn't get anybody to listen to him for ten years. One day, almost exactly in the middle of the twentieth-century, a young electrical engineer had a brainstorm while driving to work. Douglas C. Engelbart, after concentrated musing on the subject of what he should do with his life and talents, realized that humankind was creating complex problems at an accelerated rate and needed new tools for dealing with the kind of world we were moving toward. It occurred to him that if we could use the power of computers to perform the mechanical part of thinking and sharing ideas, people would be able to increase their ability to do the hardest part of thinking and solving problems together. It took decades for the rest of the world to understand the importance of Engelbart's revelation, and much of what we know as personal computing today grew directly from Engelbart's Augmentation Research Center (ARC) at Stanford Research Institute in the 1960s. Several of the strongest taproots of VR go back to ARC.

It doesn't require any knowledge of computer science to grasp the most important notion at the core of Engelbart's vision: If we can find ways to use audiovisual media to match human perceptual and cognitive capabilities with computers' representational and computational capabilities, humans will be able to increase the power of our most important innate tools for dealing with the world—our ability to perceive, think, analyze, reason, communicate.

Douglas Engelbart, now in his sixties, still derives inspiration from the vision that came to him in 1950, even though computer technology has not evolved exactly the way he envisioned. I talked to him on the telephone this morning and he was excited about the way the latest Engelbartian incarnation, the Bootstrap Institute, was getting off the ground. I had first met him seven years ago, and that was already very late in his story. One spring day in 1983, I talked with Engelbart at his offices in Cupertino about where personal computers originated and where they might be going. I had tracked him down because I was looking for the origins of the idea that computers could be used as mind amplifiers. All the bibliographies led to Engelbart's decades-old papers. Then somebody told me he was still working on his original vision, so I called him and he invited me over to his office. At that time, Engelbart and the system he had been building for more than twenty years were part of Tymshare, a computer communications service company that was owned by McDonnell-Douglas. Ironically, the man who had made possible the computer interface used by Apple's Macintosh computer was working in a cubicle in a building that was itself surrounded by the buildings of Apple's Silicon Valley "campus."

With his white hair, gentle manner, and far-focused blue eyes, Douglas Engelbart in 1983 was the image of the technological patriarch. "Moses parting the Red Sea and pointing the way to the promised land," is the phrase another computer pioneer, Alan Kay, uses to describe Douglas Engelbart's charisma. It isn't the kind of power that radiates from balconies. You have to move close to Douglas to hear what he is saying, but the sheer force of his feelings about the ideas he is describing still enlivens his tale.

It began for Engelbart in the aftermath of World War II. As a naval radar technician awaiting demobilization, he came across Vannevar Bush's 1945 *Atlantic* article, "As We May Think," in a Red Cross library in the Philippines. Back home, he followed up his radar experience with a degree in electrical engineering. Years later, his hours of trying to discern the very real threats represented by the virtual blips on radar screens would pay off when he tried to think of new ways for computers to display information. He married and found a job with a small electronics firm in Mountain View, California. Over the next three decades, Mountain View would change from a fruit orchard to the heart of Silicon Valley, and Ames, Engelbart's first employer, would become a major research center for NASA—the same one in which the human factors laboratory helped trigger the current VR research revolution. Engelbart's vision, however, came at a time when there were no more than a dozen computers in the United States.

In his office at Tymshare, amid shelves of binders documenting thirty years of research, Engelbart recalled the day in December, 1950, when he confronted his future. He looked out at the horizon, as if he could see a distinct picture in the air somewhere behind me and above my head. "At the age of thirty-five," he said, "I realized that I had achieved everything I had set out to achieve in life. I had survived the war and gained a wife, a home, an education, and a challenging profession. I realized that the question of what to do next was going to be an important decision." Engelbart has never been the kind of person who could be comfortable without a goal. It took him about a month to come up with the plan that was to occupy the rest of his career and shape the future of computer technology.

Back in 1950, his daily commute through the sunny orchards of the Santa Clara Valley gave him time to think. He had what he wanted for himself, so what he wanted was to have a positive effect on the world. As he pondered the possible futures that would give him the most leverage in making the world a better place, Engelbart discovered that he was coming up against the same problem in every scenario he imagined: our civilization's problems are getting more complicated and urgent, but our problem-solving tools have not grown more powerful. What, then, could be done to help people tackle complicated problems? And how could an electrical engineer find a way to contribute to solutions? That's when the mental picture of the system he wanted to build came to him.

Although the details took decades to work out, the main elements of his vision occurred to him all at once as a mental picture of a group of people working together in a new way: "When I first heard about computers, I understood from my radar experience that if these machines can show you information on printouts, they could show that information on a screen. When I saw the connection between a televisionlike screen, an information processor, and a medium for representing symbols to a person, it all tumbled together. I went home and sketched a system in which computers would draw symbols on the screen and I could steer through different information spaces with knobs and levers and look at words and data and graphics in different ways. I imagined ways you could expand it to a theaterlike environment where you could sit with colleagues and exchange information on many levels simultaneously. God! Think of how that would let you cut loose in solving problems!"

In the 1990s, VR technology is taking people beyond and through the display screen into virtual worlds; VR researchers today are beginning to climb through the window Engelbart and his colleagues created. In 1950, however, Douglas Engelbart seemed to be the only person in the world who thought computers could or should display information on screens. To understand why everybody thought his idea was so strange in 1950, it helps to recall that there were only a handful of computers in existence at the time, and television was an infant technology.

The idea of using these technologies together as problem-solving devices—especially if the people who used them weren't computer programmers—remained a gleam in Engelbart's eye for more than another decade. He quit his job and went to the University of California, where they had one of the two computers in California. Nobody at the university was interested in experimenting with their valuable computer time on something as fanciful as a tool to help people think. So Engelbart left academia and started his own company, then folded it when he realized he didn't want to be an entrepreneur. Having failed to interest psychologists, computer scientists, librarians, or electronic companies in investing in the research he wanted to pursue, in 1957 he took a job as a more orthodox computer researcher with the Stanford Research Institute (SRI) in Menlo Park, California.

Around the time he joined SRI, Engelbart decided to find a way for people to understand what he was talking about from the perspective of traditional computer technology. So he spent a couple of years performing conventional computer research while using his spare time to write the conceptual framework he needed.

Engelbart didn't know it at the time, but he was on his way to a convergence that would sweep him from obscurity into the mainstream of computer design—and open the portal to virtual reality. The conceptual framework he wrote in the late 1950s and early 1960s still reads like a blueprint for twenty-first-century technology. And it gave him something to do while others slowly started moving toward his way of thinking. Like most participants in such a dramatic technological convergence, he wasn't aware of the people and forces that were already launched in the direction of his dream. It took another visionary on the other side of the continent, the launching of Sputnik in October, 1957, and a few more years of growth in key enabling technologies to get Engelbart the laboratory he had been trying to build for ten years.

MACHINES TO THINK WITH

The fig tree is pollinated only by the insect Blastophaga grossorum. *The larva of the insect lives in the ovary of the fig tree, and there it gets its food. The tree and the insect are thus heavily interdependent: the tree cannot reproduce without the insect; the insect cannot eat without the tree; together, they constitute not only a viable but a productive and thriving partnership. This cooperative "living together in intimate association, or even close union, of two dissimilar organisms" is called symbiosis.*

"Man-computer symbiosis" is a subclass of man-machine systems. There are many man-machine systems. At present, however, there are no man-computer symbioses. . . . The hope is that, in not too many years, human brains and computing machines will be coupled together very tightly, and that the resulting partnership will think as no human being has ever thought and process data in a way not approached by the information-handling machines we know today.

J. C. R. Licklider,
"Man-Computer Symbiosis," 1960

When everything being dealt with in a computer system is visible, the display screen relieves the load on the short-term memory by acting as a sort of "visual cache." Thinking becomes easier and more productive. A well-designed computer system can actually improve the quality of your thinking. . . . A subtle thing happens when everything is visible: the display becomes reality. The user model becomes identical with what is on the screen. Objects can be understood purely in terms of their visible characteristics. . . . One way to get consistency into a system is to adhere to paradigms for operations. By applying a successful way of working in one area to other areas, a system acquires a unity that is both apparent and real. . . . These paradigms change the very way you think. They lead to new habits and models of behavior that are more powerful and productive. They can lead to a human-machine synergism.

Smith et al.,
"The Star User Interface," 1982

Sputnik demonstrated suddenly and dramatically that the Soviets had gained the capability of propelling bomb-sized objects anywhere in the world. Something that experts had known was made obvious to the entire population: the balance of military power in the world had shifted from those countries with the largest armed forces to those with the most advanced weapons technology. And the United States, complacent in its postwar superiority, failed to notice that in at least one critical area, American know-how was no longer the most advanced in the world. The creation of the Advanced Research Projects

Agency (ARPA) was one of the most effective official responses to Sputnik. ARPA was given the mandate to short-circuit the traditional research funding process and to fund directly the far-out ideas that might help the United States regain technological superiority. Fortunately for Engelbart and for the fate of personal computing, ARPA happened to hire another man with a vision, J.C.R. Licklider.

Licklider also had a life-transforming realization, similar to Engelbart's vision, that he came to call "a kind of conversion experience." Shortly before the launching of Sputnik, at about the same time that Douglas Engelbart, then unknown to him, started to work at Stanford Research Institute, Licklider was a researcher and professor at the Massachusetts Institute of Technology. Licklider, a psychoacoustician who was using mathematical models to understand the bases of human hearing, started getting tangled in his own data. The models had grown so complex that it took far more time to plot the data and create the models than to think about what they meant. One day, surrounded by graphs and files and stacks of research data, he decided to look into the matter of how scientists spent their time. Failing to find any existing time-and-motion studies on information-shuffling researchers like himself, Licklider decided to keep track of his own activities as he went through a normal workday.

His observations revealed, to his astonishment, that about 85 percent of his "thinking" time was actually spent "getting into a position to think, to make a decision, to learn something I needed to know. Much more time went into finding or obtaining information than into digesting it." At that time, he didn't know a great deal about computer technology, but he began to think about something very similar to the notion that had occurred to Engelbart. Perhaps some of the mechanical burden of information shuffling that seemed to have become such a large part of the scientist's job could be shifted to specially redesigned computers. Certainly, no computers at that time were usable for such a task. But the capabilities of computers, if not their uses, were changing more rapidly than anyone had expected.

Computers at that time were used to perform the kind of enormous calculations that were required for nuclear physics—the kind of computation task that is still known as "number crunching." The other primary use for computers, data processing, wasn't what Licklider or Engelbart needed, either. If you were the Census Office, overflowing with information on several hundred million people, and for some reason you wanted to find out how many divorced people over sixty lived on farms, you could use a UNIVAC to perform the sorting and calculating needed to tell you what you wanted to know. That was data processing.

Data processing involved constraints on what could be done with computers and constraints on how one went about doing these things. A process known as "batch processing" was the proper way to deal with payrolls, scientific calculations, or census data. If you had a problem to solve, the first step was to encode your problem and the data the program was meant to manipulate, usually in one of the two major computer languages, FORTRAN or COBOL. The encoded program and data were converted into boxes full of punched cards—the input devices that had become universally known as "IBM cards," the kind you weren't supposed to spindle, fold, or mutilate. Then you delivered the cards to a system administrator at the campus "computer center" or corporate "data-processing center." This specialist—the high priest who mediated between users and the mainframe computer housed in its air-conditioned sanctum sanctorum—was the only one allowed to submit a program to the machine, and the person from whom you would retrieve your printout hours or days later. If your program had an error, which could be as trivial as a misplaced punctuation mark, you would have to go through the entire process again.

If you wanted to turn a list of numbers generated by aerodynamic equations into a graphic model of airflow patterns over a wing, you wouldn't want data processing or batch processing. You would want *modeling*—an exotic new use for computers that the aircraft designers were then pioneering. Today, the field of scientific visualization is the highly evolved descendant of those first aerodynamic modeling systems. Licklider had started looking for a kind of mechanical file clerk that could help with his scientific model building. Not long after he started looking at available technology, he began to wonder if computers could help formulate models as well as calculate them, help him think about the meaning of data as well as keep track of it. When he attained tenure, later that same year, Licklider decided to join a consulting firm near Cambridge named Bolt, Beranek & Newman (BB & N). They offered him an opportunity to pursue his psychoacoustic research and a chance to learn about digital computers.

"BB & N had the first machine that Digital Equipment Company made, the PDP-1," Licklider recalled, when I talked to him in 1983. The quarter-million-dollar machine was the first of a continuing line of what came to be called, in the style of the 1960s, "minicomputers." Instead of costing millions of dollars and occupying a large room, these new computers only cost hundreds of thousands of dollars, and took up about the same amount of space as a refrigerator. Instead of programming via boxes of punchcards over a period of days, it became possible to feed the programs and data to the machine via a high-speed paper tape. Programmers could interact directly with the machine for the first time, and for the hardcore programmers of the era, that was a taste of forbidden fruit. The PDP-1 was primitive in comparison with today's computers, but it was a breakthrough at the time. Here was the candidate for the model builder that Licklider had envisioned, and the kind of real-time instrument that hotshot programmers were itching to snatch from the jealous grasp of the high priesthood and put on their own desks.

As he had suspected, Licklider found it was indeed possible to use computers to help build models from experimental data and help make sense of any complicated collection of information. Although he was convinced by his "religious conversion" to interactive computing—a phrase that turned up over and over when I talked to others who participated in the events that followed—Licklider did not yet know enough about the economics of computer technology to realize that personal computers might be possible someday. But he grasped the core issue: humans and computers would work together in new ways, if the right kind of human interface could be devised.

Over at Building 26 at MIT, the first artificial intelligence researchers were at work on computer technology that might someday replace humans as the only thinking beings on the planet. Nobody knew how long that might take. In the interim, which might last decades or might take centuries, Licklider saw the possibility of a cooperative arrangement between human wetware and computer hardware and software. He proposed a biological metaphor to describe the development he envisioned, citing the symbiotic partnerships of organisms that worked together to provide powerful mutual advantages. In 1960, Licklider wrote "Man-Computer Symbiosis," in which he predicted that "in not too many years, human brains and computing machines will be coupled together very tightly, and that resulting partnership will think as no human being has ever thought and process data in a way not approached by the information-handling machines we know today."

The way in which computer designers of the 1960s and 1970s designed the devices by which brains and computers could be "coupled very tightly" created the hardware foundation for later developments in VR technology. Indeed, VR might be described as an environment in which the brain is coupled so tightly with the computer that the awareness of the computer user seems to be moving around inside the computer-created world the way people move around the natural environment.

Fortunately for his dream of a future very similar to the one Engelbart envisioned, Licklider was connected to a force that could empower that dream—the military-industrial complex. If necessity is the mother of invention, it must be added that the Defense Department is the father of technology; from the Army's first electronic digital computer in the 1940s to the Air Force research on head-mounted displays in the 1980s, the US military has always been the prime contractor for the most significant innovations in computer technology. In the late 1950s and early 1960s, some of the best minds at MIT were called upon to help build computerized ground defense systems to protect America from nuclear attack—the "Semi-Automatic Ground Environment" (SAGE). A top-secret facility associated with MIT, Lincoln Laboratory in Lexington, Massachusetts, employed Licklider and others to work on the "human factors" side of the new computerized radar network.

Some of the thorniest problems encountered on this project had to do with devising ways to make large amounts of information available in a form that humans can recognize quickly enough for them to make fast decisions. It just wouldn't do for your computers and operators to take three days to evaluate all the radar data before the Air Defense Command could decide whether or not an air attack was under way.

Some of the answers to these problems were formulated in the "Whirlwind" project at the MIT computing center, where high-speed calculations were combined with computer controls that resembled aircraft controls, and even a primitive graphic display. The aerodynamic processes depicted on the computer screen were computed and displayed at the same speed they took place naturally—in "real time," as computer jargon has called it ever since. Whirlwind was one of the direct ancestors of both the field of simulation and the field of computer graphics and thus was a key predecessor of virtual reality technologies. Whirlwind and SAGE operators were the first computer users who were able to see information on visual display screens; moreover, SAGE operators were able to use devices called "lightpens" to alter the graphic displays by touching the screens. The matter of display screens began to stray away from electronics and into the area of human perception, which was Licklider's cue to join the computer builders. A future historian might look at this historical inflection point, the moment in history when someone who actually knew something about the way human minds operate became an architect of computer technology.

Because of his work on SAGE, Licklider was acquainted with Jack Ruina, director of ARPA, who wanted to do something about computerizing military command and control systems on all levels, not just air defense systems, and wanted to set up a special office within ARPA to develop new information-processing techniques. Licklider's notion of creating a computer capable of directly interacting with human operators via keyboards and display screens convinced Ruina that the wild-eyed minority of computer researchers Licklider seemed to know might just lead to an important breakthrough.

It was exactly the kind of breakthrough that ARPA was designed to nurture, and Licklider's plan was clearly applicable far beyond the military. "I convinced Jack Ruina of the pertinence of interactive computing, not only to military command and control, but to the whole world of day-to-day business," Licklider recalled when he and I discussed this history, twenty years later. In October, 1962, Licklider became the director of the Information Processing Techniques Office (IPTO). That particular event marked the beginning of the beginning of the age of personal computing, and set the stage for virtual reality technology in more than one way.

The computers, the input devices, and the display devices required for cyberspace systems grew directly from the technology funded by IPTO in the 1960s. In particular, Ivan Sutherland, one of the young enthusiasts Licklider found at Lincoln Laboratory, more or less single-handedly created the field of interactive computer graphics. Sutherland also succeeded Licklider as IPTO director. In 1965, Sutherland created the first head-mounted computer display. Licklider, Sutherland, and Engelbart, through a combination of vision and ingenuity, diverted the course of computer technology to the direction of human-centered computer interfaces; each of them issued part of the challenge to future VR designers, and built the first tools for building the enabling technologies of personal simulators a quarter century later.

Licklider wasn't given a laboratory; rather, he worked with an office, a budget, and a mandate to raise the state of the art of information processing. Licklider looked for start-up companies and turned to eager young programmers, some of them dropouts, at MIT, the University of California, Rand Corporation, the University of Utah, and a dozen different research groups around the country. One of Licklider's band of converts to the interactive computing crusade was a young NASA research administrator by the name of Bob Taylor. Taylor had been funding some of the theoretical work of another young fellow with wild ideas—Douglas Engelbart. After years of laboring in obscurity, Engelbart was visited in 1964 by a team of ARPA funders who promised him state-of-the-art computer equipment and a million dollars a year to create the mind-amplifying computers he had been describing on paper for years.

FROM ARC TO PARC: THE BIRTH OF THE PERSONAL COMPUTER

In this stage, the symbols with which the human represents the concepts he is manipulating can be arranged before his eyes, moved, stored, recalled, operated

*upon according to extremely complex rules—all in very rapid response to a
minimum amount of information supplied by the human, by means of special
cooperative technological devices. In the limit of what we might now imagine,
this could be a computer, with which individuals could communicate rapidly and
easily, coupled to a three-dimensional color display within which* extremely
sophisticated images *could be constructed, the computer being able to execute
a wide variety of processes on parts or all of these images in automatic response
to human direction. The displays and processes could provide helpful services
and could involve concepts not hitherto imagined (e.g., the pregraphic thinker
would have been unable to predict the bar graph, the process of long division,
or card file systems.)*

D. Engelbart,
"A Conceptual Framework for
Augmenting Man's Intellect," 1963

*The evolution of the personal computer has followed a path similar to that of the
printed book, but in 40 years rather than 600. Like the handmade books of the
Middle Ages, the massive computers built in the two decades before 1960 were
scarce, expensive and available to only a few. Just as the invention of printing
led to the community use of books chained in a library, the introduction of computer
time-sharing in the 1960s partitioned the capacity of expensive computers in order
to lower their access cost and allow community use. And just as the Industrial
Revolution made possible the personal book by providing inexpensive paper and
mechanized printing and binding, the microelectronic revolution of the 1970s
will bring about the personal computer of the 1980s, with sufficient storage and
speed to support high-level computer languages and interactive graphic displays.*

Alan Kay,
"Microelectronics and the
Personal Computer," 1977

When the ARPA funders came to visit, Engelbart had his long-con-
templated conceptual framework ready for them, along with a complete plan for a laboratory designed to "bootstrap" computer
technology into an entirely new realm. His team of researchers would
build interactive computer tools—sophisticated input and output
hardware, new kinds of communication software, advanced text editors, graphics systems—then use those tools to build even better tools,
thus "bootstrapping" the computers of the mainframe age into the
interactive computers of the augmentation age. Engelbart decided to
name his laboratory the "Augmentation Research Center" (ARC). He
deliberately chose the word "augmentation" to contrast ,with "automation," the word most often used to describe the application of com-

puters to human work. Automation means using computers to replace human labor. Augmentation means amplifying the power of intellectual labor by removing low-level barriers to high-level thinking.

In 1963, Engelbart proposed a kind of computerized writing device as an example of the kind of intellectual augmentation he wanted to create. More than a decade before the first word-processors became economically feasible, he proposed in "A Conceptual Framework for Augmenting Man's Intellect," that by using a computer and a video display screen to compose documents, it would be possible to enhance the entire process of written composition: "This hypothetical writing machine thus permits you to use a new process of composing text. . . . If the tangle of thoughts represented by the draft becomes too complex, you can compile a reordered draft quickly. It would be practical for you to accommodate more complexity in the trails of thought you might build in search of the path that suits your needs. . . . The important thing to appreciate here is that a direct new innovation in one particular capability can have far-reaching effects throughout the rest of your capability hierarchy." This example turned out to be precisely prophetic, but far ahead of the enabling technologies that would make it practical: at the beginning, the ARC researchers used consoles that looked like huge round screens that received signals from large video cameras aimed at radarlike display screens.

Word processing was only one of the inventions that flowed from ARC during the twelve years it existed at SRI. The "mouse" pointing device, now ubiquitous with personal computers of all kinds, was invented in the 1960s, although it was not commercially available until the 1980s. "Hypertext" which enabled readers to jump from one document to another by pointing at specified spots on the screen, multiple "windows" of text on display screens, computer conferences which enabled different computer users to share a written communication system, "outline processing" which enabled computer users to collapse and expand multiple views of documents, the use of video imagery in conjunction with computer graphics to convey information, the use of text and graphic information in the same computer document—most of the key characteristics that define personal computing as it exists today—all were invented at ARC.

The use of the mouse as a pointing device signaled a concrete breakthrough in the human-computer interface, one that moved directly toward the core of VR: 3D gestural input as a command language. Instead of specifying a document or a program by typing in an arcane command code, it became possible to interact with a computer by using a natural gesture; when a user moves the mouse on the desk next to the keyboard, a cursor moves in an analogous manner on the screen. It became possible to issue commands by "pointing and clicking." In virtual realities of the future, gestural input using gloves and "wireless" gestural sensors could continue this line of development to its natural conclusion by enabling people to use their most common pointing device—their fingers. Twenty years after Engelbart predicted it, the use of gestural input has been coupled with the three-dimensional graphical models he foresaw in 1962.

In 1968, Engelbart and his crew decided to gamble everything by demonstrating their radical new way of operating computers to a major gathering of the computer clans, known as the Fall Joint Computer Conference. It turned out to be a seminal event in the history of computing. Sitting on stage with a keyboard, screen, mouse, and the kind of earphone-microphone setup pilots and switchboard operators wear, Engelbart orchestrated a demonstration of the kind of navigation through information space Vannevar Bush had imagined in 1945. A true infonaut, Engelbart seized the attention of his audience and plunged it directly into a working version of the augmentation system he had dreamed about since that day in 1950. He called up documents from the computer's memory and displayed them on the big screen at the front of the auditorium, collapsed the documents to a series of descriptive one-line headings, clicked a button on his mouse and expanded a heading to reveal a document, typed in a command and summoned a video image and a computer graphic to the screen. He typed in words, deleted them, cut and pasted paragraphs and documents from one place to another. Computer conventions can be dull affairs, even to most of the people who attend them, but the 1968 Fall Joint Computer Conference was electrifying to everyone who attended. The assembled engineers, programmers, and computer scientists had never seen anything like it. It was, as scientists say, an "existence proof" of Engelbart's and Licklider's dreams. Here was a working model of the future of computers.

One of the young computer wizards in the audience was Alan Kay, a junior member of the ARPA generation who had helped create time-sharing and interactive computing. Kay became one of the key architects of the "personal computer," the evolutionary stage beyond interactive computing. It was more than the next incremental improvement in easy-to-use computers, however. In 1990, Kay recalled the excitement of the personal computer revolution that began twenty years ago, emerged in the nonspecialist population ten years ago, and is only now beginning to spread to the general population: "The actual dawn of user interface design first happened when computer designers finally noticed, not just that end users had functioning minds, but that a better understanding of how those minds worked would completely shift the paradigm of interaction."

One of Alan Kay's continuing contributions is his insistence on fitting together the worlds of psychology and computer interface design; he was deeply interested in the theories of Jean Piaget, Jerome Bruner, and other psychologists who were modeling the learning process on the idea of *exploration:* our minds are scientists, our senses are our instruments, the world is our experiment. The hypothesis held in common by these psychologists was that we discover the world by feeling our way around in it with all our senses, manipulating it with our hands, tracking it with our eyes and ears. Kay and one of his mentors, Seymour Papert of MIT's AI lab, enthusiastically married Piagetian ideas to computer technology. The sight of six-year-old children delighting in the computer language invented by Papert and his colleagues was one of the "conversion experiences" in Alan Kay's career.

Kay was a computer science student at the University of Utah in the late 1960s, when Ivan Sutherland and David Evans spearheaded the computer graphics part of the ARPA plan. Kay had been influenced by Marshall McLuhan (who inspired him to think of the computer as a medium rather than a tool), Papert (who showed him that computer languages can be thinking tools, that graphical communication is a powerful means of human-computer interaction, and that children can and should be able to use computers), and Ivan Sutherland (whose thesis, the computer code for Sketchpad, opened the possibility of the computer as interactive simulator). In 1970, Kay was one of the most brilliant of hundreds of former ARPA superstars who migrated to Palo Alto Research Center (PARC), a new computer research facility Xerox Corporation had built. For a number of reasons, the momentum of the personal computing revolution shifted in the early 1970s from ARC and the other ARPA research sites to the new Xerox facility.

The Mansfield amendment, drafted during the height of the Vietnam War, effectively prevented ARPA from funding anything but weapons-related research. And the war had a galvanizing effect on both research managers and programmers, many of whom were no longer comfortable working under the aegis of the Department of Defense. The exodus from Defense-sponsored research in the early 1970s by some of its best and brightest lights was perhaps the most important single driving force behind the personal computer revolution. A similar shift from military to civilian-based technological applications seems to be under way again in the early 1990s, which could again act as a driving force for the next evolutionary stage of information technology.

Bob Taylor, the young NASA research manager who had "discovered" Engelbart, had succeeded Licklider and Sutherland as director of IPTO. Around the time of the Mansfield amendment, he moved to PARC and started assembling the best of the researchers who had

worked on ARPA's scattered computing research programs. The
PARC researchers were the best of the legendary young programmers
who had made the ARPA interactive computing revolution a success.
The core of PARC's research staff had known each other through
ARPA meetings and ARPAnet, the computer communication network
they created; now they were gathered in one well-equipped and beautifully sited dream lab on Coyote Hill overlooking Palo Alto, managed
by their old ARPA team leaders, given generous funds and a mandate
to create "the architecture of information for the future," as the president of Xerox called it.

PARC was truly one of the shining success stories of technological
foresight, for the PARC researchers took Engelbart's innovations,
added their own, and built their own prototypes with the ever-more-
powerful enabling technologies that were becoming available in the
early 1970s. The building blocks for personal computers had been
contributed by other teams of infonauts elsewhere who had produced
the microcircuitry and display screen technologies PARC researchers
used.

Enabling technologies took startling leaps at the same time the interactive computing revolution was emerging as a clear success. The
computers at ARC were designed to extend intellectual capabilities,
but it wasn't possible in the 1960s to build a computer cheaply enough
to devote to the use of a single person—until the miniaturization
revolution went into high gear for the aerospace boom. In the early
1970s, PARC designers, many of them veterans of Engelbart's laboratory, built the first personal computer, known as the "Alto," which
I was to try out a decade later. Alan Kay's team spearheaded the effort
to create a new interface from the newly available enabling technologies. And the ARPAnet idea was the spur to create the first "local
area network," the poetically named "Ethernet," a smaller cousin and
high-speed neighborhood node of the network of networks now
known as Worldnet, which is mutating toward the infrastructure of
Gibson's Matrix. PARC in the 1970s became a utopia for the growing
army of infonauts who had joined the quest for computer-assisted
thinking tools. In the early 1980s, I interviewed several of the PARC
veterans who had been in on the design of the Alto interface, and
they recalled that the use of graphics to mediate between the human
mind and the computer was at the core of their design philosophy.

The PARC computer builders had something the earlier ARC team
lacked, a new technology known as "bit-mapped graphics." Every picture element on the screen is represented by a specific bit in the computer's memory—thus, the computer's memory contains a "bit-map"
that corresponds to the pattern of pixels on the screen. The communication between bit and pixel goes both ways: you can tweak the

bits in the computer and watch the pixels jump on the screen, and you can tweak the pixels on the screen with a pointing device and watch the computer jump. Bit-mapped screens were the enabling technology for a human-computer relationship Ivan Sutherland had pioneered on the other coast of America, a decade before PARC. Clicking on a mouse and making a computer do tricks for you with an electron beam and a phosphorescent screen was an early important step toward virtual worlds where the screen is everywhere and your gestures, gaze, and voice substitute for clicking the mouse.

The direct manipulation interface pioneered by Engelbart and developed at PARC was known in the rarefied heights of computer science for years, but the idea wasn't unleashed on the world until somebody gave Steve Jobs of Apple Computer a tour of PARC. Jobs and an even younger generation of wizards took the technology created by the previous decade's best and brightest, designed a new version for even newer generations of enabling technologies, and made mind amplifiers into an appliance—the first computer people began to feel comfortable about bringing into their homes and offices. In 1984, the Apple Macintosh computer brought the graphic interface to millions. Interactive, graphical interfaces began to diffuse through the population because most prefer to interact with computers in this way. By 1990, even the rival IBM personal computers had adopted point-and-click interfaces.

The possibilities of interactive computer graphics, the crafting of bits and pixels into visual thinking tools, had been dramatically demonstrated in a technical masterstroke by Ivan Sutherland, a graduate student in his early twenties in the early 1960s. It took another decade of working with the new enabling technologies and the tools they made possible in order for computer scientists to build Altos and Ethernets, and another decade after that for college freshmen to afford a Macintosh. Interactive computer graphics was one of the most important enabling technologies that made personal computers possible. The creation of the tools for interacting with computer graphics was also the beginning of the journey toward three-dimensional graphics and reality engines. If the ten-year rule of thumb holds true, personal computer enthusiasts by the millions a decade from now will be interacting directly with virtual worlds through their desktop reality engines.

The human relationship with images on screens, an electronic age echo of the cave paintings of Lascaux and the Eleusinian initiations, is a central theme in the history of cyberspace. It is worth backtracking a few decades at this point, focusing attention on the emergence of computer power onto screens.

SKETCHPAD: "THE MOST IMPORTANT COMPUTER PROGRAM EVER WRITTEN"

Appropriately enough, the first great conversational human/computer interface was also the first great direct manipulation one. Ivan Sutherland's Sketchpad enabled a user and a computer "to converse rapidly through the medium of line drawings." This style of interaction was primarily graphical, yet it exhibited some of the important features of human conversation. A human conversed with Sketchpad by pointing. The system responded by updating the drawing immediately, so that the relationship between the user's action and the graphical display was clear. In fact, because the feedback was so timely and relevant, it could be considered analogous to backchannels, or secondary speech in human/human communication.

SUSAN BRENNAN,
"Conversation as Direct Manipulation,"
1990

The invention of time-sharing computers in the 1960s not only enabled many people to use the same central computer by exchanging commands and results with the computer interactively, it also provided a channel of communication between humans and humans—people on the earliest time-sharing systems built electronic mail systems for sending each other messages. As the time-sharing systems evolved, so did the mail systems. Eventually, what we will send one another over the telecommunication lines will not be restricted to text, but will include voice, images, gestures, facial expressions, virtual objects, cybernetic architectures—everything that contributes to a sense of presence. We will be sending worlds, and ways to be in them. The advent of interactive computing and e-mail in the late 1960s was a twin milestone in the making of the Matrix, the network of networks where future cyberspaces will evolve. This series of developments can now be seen as the removal of a series of obstacles that separated human cognitive power from electronic computation power.

A major obstacle to direct access to computers in the early 1960s was the way computers were set up to display information to human operators. Although Engelbart had dreamed of graphical information displays in the early 1950s, it took the cold war and the space race to drive the development of interactive computer graphics. Whirlwind and the "presentation group" at Lincoln Laboratory who built the information displays for SAGE had demonstrated the possibilities of using cathode-ray tube (CRT) technology, similar to that used for television screens, in place of the teletype printers. The lightpen came out of that project, a simple but revolutionary tool that enabled the operator to manipulate the computer by touching the display screen with a penlike device. The researchers at Lincoln Laboratory in the early 1960s had the most advanced computer of its day, the TX-2, the first computer based on transistors rather than vacuum tubes as switching elements. The right kind of mind stepped into this situation and opened the possibility of virtual reality, among other things.

In 1983, Licklider and I talked about the day which he still remembered vividly two decades later, when computer graphics emerged as a distinct discipline. The presentation group veterans at Lincoln Laboratory and other ARPA researchers at the University of Utah were beginning to work intensively on the problem of CRT-based display devices. At the first meeting on the subject of interactive computer graphics, the first wave of preliminary research was presented and discussed in order to plan the assault on the main problem of getting information from the innards of the new computers to the surface of various kinds of display screens. In order to connect the bits and pixels in a functional manner, special hardware had to be designed to get the information from the computer to the screen; more difficult was the problem of creating software that would enable people to use the values of bits to control the appearance of pixels. Ivan Sutherland solved the most vexing problems with one program. He called it "Sketchpad," and every person I've met who was involved in the personal computing revolution talks about their first encounter with it in terms of the now-familiar "conversion experience."

Sketchpad was Lascaux, thirty thousand years later; instead of pigments on limestone cave walls, Sutherland used electrons and glowing phosphors on the surface of a glass bottle. The cave painting and the first interactive computer graphics were both attempts to influence the consciousness of the viewer, as a means of imparting important cultural information. A grainy black and white filmclip on Sketchpad, no more than ten minutes long, still exists. It shows Sutherland demonstrating the graphical capabilities of his computer program on a small glowing screen in a darkened room, reminding me in more ways than one of his antecedents with their subterranean light shows.

"Sutherland was a graduate student at the time," Licklider recalled, "and he had not been asked to give a paper." He was invited to the meeting, however, for several reasons: Sutherland was the Ph.D. student of Claude Shannon, who invented information theory. Claude Shannon did not sponsor ordinary graduate students. Sutherland was creating a graphics program as part of his Ph.D. research. And he was said to be just the kind of prodigy ARPA was seeking.

"Toward the end of one of the last sessions," Licklider recounted, "Sutherland stood up and asked a question of one of the speakers." It was the kind of question that indicated that this unknown young fellow might have something interesting to say to this high-powered assemblage. So Licklider arranged for him to speak to the group the next day: "Of course, he brought some slides, and when we saw them, everyone in the room recognized his work to be far better than what had been described in the formal sessions." Sutherland's thesis project, a program developed on Lincoln's TX-2, demonstrated an innovative way to handle computer graphics—and a new way of commanding the operations of computers. Sketchpad made it evident to the assembled experts that Sutherland had leaped over their research to create something that even the most ambitious of them had not dared to hypothesize.

Sketchpad allowed a computer operator to use the computer to create sophisticated visual models on a display screen that resembled a television set. The visual patterns could be stored in the computer's memory like any other data, and could be manipulated by the computer's processor. People could create images in the most natural way possible, by using their hands and eyes and a penlike device to *draw* them. In a way, this was a dramatic answer to Licklider's quest for a fast model-builder. But Sketchpad was much more than a tool for creating visual displays. It was a kind of simulation language that enabled computers to translate abstractions into perceptually concrete forms. And Sketchpad was a powerful model of a totally new way to operate computers; by changing something on the display screen, it was possible to change something in the computer's memory. It wasn't yet a bit-map, but Ivan had devised a way to use the TX-2, a cathode-ray tube display screen, and the lightpen (all developed by others) to command a computer by drawing on a screen.

Even programmer demigods (programmers who "were used to dealing lightning with both hands," in the words of Alan Kay) were impressed. "If I had known how hard it was to do, I probably wouldn't have done it," Alan Kay remembers Sutherland remarking about his now-legendary program. With a lightpen, a keyboard, a display screen, and the Sketchpad program running on the relatively crude real-time computers available in 1962, anyone could see for themselves that computers could be used for something else besides data processing. And in the case of Sketchpad, seeing was truly believing. The magic of the breakthrough is still evident to anyone who watches a skilled operator at work with it. The field of computer-aided design, now known as CAD, grew out of that Ph.D. thesis, and is one of the most powerful industrial drivers for VR development in the 1990s.

Nearly thirty years ago, the same visionaries and wizards who had foreseen and created the prototypes of a new kind of computing, became influential in the direction of future research. Sutherland, at the age of twenty-six, was recommended by Licklider to succeed him as director of IPTO in 1964; when Sutherland went back to Cambridge to build a head-mounted display the following year, he recommended Robert Taylor, also in his twenties, to be the IPTO director. The post was handed down from one true believer in the interactive computing crusade to another true believer for years. It was a kind of good-old-boy network that was linked not by a bond of social class, or college fraternity, but by belief in a specific vision of the future of the human-computer relationship.

Alan Kay was heavily influenced by Sketchpad. When Kay arrived at the University of Utah in 1966, the first job his research advisor gave him was the task of studying Sketchpad's programming code. Seeing Sutherland's brainchild in operation was one of Kay's conversion experiences. He was still excited when he described it to me, in 1983: "Sketchpad was not just a tool to draw things. It was a program that obeyed laws that you wanted to be held true. To draw a square, you used the lightpen to draw a line and gave the computer a few commands like 'Copy-copy-copy, attach-attach-attach. That angle is 90 degrees, these four things are equal.' Sketchpad would take your line and your instructions and go zap! and a square would appear on the screen."

Another computer prophet who saw the implications of Sketchpad and other heretofore esoteric wonders of personal computing was an irreverent, unorthodox, countercultural fellow by the name of Ted Nelson, who has long been in the habit of self-publishing quirky, cranky, amazingly accurate commentaries on the future of computing. In *The Home Computer Revolution*, in 1977, Nelson had this to say about Sutherland's pioneering program, in a chapter entitled, "The Most Important Computer Program Ever Written":

You could draw a picture on the screen with the lightpen—and then file the picture away in the computer's memory. You could, indeed, save numerous pictures this way.

You could then combine the pictures, pulling out copies from memory and putting them amongst one another.

For example, you could make a picture of a rabbit and a picture of a rocket, and then put little rabbits all over a large rocket. Or little rockets all over a large rabbit.

The screen on which the picture appeared did not necessarily show all the details; the important thing was that the details were *in* the computer, when you magnified a picture sufficiently, they would come into view.

You could magnify and shrink the picture to a spectacular degree. You could fill a rocket picture with rabbit pictures, then shrink that until all that was visible was a tiny rocket; then you could make copies of that, and dot them all over a large copy of the rabbit picture. So when you expanded the big rabbit till only a small part showed (so it would be the size of a house, if the screen were large enough), then the foot-long rockets on the screen would each have rabbits the size of a dime.

Finally, if you changed the master picture—say, by putting a third ear on the big rabbit—all the copies would change correspondingly.

Thus Sketchpad let you try things out before deciding. Instead of making you position a line in one specific way, it was set up to allow you to try a number of different positions and arrangements, with the ease of moving cut-outs around on a table.

It allowed room for human vagueness and judgement. Instead of forcing the user to divide things into sharp categories, or requiring the data to be precise from the beginning—all those stiff restrictions people say "the computer requires"—it let you slide things around to your heart's content. You could rearrange till you got what you wanted, no matter for what reason you wanted it.

There had been lightpens and graphical computer screens before, used in the military, but Sketchpad was historic in its simplicity—a simplicity, it must be added, that had been deliberately crafted by cunning intellect—and its lack of involvement with any particular field. Indeed, it lacked any complications normally tangled with what people actually do. It was, in short, an innocent program, showing how easy human work could be if a computer were set up to be really helpful.

As described here, this may not seem very useful, and that has been part of the problem. Sketchpad was a very imaginative, novel program, in which Sutherland invented a lot of new techniques; and it takes imaginative people to see its meaning.

Admittedly the rabbits and rockets are a frivolous example, suited only to a science-fiction convention at Easter. But many other applications are obvious: this would do so much for blueprints, or electronic diagrams, or all the other areas where large and precise drafting is needed. Not that drawings of rabbits, or even drawings of transistors, mean the millennium; but that a new way of working and seeing was possible.

The techniques of the computer screen are general and applicable to *every*thing—but only if you can adapt your mind to thinking in terms of computer screens.

In 1989, twenty-seven years after Sketchpad's debut, I attended the annual convention of the Association for Computing Machinery's Special Interest Group for Graphics at Boston's Foley Convention Center. I was one of sixty thousand people who wandered, dazed, for four days, through three huge convention floors full of devices, tools, and industries devoted on computer screens. The annual SIGGRAPH film show is the premier event for presenting the state of the art in computer graphics, a field in which the final results are displayed on screens as well as reported in journal articles; this year, 3D glasses were handed out to each of several thousand audience members for each screening. Over at the Autodesk booth, I played some virtual racquetball in their latest cyberspace demo; upstairs, at the Silicon Graphics booth, I test-zoomed Jaron Lanier's latest virtual world, an aquarium connected with a space that looked like a Greek temple. I wondered whether Sutherland might possibly be wandering through this pixelated wonderland himself, and whether he felt like Prometheus—or Pandora.

Sutherland opened a window, and pointed to the day when people would be able to go through that window and enter the abstract territory that computer simulations could create. But the idea of coupling human perceptions to computer capabilities is a big one, with many facets, and implementing it took many people in many places. Sutherland was not the only visionary who realized that computer technology was approaching a threshold.

Chapter Four

THE THRESHOLD OF VIRTUAL EXPLORATION

Though much of what McLuhan wrote was obscure and arguable, the sum total to me was a shock that reverberates even now. The computer is a medium! I had always thought of it as a tool, perhaps a vehicle—a much weaker conception. What McLuhan was saying is that if the personal computer is a truly new medium then the very use of it would actually change the thought patterns of an entire civilization. He had certainly been right about the effect of the electronic stained-glass window that was television—a remedievalizing tribal influence at best. The intensely interactive and involving nature of the personal computer seemed an antiparticle that could annihilate the passive boredom invoked by television. But it also promised to surpass the book to bring about a new kind of renaissance by going beyond static representations to dynamic simulation. What kind of a thinker would you become if you grew up with an active simulator connected, not just to one point of view, but to all the points of view of the ages represented so they could be dynamically tried out and compared? I named the notebook-sized computer idea the Dynabook to capture McLuhan's metaphor in the silicon to come.

ALAN KAY,
"User Interface: A Personal View,"
1990

As an environmental simulator, the Sensorama display was one of the first steps toward duplicating the viewer's act of confronting a real scene. The user is totally immersed in an information booth designed to imitate the mode of exploration while the scene is imaged simultaneously through several senses. The next step is to allow the viewer to control his own path through available information to create a highly personalized interaction capability bordering on the threshold of virtual exploration.

SCOTT FISHER,
"Viewpoint Dependent Imaging,"
1981

Putting your consciousness into a virtual world is only part of the task—in order to feel present, there must be a way to move around in the simulation. So the idea of "navigating" an information space in order to learn something began to guide the way people designed applications for new enabling technologies. The idea that computer technology could be merged with other audiovisual media to make "exploratory computer environments" was developed in the 1970s at MIT by researchers who would become important VR pioneers. They were also looking at ways to "tightly couple" human minds and computing devices.

Some of the roots of today's VR trace the history of a device (such as stereographic displays), a discipline (computer graphics), or an idea (exploring an artificial world), and thus cross the boundaries of scientific disciplines and academic institutions. Separate inventors or scientists, often unaware of each other, contributed puzzle pieces that have recently begun to fit together. Other roots are derived from specific places and times, from groups of people who built on one another's work, who worked together in various institutions over the years, on projects that were designed to lead to something very much like today's VR. Atari's Sunnyvale Laboratory in the early 1980s was one such site. I happened to be there in 1983 in quest of various infonauts who later turned out to be cybernauts. As I became acquainted with them, I realized that the Atari Laboratory faction of young media-technologists knew each other from specific places and times. Just as the ARC, ARPA, and PARC infonauts knew each other from joint efforts that had succeeded or failed in their past, the strong faction of today's cybernauts who are Atari alumni from the early 1980s had known each other even before Nolan Bushnell founded Atari, marketed *Pong* and sold the company to Warner's. As I got to know them, I discovered that Atari Research was packed with veterans of an even earlier institution that started at MIT in the 1970s, the "Architecture Machine Group," known as "Arch-Mac" (pronounced "Ark-Mac"), led by Nicholas Negroponte and Richard Bolt.

They had a strange name—what is an "architecture machine"?— and they were using exotic technologies that computer jockeys had never been known to mess with, like gaze-trackers and videodisks. Negroponte was the driving force and guiding inspiration for Arch-Mac and the institution that grew from it, the Media Lab. Although his name is not often mentioned in regard to the cutting edge of VR research, his vision of multisensory, human-matched computer environments set people thinking and working on real systems fifteen and twenty years ago. "How can you adapt technology to the characteristics of humans?" was the question Negroponte and his colleagues posed. Fifteen years before Stewart Brand's *The Media Lab: Inventing the Future at MIT* highlighted the role of this group, it was known only to the media-tech cognoscenti. But Arch-Mac had strong backing from ARPA.

In 1970, Nicholas Negroponte had first presented his own vision of how computers could enhance human thought and imagination by combining the presentation capabilities of cinema with the information-manipulation power of computers. He forecast the day when today's separate media will combine, when a digital-optical-audiovisual-broadcast-networked hybrid will fuse into an integrated "media technology." If all the means of encoding information (sounds, images, words, numbers, computer data) are becoming digital, and all the means of communicating information (broadcast and cable, diskettes and telecommunication networks) are being linked, then the resulting technology will become more than just an intersection of different areas; the new "metamedium," as Alan Kay called it, will constitute a world of its own. The Media Lab crew wanted to do more than explore the potential of enabling technologies. They wanted to look at what media technology might become in the future, and then materialize those visions in the form of prototypes. "Demo or die," was the in-house slogan revealed by Stewart Brand in his book, *The Media Lab*.

Just as the ARPA, ARC, and PARC infonauts had leapfrogged over existing computer technology, the Arch-Mac crowd had their eyes on the *next* breakthroughs, from voice-commanded computers to holographic films. Together with human interface researcher Richard Bolt and others, Negroponte started constructing a new kind of research institution, where the cognitive sciences and the computer sciences would deliberately mesh with cinematic and telecommunication technologies and together set their sights on the long-term future. Arch-Mac eventually became the Media Lab, the place where, as Stewart Brand pointed out, the mission is "inventing the future."

The Media Lab has never been enamored of encumbering devices like "goggles and gloves," as some people call VR peripheral equipment. In the late 1970s and early 1980s, however, several key experiments created the foundations for later VR systems. Christopher Schmandt and Eric Hulteen, working under Bolt as principal investigator (known in researchese as "the PI"), combined a wall-sized display, a gestural (finger-pointing) input device, and a voice-recognition command system into the demonstration that has been known ever since by its paradigmatic voice-gesture commmand, "Put That There." An operator could sit in a chair, facing the screen, upon which a computer-generated map of an ocean could be projected. The prototype had been disassembled years before I ever heard of it, but in 1983 I watched a video of Hulteen in the Media Room; he pointed his finger at a ship on that ocean, said aloud, "Put that," then moved his finger to another spot and said aloud, . . . there," and the computer, just like the fabled jinn, implemented the command. Objects on the screen could be scaled, shaped, manipulated, moved, through voice commands and pointing gestures. The magnetic hand-position sensor used in "Put That There," a pair of small but expensive cubes about the size of a pair of dice, manufactured by Polhemus Navigation Systems, ended up playing a central role in other parts of the VR story over the years.

Several of the research directions that were pioneered if not ultimately followed up by the Media Lab have served as blueprints for today's renewed VR research effort worldwide. Other Arch-Mac/ Media Lab research had demonstrated uses for three-dimensional displays and techniques for transmitting facial expressions and gaze direction via telecommunications; the Arch-Mac notion of transmitting "presence" was included in a vision of a "shared communication animation" in which your conversational partners would appear as virtual objects that animated their faces, cast and recast their gaze, in synchrony with their physical movements. Richard Bolt long ago stressed his strong belief that watching your conversational partner's direction of gaze is an essential element in human communication; a decade later, I found researchers in laboratories near Yokohama and Kyoto experimenting with eye-tracking technology and pattern recognition of facial expressions. Arch-Mac researchers also extended some of the work Kenneth Knowlton did in the 1970s when he was working for Bell Labs, research that was also seminal in the eventual emergence of this notion of "virtual environments." Knowlton developed a virtual desktop workspace for some of the telephone company operators who had to perform complex sets of tasks on keyboard configurations that often changed.

A half-silvered mirror, arranged at the proper angle, can create a kind of virtual image that seems to float in space over the nonvirtual world. Ken Knowlton juxtaposed a physical control device with a virtual image to create a kind of virtual visual workspace. Knowlton positioned a half-silvered mirror so the image from the mirror would overlay the sight of a blank keyboard. A computer graphics monitor displayed graphical overlays that could be used as "virtual keyboards" when aligned with the blank physical keyboard. Fisher and others were interested in the way Knowlton's experiment used a virtual space that the operators could move their hands around in. In the early 1980s, Christopher Schmandt and Scott Fisher aimed a color TV monitor down at a horizontal half-silvered mirror. The Arch-Mac crew built a small virtual environment that the operator could reach into, interact with, draw objects in three-space and pick them up, move them around. "We could literally configure the entire virtual world," Fisher recalled years later. They weren't using gloves, but the Arch-Mac cybernauts were taking a step down an important path when they followed the lead of Dan Vickers, Ivan Sutherland's assistant at the University of Utah, and added hands-on manipulation to their explorations of virtual environments, along with look-but-don't-touch navigation.

One seminal concept from Arch-Mac in the late 1970s that has taken a long time to become practical was the idea that the vast worlds of data stored in computers could be represented in some visible form ("visualized," as researchers have come to say), and somehow explored cognitively by performing physical navigation through the "data space," a concept that led to the Arch-Mac prototype of "Dataland." Mathematicians talk about "multidimensional spaces," and programmers talk about "combinatorial explosions in search space," referring to a kind of abstract place where certain formal operations take place. What if these spaces could be made visible, navigable, manipulable? Could we bring our innate neural space-exploration systems to bear on the worlds of data we now use our computers to crunch and bend, more or less blindly? Dataland was a materialization of this notion in the Media Room.

The original Media Room had wall-sized screens, stand-alone color monitors, eye-tracking, voice-input, and gestural tracking devices. But the Media Room was just the infrastructure—a medium for thinking, communicating, or experiencing something. What kind of tools could be created in such a place that could not be created elsewhere? The "Spatial Data Management System" (SDMS), a system for visually navigating through data bases, was implemented with the Media Room. "Dataland" was part of the SDMS—a visual window into the operator's personal data set (the collection of programs and files that today's personal computer users would call their "electronic desktop"). An operator sitting in a chair could use a touchpad to issue fingertip commands or grasp small joysticks to literally fly through a two-dimensional representation of a three-dimensional data structure.

The notion of an SDMS took a turn through the realm of science fiction before it resurfaced in VR research: When William Gibson coined the term "cyberspace" in his novel *Neuromancer*, he described huge virtual structures of data in the "consensual hallucination" millions of people directly connected to by "jacking in" their nervous systems. These blue pyramids of financial data or red hemispheres of corporate records were protected from eavesdroppers by visible walls of protection Gibson called "ice." When VR systems began to proliferate, the idea of flying through data space emerged again in computer science, simply because the monstrously accelerating stream of data that is generated by all the information processors we've sent out into the world has challenged the data base designers who try to build ways for businesses, scientific institutions, government, or individuals to deal with the data they need to continue operating. As Engelbart noted in 1950, we are getting entangled by the complexity of our tools, and perhaps our only escape is to build tools for coping with that complexity. By 1990, several groups around the world were actively pursuing versions of SDMS systems with head-mounted displays and reality engines, as potential means of dealing with extraordinarily large or complex collections of information.

Another Media Room experiment that continues to influence the course of VR research was "World of Windows," in which large windows of information would open on a wall-sized screen, under eye-tracking control; one panel held text, another showed still photographs, another showed full-motion video, etc. Each of the windows would be fed by part of an information-gathering network—news wire services, satellite data, computer and videodisk data bases, live video cameras—and wherever the operator chose to gaze, the soundtrack would change and/or the information window would expand. It was a way to explore ideas by browsing through structured streams of information.

Except for a few high-ticket items for the military, such as SDMSs and multimedia systems for the US Navy, the Arch-Mac and Media Lab demonstrations of navigable information-spaces didn't change the world outside the laboratory right away. Media Lab specialists are still trying to get motion holography to work, for example. Optical disks (CD ROMS) replaced older technologies in the music business (phonograph records and analog audio tapes) before they began to diffuse to the computer-using population. Like AI and VR, the potential of optical media was confounded somewhat by the storm of hype in the popular media; as Michael Naimark remarked recently at yet another conference on the future of VR, "interactive videodisk technology was heralded with visions of 'libraries on a disk' but it took more than 15 years from that prophecy for Sony to market its 'Data DiscMan' in 1990."

The early Cambridge demos of crucial bits and pieces of what has since become VR technology did not set off immediate waves of industrial R & D worldwide. But those experiments did plant some ideas in the minds of a few key people who would start making their own waves a few years later. Media Lab continues to nurture a certain amount of VR-related research: In the 1990s, two of the most important VR frontiers—autonomous computer-graphics "characters" that can inhabit virtual worlds, and devices that transmit the human tactile and kinesthetic senses—are actively pursued in Media Lab's subterranean "Snake Pit" by David Zeltzer, Margaret Minsky, and their colleagues.

VR-related projects continue to emerge from their sleek I. M. Pei building in Cambridge, but I wouldn't call Media Lab today a "VR center." Telepresence is only part of the Media Lab vision, and not a central one. When I visited Media Lab in 1989 and asked their flack, Tim Browne, he told me in a tone of voice that indicated he had explained this very patiently many times before that they were "not interested in anything that might come between the human mind and the computer." Despite that disclaimer, the Cambridge flavor to the VR brew, even in Chapel Hill and Silicon Valley, remains strong. About a dozen Arch-Mac/Media Lab/Atari alumni, with their knowledge of media rooms and eye-tracking, videodisks, head-mounted displays, position sensors, surrogate travel, voice recognition, stereoscopy, computer graphics, have become significant VR cadres in scientific and commercial institutions today. Moreover, the shared vocabulary of concepts they developed during failed or successful experiments has influenced the entire present generation of VR researchers.

The night I observed the Atari cybernauts, years ago, they used theatrical improvisation as a way of thinking about the nature of human-computer dialogue in the future. All of the participants, except for Scott Fisher, placed their different vignettes within the context of a media room. When his turn came to improvise an interaction, Scott didn't mime walking into an imaginary room. Instead, he put on an imaginary head-mounted display. Everybody laughed. His interest in stereoscopic displays appeared to be sufficiently well known to his peers to constitute a wordless cue for amusement. And while Scott's interest in head-mounted displays was one of many paths the Atari researchers intended to pursue, I got the feeling from the other researchers (which I shared at the time) that the media room approach was considered sexier. Michael Naimark once remarked that even before Arch-Mac, when Fisher was teaching at the Center for Advanced Visual Studies at MIT, "we used to call Scott 'the 3D guy.' "

Fisher had a strong interest in using technologies for artistic representation. He was drawn to MIT's Center for Advanced Visual Studies, an arts-oriented if technologically sophisticated enclave in an engineering-oriented environment, where he lectured on 3D imaging and held a research fellowship from 1974 to 1976. When Arch-Mac started experimenting with "surrogate travel" in 1978, Fisher moved there to help principal investigator Andrew Lippman and others create a new information-tool they called the "Movie Map." The "Aspen Map," as it sometimes is called, was an important predecessor to VR technology. The "user" began to become the "operator." The job of creating the feeling that an operator is inside a simulated space has two aspects: First, the perceptual technology must convince the operator that the simulation is a three-dimensional environment that surrounds him or her; this aspect has become known as "immersion." There is another key idea, however—the question of whether the operator is a passive observer of this environment (as in Sensorama) or has the power to actively navigate and explore it. Together, immersion and navigation constitute the elements of a new kind of beast, a "personal simulator." Arch-Mac was another inflection point—several inflection points—in VR history, as significant for its effects on the minds of the people who were there as for the research itself.

As Fisher wrote years later, the Arch-Mac prototype demonstrated part of the capability that would later be important to VR technology:

> The technology has been moving gradually toward lower cost "personal simulation" environments in which the viewer is also able to control his own viewpoint or motion through a virtual environment— an important capability missing from the Sensorama prototype. An early example of this is the Aspen Movie Map. . . . Imagery of the town of Aspen, Colorado, was shot with a special camera system mounted on top of a car, filming down every street and around every corner in town, combined with shots above town from cranes, helicopters and airplanes and also with shots inside buildings. The Movie Map gave the operators the capability of sitting in front of a touch-sensitive display screen and driving through the town of Aspen at their own rate, taking any route they chose, by touching the screen, indicating what turns they wanted to make, and what buildings they wanted to enter.

In one configuration, this was set up so that the operator was surrounded by front, back, and side-looking camera imagery so that he was completely immersed in a virtual representation of the town. There was no head-mounted display or glove or 3D technique, but the Movie Map was a kind of virtual world.

You could sit inside a room anywhere, and a photographic map would surround you. Look straight ahead and sight down a street in Aspen. Make a gesture, and your point of view starts moving down the street. Stop the motion at any point and freeze the frame. Look to your right and decide to go down that street, and the points of view on the four screens around you change in order to keep you headed in the direction you choose. If that house over there looks attractive, zoom in. Text information about the history of that house appears at the top of the frame in front of you. Move forward or backward in time and see that house in winter or summer. Enter the house, if you wish, and peek at the interior. It isn't hard to see how movie maps might interest the military, who would be willing to build extremely detailed videodisk maps of certain key installations in order to realistically rehearse an operation.

The technology was different from what is used in most VR research today. A videodisk is a way of encoding text and grahic visual information in a form that can be stored digitally in the form of on-off (binary) "bits" of information that can be read by a laser, computationally reconstructed from the bits as an image and played back through a screen-based display device. It is possible to use a videodisk system to retrieve very quickly a specific chunk of text, sound, or image from a fairly large collection stored in a compact, inexpensive medium. A certain amount of information can be stored on each "platter"— 54,000 still photographs or thirty minutes of full-motion video—and the use of various strategies for storing the information and for retrieving it makes it possible for human operators to "navigate" pathways of their choosing through the images, thus making it an "interactive videodisk system." The information can be as abstract and nonspatial as an illustrated history of a country or an epoch; it can be as concretely spatial as the Aspen Map. The concept of navigation transcends the particular kind of technology used to manipulate information and the particular form of the information—it is possible to navigate through a text data base (as Engelbart and Nelson envisioned when they started dreaming of "hypertext"), a library of still or moving images, a virtual simulation of the physical world, or via telerobotic control, through a remote part of the physical world. The operators can interact with what they see by choosing how they see it, rather than simply perceiving whatever the computer conveys.

There's an indefinable distance between a technology's potential and the circumstances in which it becomes practical to use that potential, as Scott Fisher learned. In the early 1980s, the amount of computing power necessary to create a completely computer-simulated virtual world was too expensive for a master's degree project. But it was possible to explore immersion and navigation by combining a stereoscopic display with a cleverly designed interactive videodisk system. Fisher's master's degree in media technology thesis project combined stereographic 3D display technology with videodisk image storage and retrieval technology, in a way that enabled the operator to explore a three-dimensional environment. Two videodisks contained two precisely different and precisely matched sets of images, one for display to the right eye, the other for display to the left eye. The images in these pairs were photographs of the same scene, taken 65 millimeters apart—the "interocular distance" that separates human eyes. The image data base was structured to provide a rich series of pathways for exploring the depicted environment; wherever an operator is likely to move or look, there is a corresponding pair of images taken from that point of view.

The images in the data base were linked to a computer model of the environment so that the displayed image could match the operator's position within a small radius of movement. The operator's position was communicated to the computer data base by the magnetic tracking device used in "Put That There," the sensor provided to the Air Force and others by Polhemus Navigational Systems. To this day, the device itself is known informally in the VR trade as "the Polhemus." The correct pair of images for the operator's position were tracked by the computer and retrieved from the videodisk, then displayed to a user via a 3D display known as a PLZT display because of the piezoceramic viewing glasses worn by the operator.

PLZT glasses were "electronic shutters" that switched the visibility of the screen on and off very rapidly, alternating the right- and left-eye views. PLZT glasses have been supplanted by liquid crystal display (LCD) glasses; they are similar insofar as both create a 3D effect through "time-multiplexing." PLZT and LCD glasses don't show the right and left images at the same time; they slice up and shuffle together the images in time, rather than space. The two videodisk frames for every viewing position are "interlaced" during alternate scans of a CRT monitor: Each 1/60 of a second the right-eye view is displayed, and each 1/60 of a second the left-eye view is displayed. The operator looks at the display, and the shutters blank the view alternately so that each eye sees only the display meant for it, at such rapid intervals that the view fuses into a single, three-dimensional representation. Look at an object, and it seems to stand out in space. Test your perception by moving your position slightly, and the Polhemus sensor signals the change to the computer, which retrieves the proper image pairs from the videodisk and projects them through the 3D display.

There is a fundamental limitation to this method of chopping up the information into slices that are displayed in rapid sequence: it is impossible to photograph and store every possible view of any space over a certain size or of more than a very low level of complexity. But the constraints of the medium weren't Fisher's primary concern in the early 1980s. Fisher wanted to find out if you could use these new information media to create an authentic feeling of navigating an artificial space. He used the image pairs to create the 3D effect of "binocular parallax," mimicking the way humans create a sense of three-dimensionality by stringing together parallel 2D views of the world; our eyes see everything from slightly different angles, and our brain does all sorts of sophisticated depth-calculations on the differences between each eye's view. He used the Polhemus sensor and data base system to create the 3D effect of "motion parallax" that shows how the world you are seeing changes position as you move your own point of regard, the way Sutherland's first system had done.

Scott Fisher and his colleagues were learning to build virtual space probes. Stereography, parallax, navigation, position-sensing have turned out to be fundamental building blocks of VR systems. The capabilities of the enabling technologies had to become an order of magnitude more powerful before the first experimental gear of the 1970s could fit together into a reality system worth tinkering with. Computer graphics and modeling techniques that take advantage of the power of these future computers would have to replace the videodisk as the medium for storing and retrieving images. Another crucial element—the position-sensing glove that enabled the operator to reach out and manipulate the virtual world as well as navigate it—still had to come along. The long hard work of finding out how to glue all these devices together with the right kind of software would remain to be done even after hardware of sufficient power arrived and the software algorithms for manipulating 3D graphics were worked out by hundreds of programmers at universities and commercial research laboratories. But Scott Fisher and others he would meet again at NASA were already learning what they needed to know to build the first generation of affordable cyberspace engines.

From Sensorama to ARC to PARC to Arch-Mac, there seem to be many paths converging on today's VR technology. The point at which the idea of virtual reality as it is understood today was most closely articulated came surprisingly long ago, when Ivan Sutherland first used the combination of head-mounted displays, head position track-ing, and real-time interactive computer graphics to actually put a human being inside a computer-generated world.

Moving things around on computer screens was only the beginning of Ivan Sutherland's bright ideas. Almost immediately after getting computer graphics off the ground, he started thinking about ways to get *inside* the computer screen. As Theodor Nelson noted, Sutherland had hit upon the importance of using human eye-hand coordination. If Sketchpad "let you slide things around to your heart's content," how much more power to formulate, model, design, and think might computer users gain if they could actually walk around a graphic object in three-space, rotate it, mold it, and deal with it the way we handle objects in the physical world? While inventors of stereoscopic devices, particularly Morton Heilig, had been experimenting with various kinds of glasses and other head-mounted displays, Sutherland was the first to propose mounting small computer screens in binocular glasses—far from an easy hardware task in the early 1960s—and thus immerse the user's point of view inside the computer graphic world.

Chapter Five

ENTERING CYBERSPACE

IVAN AND THE SWORD OF DAMOCLES: THE FIRST HEAD-MOUNTED DISPLAY

The fundamental idea behind the three-dimensional display is to present the user with a perspective image which changes as he moves. The retinal image of the real objects which we see is, after all, only two-dimensional. Thus if we can place suitable two-dimensional images on the observer's retinas, we can create the illusion that he is seeing a three-dimensional object. Although stereo presentation is important to the three-dimensional illusion, it is less important than the change that takes place in the image when the observer moves his head. The image presented by the three-dimensional display must change in exactly the way that the image of a real object would change for similar motions of the user's head.

IVAN SUTHERLAND,
"A Head-Mounted Three-Dimensional
Display," 1968

By 1966, having created an interactive system on a flat screen, Ivan Sutherland wanted to take Licklider's notion of "intimate contact" between human mind and computer to the extreme of putting the user inside a computer-generated, three-dimensional graphic world. There are many roots to VR technology, especially Morton Heilig and Myron Krueger (who appears later in this chapter). Ivan Sutherland's research in the late 1960s and early 1970s, however, turned out to be a fundamental event in the genesis of cyberspace technology. Heilig, a cinematography and multisensory media specialist, couldn't find the funding for his head-mounted display and audiovisual-olfactory-tactile experience booth, and he was not a computer expert. Krueger, as much artist as technologist, ran into similar funding difficulties. They fell into the "Engelbart gap"—the gulf between different conceptual frameworks, the dead zone in the visual fields of those who can't see the future when it is pointed out to them.

Funding from understanding sponsors wasn't a problem for Ivan Sutherland, who had just vacated the post of IPTO director, the most important single funding source for leading-edge computer science. Beginning in 1966 at MIT's Lincoln Laboratory, Sutherland and several colleagues conducted the first experiments with head-mounted displays (HMDs) of different kinds. Just four years after Sketchpad, Ivan Sutherland's ideas wielded a significant amount of clout. Harvard was involved as well as MIT. ARPA and the Office of Naval Research cosponsored the first HMD research.

Sutherland approached the problem in characteristic ARPA fashion. He started with what he wanted to do—to put humans inside computer-generated graphic simulations—then collected the technologies he would need and tackled the technologies he would have to invent. He decided not to add stereoscopic graphics until later. Although the first HMDs were binocular, each eye saw the same image. The illusion of three-dimensionality created by Sutherland's first display took advantage of the way we are accustomed to seeing our view of the world change when we move our head. Look at an object, then move your head to the left and notice how the object moves to the right side of your field of view. In order to change the appearance of the computer-generated graphics when the user moves, some kind of gaze-tracking tool is needed. Because the direction of the user's gaze was most economically and accurately measured at that time by means of a mechanical apparatus, and because the HMD itself was so heavy, the users of Sutherland's early HMD systems found their head locked into machinery suspended from the ceiling. The user put his or her head in a metal contraption that was known as the "Sword of Damocles" display. I know from my recent personal experience with telerobotic manipulators the uneasy feeling that comes over you when you lock a piece of heavy metal experimental machinery onto your cranium.

The HMD work continued when Sutherland moved to the University of Utah. The 1966–67 experiments involved the use of partial systems that were constructed to solve key problems and test concepts; the computer hardware was last-generation technology. The next stage was to use state-of-the-art hardware to build a real laboratory instrument with all the debugged systems working together. The first fully functional HMD system was fired up in a laboratory in Salt Lake City on January 1, 1970. Daniel Vickers, who was a student at the University of Utah at that point, has been at Lawrence Livermore National Laboratory ever since. He had the job of getting the systems running together and creating the software that would integrate them. I called him at Livermore and asked him if there was a specific day that he can remember when the first fully functional HMD system with the first virtual world software was working. He laughed and recalled: "I remember getting the software up and running and making the first successful test on January 1, 1970, because the system consisted of several components that were always being used, and there weren't too many time slots available to debug the software. It was easy to get hold of the systems New Year's morning because everybody else who used the equipment had been partying the night before. The first image was a wire-frame cube, two inches on a side. I called home and told my wife and brought my family over to see it. We felt it was a breakthrough. We could see real possibilities for the future."

The system that was running at Utah in the late 1960s and early 1970s consisted of six interconnected subsystems, many of which Sutherland and his colleagues had invented previously in Massachusetts: a *clipping divider,* a *matrix multiplier,* a *vector generator,* a *headset,* a *head position sensor,* and a *general purpose computer* constituted the world's first reality engine. These components were the first specific machines for creating a virtual reality—the software cyberdozers that laid out the landscape of the virtual world. The *system* that constituted the MIT-Utah HMD was another inflection point in the history of VR: Using binocular displays to create a three-dimensional visual perspective is one distinct element of VR; using computers to create the graphics that are seen in 3D is another distinct element; tracking the user's direction of gaze to immerse the user in the virtual world is yet another distinct but essential element. Putting the elements together into an integrated system that enabled the user to walk around inside a computer was Sutherland's brilliant contribution to the birth of VR, and not his only one. Knowing what these devices might be used for, and how they might affect everyone's life when the technology matures, were also crucial insights Sutherland communicated early in his research. In his earliest papers, more than twenty years ago, Sutherland laid out the explicit path from his crude prototype to the virtual realities today's researchers pursue.

In order to create the three-dimensional graphics and manipulate them in real time as the users moved their heads, Sutherland, his colleagues, and his students created the "clipping divider," a special purpose computer that makes it possible to eliminate all the lines in the world model that would be behind the user's head or outside his field of view. That takes care of the "clipping" part. The "divider" refers to the capability of this task-optimized hardware to convert the data description of a three-dimensional object into a two dimensional description for display on a graphics screen. General purpose digital computers are enormously flexible, but they are not as fast as special purpose electronic circuits that are built to optimize the performance of a specific kind of calculation; Sutherland's group and every VR team thereafter have assembled systems that orchestrate special purpose and general purpose computers.

Another technical problem of three-dimensional graphics displays that Sutherland's colleagues attacked was the "hidden line" problem— removing all those lines portraying the skeleton of an object in a three-dimensional depiction that are hidden by parts of the object itself or by another object. When I look at a chair from the front, I can't see the back. When a person is sitting in the chair, I can see only part of the chair. When a computer reconstructs a depiction of a chair on the screen, and a user is looking at it from the front, the computer system must perform an enormous number of calculations to continually hide the lines that would not be visible from the user's point of view. John Warnock, then a graduate student at the University of Utah, came up with an algorithm (a way of programming the problem so it could be solved efficiently) that was usefully paired with Sutherland's clipping divider. Warnock was later one of the principal researchers at Xerox PARC, and later founded Adobe Corporation.

The clipping divider takes care of the "room coordinates" that link the user and the virtual objects to the virtual space. Seeing synthetic objects from different angles within a virtual space requires performing a calculation-intensive mathematical translation between the user's position and room coordinates. If a large number of points in each virtual object and the virtual room must contribute data to such a calculation, the cascade of secondary calculations can require literally millions of separate mathematical operations per second—something that is inexpensive today but was impossible in the 1960s, when the first HMDs were created. Sutherland and his group created the "matrix multiplier," which solved this problem in a clever way.

The wire-frame worlds on Sutherland's display screens were produced by the kind of graphic display technology known as "vector graphics." The visible light on these screens, like the light on oscilloscopes or televisions, is caused by the property of certain materials to emit light when stimulated by an electron beam. The position of the electron beam in a CRT is controlled by electromagnets that deflect the electron beam to specific locations on the surface of a glass vacuum tube that is coated with a film of material that glows (emits photons) when activated by electrons. In a television "raster" display, the electron beam is systematically scanned back and forth, up and down the screen, refreshing specific activated points during each scan. In the United States, every standard television set has 480 horizontal scan lines, regardless of the size of the display, with 640 separate picture pixel positions on each line. The pattern of activated points creates the mosaic that a viewer sees as a television image; changes in the patterns of activated phosphors are interpreted by the human visual system in terms of movements of the image on the screen.

A mosaic, however, is not the only way to draw with electrons on phosphors. In a *vector* display, the electron beam is moved directly between different points on the screen, creating visible "vectors" similar to the kind of lines you can see in the dark by waving a flashlight quickly. The matrix multiplier took the endpoints of each 3D vector that formed the edge of a virtual object, and multiplied them by the set of numbers provided by the head position indicator, and automatically updated the positions of the vector endpoints, effectively moving the view of the displayed objects in synchrony with the user's head motion. Robert Sproull, at that time an undergraduate at Harvard, designed most of the clipping divider; Sproull later was one of the cohort of infonauts who carried the interactive graphics part of the computer revolution into the era of the personal computer.

The clipping divider and matrix multiplier took information from the head position sensor, transformed it, and fed the output to the vector generator, which was responsible for "painting" the right vectors on the display screen. The display screens were part of the headset worn by the user. The 6-inch-long ¹³⁄₁₆-inch-diameter CRTs were mounted next to the temples, like a pair of small flashlights projecting forward from either ear. The light emitted by the CRTs was shunted through a series of lenses and half-silvered mirrors that projected a virtual image about 14 inches in front of the user. The image appeared as an overlay on the physical world. ("Thus," Sutherland wrote in 1968, "displayed material can be made either to hang disembodied in space

or to coincide with maps, desk tops, walls, or the keys of a typewriter.")
The first displays were capable of displaying over 3000 lines at 30
frames per second—impressive even by today's standards. The users
of the Sword of Damocles systems had a 40-degree field of view—
better than the 4- to 6-degree field of view afforded by a television-
sized screen on a desk top, but not as good as the 120 degrees of the
NASA HMD systems twenty years later, or the 300 by 150 degrees of
a multiple-screen F-15 flight simulator of the mid-1980s.

Sutherland experimented with ultrasonic as well as mechanical sensors from the beginning as head-tracking devices. Ultrasonic sensors
work on the principle that the propagation of ultrasonic signals occurs
at a constant and measurable rate, and the wavelengths of the signals
from fixed locations can be adjusted so that very fine changes in position of a mobile sound source can be located in three-space. If the
user wears an ultrasonic sound source as part of the HMD, and there
are four ultrasonic sensors mounted in the four ceiling corners of the
laboratory, it is possible to detect changes in the relationship of the
sound sources as the user's head moves, and to perform calculations
quickly enough to convert these signals into a measure of head position
in real time. But there are problems with ultrasonic sensors, including
the fact that they don't work if objects in the room come between the
ultrasonic source and the sensors.

A mechanical position-measuring system afforded the user a fairly
small range of movement, but made accurate, real-time measurements
of head position much easier. A·pair of telescoping tubes that slide
freely along their common axis were fastened via universal joints to
the headset and to tracks in the ceiling—a classic analog device that
outperformed the digital technologies of its time. The user had a
volume of possible head motion that was about 6 feet in diameter and
3 feet high. The user was free to move, turn around, tilt his or her
gaze up or down as much as 40 degrees. The fact that it made the
user a captive of the machine in a physical sense was understood to
be a temporary artifact of the primitive enabling technologies of the
time; by the early 1970s, most computer scientists were beginning to
get used to the idea that their basic tool would grow twice as powerful
and half as expensive every two years.

The engine for the first HMD was the TX-2, which was beginning
to age by 1967. The first virtual object was a cube, about two inches
on a side, that appeared to the person wearing the HMD as an object
of light, floating in space. Another early virtual object was a molecular
model—a skeletal perspective view of the compound cyclohexane.
(When Frederick Brooks and his students began their multidecade
development of molecular design tools at the University of North

Carolina, they took that model's particular explicit suggestion for future applications to the high degree of sophistication achieved by their Grope system in the late 1980s.) The later HMD experiments at Utah used a more powerful DEC PDP-10. At Utah, the cube that could be viewed from all sides remained as part of the repertoire, and in addition a much larger "room" was constructed, with the four walls marked N, S, E, and W, the ceiling marked C and the floor marked F. An observer saw the walls of the virtual room, painted in light, hanging in space, within the physical room housing the HMD. The first room in cyberspace was plain, square, and monochromatic.

The idea of building rooms in cyberspace has not ceased to attract attention by VR researchers. In 1988, NASA's VR lab had a model of the VR lab. I remember reaching out in virtual space to put my virtual hand on a virtual bookshelf, and feeling my physical hand touching the physical bookshelf, a strange sensation of being in two worlds at once. Autodesk's 1989 demonstration included an "open plan" office. In 1990, participants in the annual conference of the Association for Computing Machinery's Special Interest Group for Computer-Human Interfaces went to the Human Interface Laboratory in Seattle to fly through "Virtual Seattle," a colored, shaded, cartoonlike but recognizable and roughly accurate cyberspace replica of Seattle. By now, dozens of rooms and buildings, parts of cities and even crude solar systems exist within reality engines from Redwood City to Kawasaki. In Santa Barbara, I met a man who is converting the map of Tokyo's power network into a virtual space. By 2010, that first room of Sutherland's will have multiplied itself into a virtual cosmos. It is impossible to say, in today's terms, how vast that future cyberworld will be.

Sutherland, in the last sentence of his only partially tongue-in-cheek publication, "The Ultimate Display," extrapolated the future of virtual environments in a way that made it clear he was aware of the "magical" implications of the portal he had opened: "With appropriate programming, such a display could literally be the Wonderland into which Alice walked." In a report about a special input device for drawing lines of light in space, Sutherland's student Daniel Vickers acknowledged the magical flavor of even the earliest cyberspaces. "An observer within the 3D environment of the head-mounted display system," Vickers wrote, "has at his disposal a wand system by which he can reach out and 'touch' the synthetic objects he sees. A wand for creating and interacting with synthetic objects visible only to one wearing the headset relates an aura of sorcery to bystanders and is the basis for the name: Sorcerer's Apprentice."

Sorcerer's Apprentice was a pistol-grip-shaped wand, with four pushbuttons, a slide switch, and a small potentiometer (a dial that works like a dimmer switch). Two different replaceable tips afforded either an ultrasonic position-tracker or a mechanical position-tracker. In order to have full capabilities to create and manipulate three-dimensional computer graphics with only four buttons, the capabilities of the wand controls were extended by the use of a wall chart that mapped onto virtual space; by pointing the wand at a command listed on the wall chart and pushing a button, it became possible to change the mappings between buttons and commands. Then the user could reach the wand toward objects in the virtual world and perform different kinds of graphical magic—causing objects to appear, stretch, shrink, rotate, disappear, fuse together and pull apart. The wall chart mapped into cyberspace as a means of extending the power of the command language was a key antecedent of the "pull-down menus" that enable users today to select commands from a kind of "wall chart" on their computer screen.

Adding the wand to the system dramatically increased the sense of presence experienced by the user of the HMD, Vickers recalls. "We discovered that the sense of presence was enhanced when we added the wand. The more senses you involve, the more complete the illusion appears to be," said Vickers when we spoke in 1990.

It sounds like great fun, but before the era of computer-aided design (CAD), it was possible for those without sufficient foresight to wonder why it was necessary to spend taxpayers' money on it. From the first demonstration of Sketchpad, Sutherland made it clear that the use of such a tool in the interactive design of three-dimensional objects, from microwidgets to cities, could amplify designers' power the way other computer capabilities amplified the power of bookkeepers, census takers, and scientists. Designing objects is largely a matter of envisioning them clearly and conveying that vision to others in the form of drawings, sketches, models, blueprints, and other visual models. Just as the word processor augments the higher-level skills of writing by removing some of the low-level tasks involved with moving words around on a page, a CAD system, as the pioneers envisioned it, would augment the high-level skills of automobile designers, architects, product designers, city planners to perform tasks requiring moving lines around on a page. In the case of CAD, the computer industry was quick to pick up on ARPA's lead (although they were slower to catch on to interactive computing and time-sharing). In the early 1960s, not long after Sketchpad, IBM built a computer graphics design tool for General Motors. The acronym IBM and GM used was DAC ("design augmented by computer"), but the field ended up being known as CAD.

A lot more than the shape of next year's automobiles might be influenced by augmented 3D design tools. As several different branches of medical research are beginning to reveal, the human body is a matter of three-dimensional design, a characteristic that is very important to surgeons and diagnosticians who try to analyze and repair physiological dysfunctions by looking at two-dimensional X-rays. One of the first potential applications for the Utah HMD was medical. As early as 1971, a medical team working with Sutherland's group devised a method for repairing the simulated intersection of two large arteries. Not long after that, Frederick Brooks and his group at the University of North Carolina (UNC) began to extend Sutherland's work in medical imaging by bringing working doctors into the design loop. Sutherland went on to build his flight simulator company, Evans and Sutherland, then to work on robots that walk on legs. His students were caught up in the personal computer revolution. The University of Utah remains a center of research and innovation in computer graphics, and another project associated with the medical school, the development of mechanical and control systems for the electronic prosthetic "Utah Arm," converged with a different aspect of VR technology—the use of cyberspace interfaces to operate remote robots.

CAD developed and diffused through industry through the 1970s and into the 1980s as an expensive, powerful, fairly arcane command language for large, expensive, mainframe computers, and in that role it catalyzed enormous changes in many industries. But, like early computers, it remained a tool for those who could afford it and those who took the time to learn their way through human interfaces designed by engineers. And it remained a two-dimensional tool for envisioning three-dimensional objects. Instead of seeing a pillar or carburetor in three-space, one saw a schematic or a pseudo-3D rendering. Instead of reaching in and changing it, one typed a command code on a keyboard. It lacked the feeling of immersion and the power of navigation, but CAD was a powerful tool for using a computer screen and computer modeling capabilities to spur insights. In the 1980s, a programmer by the name of John Walker helped found a company called Autodesk on the belief that small businesses with personal computers would pay several hundred dollars for a personal computer CAD tool that offered a significant fraction of the functions of a mainframe CAD program costing thousands of dollars. (Walker will enter this story again later.) None of the major CAD vendors ventured into 3D viewing techniques until Autodesk came along in 1988 with their "Cyberia Project."

The idea of *immersion*—using stereoscopy, gaze-tracking, and other technologies to create the illusion of being inside a computer-generated scene—is one of the two foundations of VR technology. The idea of *navigation*—creating a computer model of a molecule or a city and enabling the user to move around, as if inside it—is the other fundamental element. Nothing about either of these key elements requires that they be implemented in one specific kind of technology. Images can be created with optics or electronics or both. Gestural input can come from gloves and keyboards and steering wheels. Head-mounted displays, gloves that sense finger position and movement, and magnetic head-position trackers can create the feeling of immersion and grant the power of navigation through simulated environments. Goggles and gloves aren't the only way to enter a computer. And sooner or later, sensibilities from outside the hardware or software engineering disciplines were bound to join the reality creation enterprise. Enter Myron Krueger, artist and engineer, iconoclast dreamer, the man who has spent twenty years using artificial reality as a medium for artistic expression in human-computer interaction.

"RESPONSE IS THE MEDIUM": MYRON KRUEGER AND ARTIFICIAL REALITY

The central idea of this book is the Responsive Environment, a paradigm like Hobbes' model of the world as a huge machine, which aids in understanding the essence of our current and future experience. This experience will be characterized by computer systems able to sense our needs and respond to them.

MYRON KRUEGER,
Artificial Reality, 1983

The responsive environment has been presented as the basis for a new aesthetic medium based on real-time interactions between men and machines. In the long range it augurs a new realm of human experience, artificial realities which seek not to simulate the physical world but to define arbitrary, abstract and otherwise impossible relationships between action and result. In addition, it has been suggested that the concepts and tools of the responsive environments can be fruitfully applied in a number of fields. . . .

We are incredibly attuned to the idea that the sole purpose of our technology is to solve problems. It also creates concepts and philosophy. We must more fully explore these aspects of our inventions, because the next generation of technology will speak to us, understand us, and perceive our behavior. It will enter every home and office and intercede between us and much of the information and experience we receive. The design of such intimate technology is an aesthetic issue as much as an engineering one. We must recognize this if we are to understand and choose what we become as a result of what we have made.

MYRON KRUEGER,
"Responsive Environments," 1977

Climbing into VR gear is a nontrivial exercise. Today, if you want to visit cyberspace, you need to put on sensor-equipped gloves and cover your face with a head-mounted display, perhaps wriggle into a full-body suit. The portal to VR-land now consists of what Jaron Lanier, who markets these devices, calls "computerized clothing." (And I've heard the cyberians at one VR software vendor refer to the head-mounted display, which was clearly built around a scuba mask, as "the face-sucker.") When certain technical problems are solved, however, virtual reality might not be anything you *wear*, but instead might be built into the space you *inhabit*. Doesn't cyberspace sound more comfortable as a place rather than a wardrobe? Instead of head-mounted CRTs, wouldn't it be more convenient if a three-dimensional screen simply wrapped around the VR room? Instead of gloves and suits, what if the room had sophisticated sensors that automatically detected your location, position, posture, even the direction of your gaze and your facial expressions?

We aren't used to thinking about places that can interact with us, perhaps even play with us. How would a place play with a person, even if it could? What would be the use of such an exercise? These lines of inquiry might become more than merely philosophical speculations. This semi-spooky notion of interacting with an intelligent space might point toward a place where human and machine intelligences could meet. We already are surrounded and supported by webs of computers. But most of us can't really communicate with them, and those who do operate computers are forced to use artificial languages rather than our familiar tools of speech, gesture, expression, posture. If we were to be able to interact more naturally with computers, a quasi-sentient room sounds like a likely place to establish communications with the new electronic global culture we've built.

Myron Krueger, the man who started me thinking about these questions, still looks too young to be considered the "grandfather" of anything—"boyish" is an adjective that always comes to mind when I see him—but he certainly qualifies to be called one of the founding fathers of cyberspace technology. He's old enough to have children in their twenties. His blond hair is showing gray at the temples, but something about his face keeps him from looking anything like a gray eminence. The corners of his eyes betray the presence of a prankster laying in wait behind the cultivated ordinariness of his appearance.

Krueger reminds me of a few other visionaries I have met, people who have stubbornly pursued strong personal visions despite a lack of recognition by the mainstream. It's hard not to feel hurt personally when you see a way to make the world a better place, but others fail to see the same possibility. And then, twenty or thirty years later, it has to be frustrating when everybody begins to catch on to the possibility you've been trying to point out and everybody starts rushing down all the alleys, blind and otherwise, you had explored so carefully yourself long ago.

I first saw Krueger when he stopped by my house in California, not long after I started writing this book. Knowing what I now know about his bad luck with journalists in the past, I can't help suspect that he had responded to my invitation just to make sure I wasn't going to ignore his place in the history of VR. It had taken him years to find a publisher for his book, *Artificial Reality*. In April, 1989, when *The New York Times* ran their first front-page article on "What Is Artificial Reality?," the piece failed to mention Krueger, who invented the term. When I called him to introduce myself, I discovered that he already knew about my book. He found himself in California a few weeks after we first talked, and took me up on my invitation to drop in. I was curious about the reasons he'd been working on his approach to artificial reality, despite setbacks, for all these years.

We sat at my kitchen table and talked for hours. Months later, when I took a train from New York to Hartford to visit him and see his artificial reality laboratory, Krueger rented a car to transport me to the University of Connecticut, because the seat belt on the passenger side of his own car wasn't working. He does things like that. And the dearth of clippings I managed to turn up about him made it clear that Myron Krueger hasn't been besieged by hordes of journalists and historians, despite his years of hard work on what now appears to be a fundamental technological breakthrough. He was genuinely happy to see anybody who was interested enough in his ideas to make the trip to Storrs, Connecticut.

Although the mainstream of today's VR technology-industry-science research involves computer-modeled illusions and three-dimensional visual displays, another vision from the past that might become increasingly important in the future is the idea that the very walls around us can be made aware of human behavior and can learn how to respond to us. The part of the system that fosters the feeling of immersion and interactivity, the sense of being in a new kind of space, might develop into directions other than face-sucking masks and sensor-laden bodysuits. To the extent that a "media room" rather than "reality gloves and goggles" is a strong if technically difficult vision for the long-term future of VR, Myron Krueger must be credited for his early work in the field that he named "artificial reality." Indeed, it's hard to find a VR frontier today that Krueger didn't explore (with the help of now-obsolete technologies) decades ago.

Like Morton Heilig and Doug Engelbart, however, Myron Krueger seems to have made a career of falling into the gaps between the conceptual frameworks of scientists and artists, technologists and academics, computer scientists and educators, performers and programmers. He and I have talked about his odd life story, and Krueger admits, "I've never gone out of my way to force myself to fit into the categories." The 1990s started bringing him the recognition he might well have been given twenty years ago. When all the hardware and software that make VR possible evolve to the point where we don't notice them, people will get down to the business of learning what human beings will do with the powers unique to the new medium. VR researchers of the future will do well to remember something Krueger has been saying for two decades: "Response is the medium."

Krueger believed from the beginning of his quest, which has by now included rooms full of homemade electronic gear and tens of thousands of hours of computer programming, that the visual and audio effects, the video imagery and computer graphics, the input technologies and display systems, and the software that controls it all, can be seen as a tool for eliciting new kinds of human behaviors. Concentrating on the tools is good and necessary, Krueger will concede, but please let's not continue to ignore the behavioral, psychological, social, artistic components of VR, he continues to argue.

At the center of every VR system is a human experience—the experience of being in an unnatural or remote world. Attention is the instrument in cyberspace. Painters, poets, therapists, teachers, shamans, and playwrights might have something to contribute to our knowledge about the experiential side of VR, and much will be learned only by inventing new art forms. Just as the Impressionists began a perceptual revolution in the medium of oil painting at the height of the Industrial Revolution partially in reaction to the mechanical worldview engendered by the introduction of the camera, future artists will have to wield VR engines as brushes or bassoons, and paint the silence with the kind of possibilities only artists can show us. If art is about seeing the world in new ways, and VR is an instrument for creating worlds, artists might furnish the clue to a key question: If our technology ever allows us to create any experience we might want, what kinds of experience should we want to create?

Like others in the VR story, Myron Krueger had a kind of conversion experience (a series of them, in fact) that has compelled him to follow his particular vision of cyberspace technology for the past two decades. I met him almost exactly twenty years after the day GLOWFLOW opened to the public. In April, 1969, visitors to the Memorial Union Gallery of the University of Wisconsin at Madison had the opportunity to witness and participate in the event that started Krueger on the road to artificial reality (a term he first started using in proposals around 1974). GLOWFLOW did not use computer graphics, but created visual effects through the use of other technologies. Hidden minicomputers and sound synthesizers and a network of tubes filled with phosphorescent colored fluids turned a darkened space into something nobody had ever seen before. Krueger, a graduate student at the time who had been invited to join the GLOWFLOW project, has been working ever since toward the ultimately responsive environment, an "artificial reality" that simply surrounds users, without encumbering their hands or head with gloves or goggles.

GLOWFLOW used the properties of a computer-controlled light-sound environment to give participants the sensation of inhabiting a space that responds to human attention and behavior. The walls of the GLOWFLOW environment were lined with opaque, vertical columns and transparent, horizontal glass tubes. Phosphorescent particles were suspended in water and pumped through the tubes at intervals, passing through the columns in which hidden lights temporarily activated the phosphor, creating instant vectors of light that floated in space and decayed into blackness. The patterns pulsed on and off according to instructions mediated by the minicomputer hidden behind a partition; the light show was accompanied by electronically synthesized sounds. The responsive part of the environment was embedded in the floor, where pressure-sensitive plates enabled members of the audience to control the displays of lights and sounds, unaware of their role in the performance. Unlike today's VR technologies, the machinery in GLOWFLOW was mostly invisible. The effect on the people in the galleries, however, was extraordinary.

In 1983, Myron Krueger wrote this about GLOWFLOW:

People had rather amazing reactions to the environment. Communities would form among strangers. Games, clapping, and chanting would arise spontaneously. The room seemed to have moods, sometimes being deathly silent, sometimes raucous and boisterous. Individuals would invent roles for themselves. One woman stood by the entrance and kissed each man coming in while he was still disoriented by the darkness. Others would act as guides explaining what phosphors were and what the computer was doing. In many ways the people in the room seemed primitive, exploring an environment they did not understand, trying to fit into what they knew and expected. Since the GLOWFLOW publicity mentioned this responsiveness, many people were prepared to experience it and would leave convinced that that room had responded to them in ways that it simply had not. The birth of such superstitions was continually observed in a sophisticated university public.

The artists who had recruited his help were committed to maintaining a contemplative quality to the piece. Krueger wanted to zero in on the individual audience members' awareness of the way their actions influenced the patterns of light and sound. The visible and audible artificial space was beautiful, but more interesting to Krueger was the way this new kind of place elicited strong human reactions. GLOWFLOW, in Krueger's opinion, could be an instrument for probing this new unknown psychological territory they had stumbled into. But the artists insisted on introducing delays and otherwise disguising the precise relationship between the movements of participants and the environment's reactions. Krueger decided to build an artificial environment that focused on the aspects that interested him.

Early in his career Krueger decided that while the use of computers to amplify traditional visual, musical, and dramatic arts is a legitimate application of the technology, none of these artistic tools took advantage of the unique properties of computation. In fact, nobody had seriously explored what these unique properties might be. He began to see the outlines of what these unique computational properties might contribute to a pure form of computer art; more important, after GLOWFLOW he began to see ways to experiment with computers' unique artistic potential, and thus test his hypotheses. Such experimentation was clearly fun in itself, but Krueger also believed from the beginning that "responsive environments," as he began to call them, were laboratories for finding out how humans might harmonize with our technological environment, instead of just reacting with shock and fear to the new tools that were entering our lives.

Krueger continues to believe that computer art is fundamentally interactive: "Other artistic uses of the computer are of interest," he concedes, "but they do not constitute a new art form based on the computer." Krueger had no interest in arguing the legitimacy of other computer-based arts; he simply wanted to show that the most appropriate computer-based arts remained to be discovered, and that the works of art themselves could be the instruments of discovery. He built these tools in engineering shops, exhibited them in galleries, and ran them as social experiments. To Krueger, responsive environments were laboratories for experimentation in education, psychology, art, and computer science, all at the same time. Those who participated in the exhibition-experiments were excited, but theorists in both the arts and the sciences reacted with confusion. Responsive environments as Krueger envisioned and built them weren't the kind of commodity or performance that artists were accustomed to seeing as art. They aren't paintings that can be reproduced, or even musical compositions that can be identified by their composers, but truly artificial realities that present invisible social constraints and communication opportunities, invite certain kinds of behaviors and discourage others, amplify some aspects of human thought and behavior and mask others.

After GLOWFLOW, Krueger concocted yet another kind of responsive environment. METAPLAY was exhibited a little more than a year after GLOWFLOW, in May, 1970, again in the Memorial Union Gallery in Madison. (One Kruegerism that seems to have persisted for the past twenty years is his use of capital letters, a reminder of the days when capital letters were the only kind of text that computers could print.) In Krueger's words, "Traditional criteria of art, beauty, and responsive subtlety were set aside. The focus was on the interaction itself and on the participant's awareness of the interaction." With METAPLAY, sponsored by the National Science Foundation and the University of Wisconsin's Computer Science Department, Krueger went beyond simple lines of light and synthesized sounds by including video cameras, back-projected video screens, computer graphics tools, and 800 pressure-sensitive switches. Krueger convinced Digital Equipment Corporation (DEC) to loan him a PDP-12 computer (the descendant, a decade later, of the PDP-1 that had triggered Licklider's conversion experience, and the ancestor of today's DEC computers used as reality engines at several research sites).

One wall of the METAPLAY room was replaced by an 8-by-10-foot rear-projection screen; a video projector was hidden behind the screen, and so was a video camera aimed at the participation area. The participants could look at the screen and see their own video images; sometimes, the video image would be overlaid with computer graphics images. The pressure-sensitive switches were concealed on the floor under a sheet of black polyethylene. The interaction between participants and the environment was mediated by a human facilitator who was monitoring the proceedings via video monitors in a control center located in another building.

Inside the METAPLAY room, a few participants looked at the screen on which their video images had been projected. Back in the control room, a quarter-mile away, a drawing tablet enabled the facilitator to create computer graphic designs and display them on one of the first special-purpose computer graphics systems, an Adage Graphic Display Computer. A video camera was aimed at the monitor of the Adage, and the video signal from Adage went through a video mixer along with the live video feed from the METAPLAY room. By this method, the computer graphics were converted to video format. The computer graphics and video feeds in METAPLAY were all under the facilitator's control, and the facilitator could overlay the computer graphics designs drawn with stylus and tablet on top of the video image of the audience captured by the camera. Various knobs and dials enabled the facilitator to shrink, expand, and apply other special effects to either the video or computer graphics displays.

Hoping to provoke a reaction in one of the first METAPLAY audiences, Krueger selected one of the participants and used the graphic tablet and pen to superimpose an outline of light over his hand. When the participant moved his hand, Krueger quickly drew a fresh outline of the hand in its new position. The participant then turned the tables on Krueger and transformed him from an invisible prankster into an artistic collaborator. The participant started using his finger like a stylus and mimed the act of "drawing" a line in midair. Krueger drew a line that followed the participant's gesture. The line appeared on the wall-sized screen in the METAPLAY exhibit space for all participants to see. From that moment, the environment itself and the behavior of the hidden facilitator became a magical device for drawing lines in video space. Participants in METAPLAY found that they could pass control of the "magic finger" from one to another by touching their fingers. A possibility inherent in the system but not explicitly designed by its creator had revealed itself through interaction. The participant played with the capabilities of the artificial space and discovered a new tool. And the facilitator metaplayed with the participant's reactions to draw attention to that inherent possibility.

His next piece, in 1971, was PSYCHIC SPACE, a "composed environment." The walls and ceiling were covered with black polyethylene, the floor contained six rows of eight 2-by-4-foot pressure-sensitive modules, and a false wall concealed a wall-sized rear projection screen opposite the phosphorescent wall. Again, a control booth in the computer center on the other side of campus received a video image from the camera along with messages from the floor sensors, manipulated that information by means of computer programs, and fed mixed video and computer graphics displays back to the participant in real time. Thus, a close relationship between the movements of participants within the physical space and the response of the audiovisual displays was made explicit in the design of the environment.

PSYCHIC SPACE was a tool for creating a variety of realities. In "the maze," Krueger managed to design an amusing interaction within the virtual space defined by the video display and position-sensing floor. Walking into PSYCHIC SPACE alone, a participant in the maze finds a graphic symbol on the screen corresponding to his or her position in the room; moving forward moves the symbol closer to the top of the screen, moving back away from the screen moves the symbol closer to the bottom. Movements to the left and right also were mapped onto the screen. After the participant seems to understand the relationship between movement within the space and the position of a graphic object, the maze automatically puts another graphic object on the screen in a different place. Invariably, the participant walks toward the part of the room depicted by the second object, to see what would happen if both objects occupied the same place. At that point, a maze appears on the screen; participants then make their way through the dark, unmarked floor space and thread their way through the virtual maze by observing the movements of their symbol on the wall-sized screen.

Simply solving the maze, however, just produced another maze: eventually, participants realized that apparently there was no way to "win." When the participant decided to "cheat," however, and the maze replied by stretching a boundary so the participant couldn't step over it, or moved it one square beyond wherever the participant stepped, the purpose of the interaction—playing with the mental boundaries we pretend into existence, like the rules of the ruleless maze—began to become more clear. The maze was a game of ideas, invented by Krueger, implemented by a computer program, using movements and visual signals as tokens, played out within the floor space of PSYCHIC SPACE, to create a virtual place that existed chiefly within the mind of the participant. The "psychic space" hidden within the darkened room was thus created by a process of experimentation and discovery.

Between 1972 and 1974, Krueger's continuing experiments with different systems for sensing position and responding to video images led to the creation of a general-purpose artificial reality laboratory instrument. After PSYCHIC SPACE, Krueger started working on VIDEOPLACE. Although a preliminary version was exhibited in October, 1975, at the Milwaukee Art Center, this environment was designed to be an open-ended laboratory that would evolve over a period of years. Now that he had a glimpse at the possibilities of mixing video, computer graphics, and gesture/position-sensing technologies, Krueger had an idea of what power he would be able to gain from the increases in computation power that were coming along by then. Computational tasks such as video pattern recognition—a computer program that could analyze a video image of you and determine whether or not you were pointing your hand at an object on the screen—had been unthinkable with the computers available to him in the 1960s, but the computers of the late 1970s made it possible to dream of a system that could not only record but recognize the positions or gestures of participants in a responsive environment.

By the mid 1970s, VIDEOPLACE, by then a full-blown artificial reality laboratory, had migrated to the University of Connecticut. This idea of a new kind of environment, created by human perceptions triggered or mediated by video and computer technologies, continued to be an important theme in his explorations. In his 1977 report to the National Computer Conference, Krueger wrote this statement about VIDEOPLACE, revealing something about his thinking regarding the nature of the "artificial realities" he was building:

> VIDEOPLACE is a conceptual environment with no physical existence. It unites people in separate locations in a common visual experience, allowing them to interact in unexpected ways through the video medium. The term VIDEOPLACE is based on the premise that the act of communication creates a place that consists of all the information that the participants share at that moment. When people are in the same room, the physical communication places are the same. When the communicants are separated by distance, as in a telephone conversation, there is still a sense of being together although sight and touch are not possible. By using television instead of telephone, VIDEOPLACE seeks to augment this sense of place by including vision, physical dimension and a new interpretation of touch.

The early versions of VIDEOPLACE consisted of two or more rooms separated from one another by various geographic distances. Video cameras, mixers, and projectors would make it possible for people in any of the different rooms to interact with the video images of others who were physically located elsewhere. Krueger had noticed by that time that people identified themselves very strongly, almost physically, with their video images, even in the form of silhouettes. In an early experiment, when Krueger and an assistant at a remote location were using video silhouettes of their own hands to point at objects in a shared video space, he accidentally moved his hand's video image so that it intersected with the video image of his assistant's hand. The assistant moved his hand away, as if he had been touched. In a visceral way, mixing people's video images together in a way that was visible to them created a new kind of communication space, complete with a sensitivity to the boundaries of one's virtual body.

What could be done with such a strange hybrid medium? "Let people play with it and find out for themselves," has always been Krueger's answer. As he wrote in a later article in *Leonardo,* a magazine about art and technology: "In VIDEOPLACE, two fundamental cultural forces—television, purveyor of passive experience, and the computer, symbol of forbidding technology—have been married to produce an expressive medium for communicating playfulness and inviting participation."

The idea of "picturephones" has been the stuff of Sunday supplement articles for decades, but experiments in video teleconferencing have never proven wildly successful. However, picturephones merely represent the video image of communicants in their separate physical spaces. Krueger made the crucial modification of putting the video images of all communicants together in a shared video space visible to all. As early as 1977, Krueger acknowledged that he was experimenting with a new form of telecommunication as well as a new way of interacting with computers, a statement that makes a new kind of sense today, with Japanese telecommunication companies sinking tens of millions of yen into VR communication research:

The responsive environment is not limited to aesthetic expression. It is a potent tool with applications in many fields. VIDEOPLACE clearly generalizes the act of telecommunication. It creates a form of communication so powerful that two people might choose to meet visually, even if it were possible for them to meet physically. While it is not immediately obvious that VIDEOPLACE is the optimum means of telecommunication, it is reasonably fair to say that it provides an infinitely richer interaction than Picturephone allows. It broadens the range of possibilities beyond current efforts at teleconferencing. Even in its fetal stage, VIDEOPLACE is far more flexible than the telephone is after one hundred years of development. At a time when the cost of transportation is increasing and fiber optics promises to reduce the cost of communication, it seems appropriate to research the act of communication in an intuitive sense as well as in the strictly scientific and problem-solving approaches that prevail today.

Physically, the public area of VIDEOPLACE is a 16-by-24-foot room, with a wall-sized video screen and a video camera. The basic hardware for capturing and displaying video images, mixed with graphics, remained the same. What Krueger concentrated on improving was the parts of the system that gave the room a greater degree of sentience regarding human behavior. This was done by a combination of special purpose hardware and software techniques for analyzing the video image. In order to simplify the task, Krueger long ago decided to work with the colorized silhouettes of participants and leave full video imagery for future research. With a silhouette, it became possible to analyze images by focusing on the pixels and scan lines in an image. And by applying a set of mathematical transformations to the information analyzed in such a way, higher-level determinations could be made automatically—edges could be detected, intersections of objects could be pinpointed, orientation of lines could be determined.

A line on a screen, visible to a participant, consists of a number of adjacent pixels in one or more scan lines that are activated by the CRT's electron beam; the position and state of each pixel can be stored and updated in the computer's memory. The computer can determine which pixels are on and off in each line, perform calculations on those endpoints, and thus submit the digitized image to certain mathematical tests. By extracting the position of the endpoints of a line, for example, and then testing the endpoints of that line after a specific number of scans, it becomes possible to determine whether the line is at rest or moving. By testing the position of the image-pixels on the screen it is possible to track the top or the bottom of an image. Curves can be distinguished from straight lines. It is possible to design a set of primitive tests like these that could become building blocks for creating on the fly a model of the activity captured by the video camera—and that would be the way to teach a computer to automatically recognize whether a person is pointing or jumping or moving, and respond to the motion appropriately in real time.

The problem with approximating human perception with technology, then putting it to the test, is that human perceptions do things that no technology can do. The problem in implementing a system capable of recognizing simple shapes on a general purpose computer is that such calculations can add up to such enormous demands on the computer's capacity that it is not possible to do them in real time; and thus interactivity, Krueger's *sine qua non,* suffers when precision in the direction of pattern recognition is increased. You can put a fuzzy or simple image on a screen quickly, and even move it around, but if you want to put a sharper or more complex image on the screen, it isn't possible to move it around as quickly.

Adding intelligence to the system without slowing it down was a supreme technical challenge. Krueger, self-taught electronic engineer, did exactly what Ivan Sutherland had done with the first head-mounted displays when he built clipping dividers and matrix multipliers to perform some special functions at very high speeds with hardwired circuitry rather than performing those functions more slowly by programming a general purpose computer. General purpose computers can do a lot of different kinds of tasks fairly swiftly—extremely swiftly from the human point of view. That doesn't mean computers can perform *all* tasks so swiftly. The computation-intensive nature of computer graphics means the more complex you make your sets of tests to try to understand what is happening in a scene, and the more often you want to perform these tests on input information, the longer it will take to process the output. If, however, you could build a specific processor for performing a specific test the way electronic circuits can do, it is possible to perform special purpose calculations at far higher speeds than general purpose computers can achieve. So Krueger built special processors, amounting to twelve by 1990, to perform subsets of critical pattern recognition functions very quickly. Each handmade processor he added to VIDEOPLACE over the years was a vision algorithm optimized in silicon, a hardwired circuit that did what a specific pattern recognition program could do, but far faster.

"There are processors to determine whether a line is long and skinny or short and fat, processors to measure curves and boundaries, processors to count all the crossings of lines, processors to find the edges of an object," Krueger explained, when we talked in 1990. He wanted to be able to examine a video silhouette and determine automatically whether the person in the video image was making a gesture; this quest landed him in the middle of one of the knottiest problems of robotics and AI.

Built into that hardware-software architecture is something that Krueger believes to be dominant at all levels of artificial reality design: *context.* That's where the world model created for the general purpose computer comes in. In the context of a human body, the highest line of a silhouette means the top of the head, the rightmost endpoint of a skinny edge means a fingertip, and in the context of accepted gestural grammar, a fingertip is an indicator to pay attention to where it is pointing. Context can be psychological or social: the human propensity for problem-solving was part of the context for the maze. Shadows, perspective, parallax, and other visual cues are part of the perceptual context for the cognitive experience of being immersed in a three-dimensional world. It's the kind of intuitive recognition of the texture of certain complexities that enables us to see key parts of the complex system in simple ways. Our eyes seem to have hardwired circuits for detecting edges, fast-moving objects, dots of red, small amounts of light, small motions in large environments. It is one way humans cut through the morass of calculations that must be involved in the kind of pattern recognition we do every second, without thinking about it.

CRITTER, a VIDEOPLACE experiment in building a context-seeking artificial creature, is a cartoon, resembling a small bug, that plays with the participant's video image. Imagine walking into a room where a bright blue image of your silhouette appears against a red background on a wall-sized screen. You are alone in the room, but on the screen, you appear to have a companion, a yellow, round, four-legged cartoon that scurries around the screen, just out of reach. If you chase it, CRITTER eludes you. If you stand still, it moves toward you; if you walk away, it follows you. If you remain still and extend your hand out from your side, CRITTER will land on it. If you move your position, CRITTER will cling to your silhouette until you stop moving again, and then continue its task of climbing up your outline. When it reaches the top of your image's head, it dances a jig—the first time. If it makes it to the summit again during this interaction, CRITTER will pace back and forth if your hands are by your sides, taking a flying somersault into your hand if it is extended at shoulder level. Depending on how you move, and the history of your interactions, CRITTER performs surprisingly sentient-seeming tricks within the VIDEO-PLACE virtual space. By now, the lines of programming code that constitute CRITTER's personality number in the tens of thousands.

When I was in Connecticut, in the summer of 1989, VIDEOPLACE was en route to Japan, where Krueger and his latest version spent a week at the opening ceremony of a Japanese "Science City." I had decided to come out to see Krueger's laboratory, even though his equipment was in transit, in order to continue the conversation that had started at my kitchen table. The equipment he uses seems alien and archaic in comparison with the head-mounted displays and cyberclothing the other VR labs use, but Krueger simply has logged more hours building artificial spaces and putting them through their paces than anybody else in the VR world. He picked me up in Hartford in his rental car. We drove to Storrs and made our way across campus. His Artificial Reality Laboratory is located in the back of the Connecticut State Museum of Natural History, in a room that smells just like every natural history museum my mother shlepped me to when I was a little boy. Rows of raptors—hawks, ospreys—"the finest collection of mounted birds in New England," stare mutely from their cubicles. High ceilings, echoing footsteps, the cool muskiness of an old museum on a summer's day.

One of the pieces of equipment that Krueger shipped off to Japan was a desktop version of a VIDEOPLACE-style artificial reality—VIDEODESK. A camera above your desk captures the image of your hands, in silhouette; a camera over your remote partners' desks captures the images of their hands. In a shared video space, you can use your hands to gesture and point at text or graphic materials. "I can give you your desk space back. There's no need for a mouse or a physical keyboard. One gesture can cause a keyboard to be projected onto the artificial desk. You could paint with your fingers, or with the use of a three-dimensional drawing system, you can sculpt 'graphic clay' with your hands." Still chasing the grail, still three steps ahead of the state of the art, still slightly out of synch with everybody else, still half a step behind his funding, Myron Krueger is charging into the 1990s with a full head of steam. The last time I talked to him, he was negotiating with an international rock star to build exhibits for a theme park, designing a gestural input system for a major personal computer manufacturer, and drawing up the preliminary design for an exhibit in Spain.

I ended up visiting Japan about six months after Krueger had been there. VIDEOPLACE had been part of an event called, in the sometimes slightly stilted Japanese style, "Wonderland of Science/Art." It was a week-long ceremony celebrating the opening of Kanagawa Science Park, a research and development complex sponsored by a prefecture outside Tokyo. VIDEOPLACE itself was installed in a laboratory in one of the four 20-story buildings that made up the brandnew research complex. The Japanese, it turns out, are not only very interested in "Science/Art," but committed to building the kind of communication tools Krueger had dreamed about for decades.

Over the years, several other related experiments began to spawn their own pre-VR media cultures. In the 1970s Kit Galloway and Sherrie Rabinowitz used a video space concept in a geographical experiment called "Hole in Space": A large video screen and camera were set up in a public space in New York City; another setup was in Los Angeles. The rest was up to whoever happened upon the experiment. People started reacting to one another. The word started to spread and people used the telephone to make appointments to meet at the Hole in Space, which stayed in existence for days at a time—a place of electrons and gestures, created by humans and technology, exploring the boundaries of a new communication medium. A similar project was conducted at Xerox PARC in the late 1980s; PARC called it "Electronic co-presence." One wall of a room at PARC showed the video view of a room at another Xerox research facility in Oregon. People would schedule meetings, or they would just wander into their respective ends of the two-way, full-time, shared video space, and strike up a water-cooler conversation. While I was conducting research for this book, I saw definite attempts to replicate Krueger's work and other early work in shared video spaces, both in Japan and in the Netherlands.

Although he took the first steps down several research paths that are being seriously pursued today from Kyoto to Eindhoven, Krueger's research was not in the mainstream of VR's early emergence. VR as we know it today did not flower directly from the work of Heilig or Krueger, but from an intersection of computer science, stereoscopy, and simulation, in academic, military, and commercial research laboratories. The science and technology of VR as it appears today emerged in several places during the late 1970s and early 1980s. Frederick Brooks, Stephen Pizer, Henry Fuchs, and other computer scientists have been advancing the state of the art since the late 1960s at the University of North Carolina at Chapel Hill—Brooks dates the beginning of the effort to 1967. Thomas Furness's "supercockpit program" in the US Air Force sponsored two decades of continuous research at Wright-Patterson Air Force Base into head-mounted displays. Scott Fisher was working at MIT in the late 1970s on an interactive stereoscopic display for what he called "virtual exploration." Negroponte and others pursued "the transmission of presence" at MIT. But it was in Mountain View, California, that the "conversion experiences" started happening again.

The field of VR began to crystallize when the right combination of sponsors, visionaries, engineers, and enabling technologies came together at NASA's Ames Research Center in Mountain View, in the mid 1980s. It was there that a human interface researcher, a cognitive scientist, an adventure-game programmer, and a small network of garage inventors put together the first affordable VR prototypes. It was there that a generation of cybernauts donned a helmet-mounted display and glove input device, pointed their fingers, flew around wireframe worlds of green light-mesh, and went back to their laboratories to dream up the VR applications of the 1990s.

For me, one of the most intriguing aspects of my encounters with the NASA group and the cyberspace machine they had assembled was the fact that I had met many of them years ago. The first time I met them, I was pursuing what I thought was a different topic, but it has turned out to be one of the key plot lines woven into the virtual reality scenario: if you set out to build a mind amplifier, you are likely to start thinking about the ultimate human interface. Seven years ago, I was one of the fairly small number of technology observers familiar enough with the leading edge of computer research to know about the notion of head-mounted displays, but like most of the others, I didn't realize that HMDs would kick up such a fuss before the decade was out. In 1988, when I saw the system that these young cybernauts had assembled, I had the conversion experience that changed my mind about the potential importance of inhabitable computer spaces and the personal allure of the VR experience. In the two years that followed my first flight through cyberspace, I've met people around the world who were fired up to start their own VR laboratories because they had experienced a similar demonstration in Mountain View.

Part Three

THE REALITY-
INDUSTRIAL COMPLEX

Chapter Six

BLASTOFF AT NASA

Although the current prototype of the Virtual Environment Workstation has been developed primarily to be used as a laboratory facility, the components have been designed to be easily replicable for relatively low cost. As the processing power and graphics frame rate on microcomputers quickly increase, portable, personal virtual environment systems will also become available. The possibilities of virtual realities, it appears, are as limitless as the possibilities of reality. They can provide a human interface that disappears—a doorway to other worlds.

SCOTT FISHER,
"Virtual Interface Environments,"
1990

The first time I peeked my head and poked my hand in a virtual world, I began my journey by slithering my right hand into a VPL DataGlove. The snug Lycra garment, lined with sensors running down the backs of my fingers, translated my finger motions into a stream of digital snapshots and piped the data through a cable to a refrigerator-sized computer in one corner of the lab. The glove was stretchy enough and the cables light enough to give me more degrees of freedom than a pair of ski mittens and less than that afforded by my naked hand; I probably couldn't thread a needle easily, but I could definitely pick up a chess piece or wield a screwdriver. NASA researcher Scott Fisher had helped me fit my face into a pair of goggles that contained two miniature television screens and optics aimed at my eyes. Like the glove, the head-mounted display was cabled to a computer that was running the software models of the world I was about to enter.

The computer and special software together constitute what is known in the VR trade as the "reality engine." A detailed three-dimensional model of a virtual world is stored in the computer's memory, encoded into microscopic lattices of bits. When a cybernaut shifts his gaze or waves her hand, the reality engine weaves the data stream from the cybernaut's sensors together with updated depictions of the digitized virtual world into the whole cloth of a three-dimensional simulation. The computer-based engine, however, contributes only part of the VR system. Cyberspace is a cooperative production of the microchip-based reality engine sitting on the floor of the laboratory and the neural reality engine riding in my cranium. The computer converts its digital model of a world into the right pattern of dots of light, viewed from the right perspective, and audible waveforms, mixed in the right ways, to more or less convince me that I am experiencing a virtual world.

When I fit the eyepiece into place, my entire field of vision was cut off from the outside world. I heard keys clicking on a keyboard somewhere. Suddenly, the world lit up around me, green on black. I found myself floating inside a luminous wire-frame depiction of a space shuttle. Precisely disparate images presented to each of my eyes combined to create a near-perfect three-dimensional stereogram, a close to true 3D image. Stereographic enthusiasts would approve of the qualifiers: computation-intensive "shaded polygons" (the building blocks of high-resolution computer graphic models) and other techniques to present secondary depth cues could produce a more perfect three-dimensional image than wire-frame. "Wire-frame" is a way of representing a solid object on a computer screen as a mesh of contoured lines that outline the object like a crosshatched, tight-fitting garment. In cyberspace, the computer has to recompute the whole world every time you move your head. It takes more computer power—and thus, more expensive computers—to create a solid, colored, properly lit and shaded object.

NASA, appropriately, was the institution that launched the first real public exploration of cyberspace. At NASA, the streams conjoined. Scott Fisher, an MIT student of Negroponte's, had on the wall of his office the Sensorama poster that led me to Morton Heilig. Fisher considers himself a scientist and an artist, and would be the last to consider himself an evangelist. All he did was show people to the goggles and gloves and let them judge for themselves. NASA was the opening of the era of the reality-industrial complex—the network of academic and commercial research and development and entrepreneurial ventures that might grow into a new technology-based industry. The work at UNC on all aspects of VR had been continuing, quietly, for years. But they didn't give a lot of demonstrations in Chapel Hill, and are growing understandably reluctant to spend research time on public relations. At NASA, especially in the "proof of concept" stage, giving demos is a way of life. Fisher and his colleagues were intent on bringing surgeons and educators and scientists into the process.

Futurists and intellectual adventurers like myself managed to slip into the lab and take a look at the brave new technology—and catch the fervor ourselves, from the inside. Whether we are for it or against it, extrapolate it or debunk it, those of us who have tried today's VR technologies for ourselves always seem to want to talk about what it all means. As I traveled around the world to different VR sites, I discovered that many who were to carry the banner of VR research to Manchester and Seattle and Tsukuba had their first head-in demonstration of VR in Mountain View, California, at the NASA/Ames Human Factors Research Division. The same Ames, in the same place, that Douglas Engelbart was working when he had his augmentation brainstorm in 1960. Like ARC, the Human Factors lab was the site of the conversion experience. The descent into the electronic cave.

My first ride was in a cheap version of cyberspace, constructed from off-the-shelf components. The NASA system I tested in 1988 was created in part to show that a reality system could be constructed relatively inexpensively; the Air Force had been flying million-dollar helmets for years. Using wire-frame rather than shaded-polygon depictions knocked a couple of zeroes off the price—always a key consideration at an impoverished bureaucracy such as NASA. And NASA's demonstration of the feasibility of near-garage-scale virtual reality research set off a wave of commercial and academic interest in the late 1980s that created the foundation for the VR industries of the 1990s, when garage-scale R & D became possible.

Because two key researchers had known each other from the time they both worked at Atari Research Laboratory, the NASA prototype was the first time a head-mounted display and a DataGlove were joined. The first time I saw the glove, in 1985, in Jaron Lanier's musical-instrument-strewn bungalow in Palo Alto, he was using it with an inexpensive, regular flat-screen 2D personal computer. In Mountain View in 1988, inside my goggles, a 3D wire-frame depiction of one gloved hand floating in virtual space mimicked every movement of my hand in real space. I waved. There was a perceptible lag. A couple of hundred milliseconds later, the hand waved. I flexed my fingers. The hand flexed in precisely the same pattern. I took a step forward. A couple of hundred milliseconds later, the hand and my point of view moved forward in virtual space. Cyberspace was everywhere I looked—above me, below me, behind me. I wasn't just watching it. I was *in* it.

"Point your forefinger in the direction you want to fly," Fisher's voice instructed me, from somewhere out there in physical reality. I pointed. My point of view started to fly. *I* started to fly. Although my real feet remained on the ground in California, my point of view and my simulated hand suddenly gained the power to fly through a cartoon version of outer space. I could point up and zoom out into space to observe the space shuttle model from afar, or point down and zoom in to inspect the loading bay. My velocity could be accelerated if I moved my hand away from my body, or slowed to a crawl by moving my hand closer to my body. I could reach out my real hand, and my simulated hand would grab simulated objects in virtual space, change their position or orientation, hurl them into the distance. During my explorations of this strange new synthetic space that few others have yet visited, I stood, squatted, crouched, walked, adopted inexplicable postures, gazed into odd corners of the room, aimed my head down at the floor or up at the ceiling, pointed my forefinger at things invisible to everyone else. I'm sure it was a humorous sight to the other people in the room. But that didn't matter to me at the time. I was in another reality.

By using computer techniques combined with electronic audio technology, NASA researchers have learned to simulate a three-dimensional acoustic environment the way head-mounted displays and stereographic images simulate a three-dimensional visual environment. In later demonstrations at NASA, 3D sound via earphones removed me even farther from contact with the nonvirtual world. But that first time, I learned that my universe was likely to change around me whenever I heard the sound of fingers on a computer keyboard. Scott Fisher was diligent in warning me about what was going to happen next. He typed a new command and, just as he promised, I found myself inside a six foot-high tinkertoy of a hemoglobin molecule. The molecule was an enigmatic symphony in three-space, folded into the shape that enables air-breathers like myself to suck oxygen out of the air and distribute it to our internal energy factories through our bloodstream. It was a stick-figure representation, very low-tech compared with the high-resolution molecular models I was to see elsewhere. But I was hooked. By the time I started to wonder whether VR was really as exciting as I started thinking it was, I found myself in Chapel Hill and Cambridge, poking my perceptions into ever-more-real virtual worlds.

I can see now that my meetings with Scott Fisher and others from the Arch-Mac gang over the past seven years can be seen as data points on a time line, a graph of my intersections with VR, the shape of which I have only recently begun to perceive. Thinking about these encounters helped me notice how virtual reality has been drawing me toward it for years, like a strange attractor in my writing career. In 1983, I met Fisher at Atari's Sunnyvale Research Laboratory, a legendary outfit, now defunct, that served as a place-de-passage for many of today's cybernauts. I can also see now that Atari Research, like Arch-Mac before it, was another kind of strange attractor that sucked in, blended and bonded, inspired and dispersed many of the reality engineers in today's VR labs.

Atari Research existed during a peculiar moment in history: video games, a technology that didn't exist a few years previously, were generating more cash receipts than Hollywood and Las Vegas combined. Put a CRT, a microprocessor, and a teenage game programmer together and you get a device that draws a portion of the world's small change like a kind of street-legal mesmerism machine. Action as hot as that attracts attention. Small companies run by enthusiasts were swallowed by or metamorphosed into media giants. The Warner managers who had taken over Atari, Inc., had hired Alan Kay to gather the best and brightest of the infonauts. Video games were just the beginning. Personal computers were ramping up as a consumer product, and who knew what mind-bending, money-making, microprocessor-based gadgetry might loom on the horizon. With Atari's generous assistance, Kay assembled the architects of the next computer revolution, the hypermedia designers and roboticists, the AI hackers and hardware wizards, the miniaturizers and interfacers, to plan the entertainment technologies of the twenty-first century. There was a buzz in the air in Sunnyvale, at their research center—the kind of buzz that used to electrify ARC and PARC.

Less than ten years ago, before word processing replaced typing pools, before spreadsheets turned business people into computer jockeys, before telecommunications linked the electronic cognoscenti into a packet-switched Worldnet, a small but fanatic subset of computer enthusiasts understood that the personal computer would shape the way businesses worked, the way scientists operated, the way students learned, over the coming decades: the *infonauts,* who helped change the way we live today. An even smaller and at least equally fanatic subset of that subculture began to think not only about what lay beyond the horizon but about what might be found beyond that: the *cybernauts,* who are changing the way we will live in the future.

Seven years ago at Atari Research, in one of those timeless, windowless, fluorescent-white rooms that characterize Silicon Valley office parks, I watched Scott Fisher and his colleagues, most of them in their twenties, brainstorm the computer interface of the late 1990s. They were taking steps toward the day when the computer interface would not be something users *look at* through the window of a small screen, but a place they *walk into*—the responsive, three-dimensional, computer-generated promised realm that Ivan Sutherland had prophesied. Brenda Laurel was leading the exercise; back then, Laurel was exploring the possibility of using AI systems to build "first-person" computer games; seven years later, the giant Japanese electronics company Fujitsu has based part of their electronic entertainment R & D on Laurel's theories of how computers can be seen as theatrical devices. Eric Hulteen, another MIT emigrant, was part of the group that met that night in Sunnyvale, too; in 1990 he was working on gestural input and other human interface frontiers for Apple computer. Now married to Brenda Laurel, Hulteen had been one of the two key implementers of Richard Bolt's "Put That There," the computer system in the late 1970s that responded to gestures and voice commands.

Michael Naimark was there the night I visited Atari; at MIT, he and Scott Fisher had been among those instrumental in creating the "Aspen Map," an interactive videodisk that made it possible to use the computer screen to take a virtual tour of the city of Aspen, Colorado. Susan Brennan became an expert on how truly difficult it is to create a computer that can participate in the simplest human natural language conversation. Kristina Hooper, another MIT-Aspen veteran, now manager of Apple's multimedia laboratory, was Fisher's, Laurel's, Hulteen's, Naimark's, and Brennan's boss. After Atari Research fell apart, clusters of the researchers who had been there that night ended up working with each other again, at NASA and Apple. Around the time Atari Research was at its height, I met another fellow who was from Missouri, not Cambridge, but who would figure in later chapters of the VR story, Warren Robinett. Then I went on to other things, for years. Brenda Laurel sent me to get a demonstration from Scott Fisher in 1988, and my subsequent exploration of the state of VR research led me back to them in 1989 and 1990. The accident of our former encounters gave me a kind of temporal perspective, a visceral sense of the pace at which VR is emerging.

Scott Fisher was one of a whole constellation of Arch-Mac veterans scooped up by Alan Kay—and attracted by the opportunity to work with the other media-tech magicians Kay had assembled. At Atari Research, Fisher found himself teamed up with some of the best young technology wizards of the early 1980s, granted a mandate to dream up the entertainment and education media of the future, and promised tens of millions of dollars to work with. It was like the early days at PARC, but they were in the fun and games business rather than the office automation business. The Atari crew decided to take their time and build a conceptual framework while they were building their hardware and software laboratory. It would take a few years to get up to speed, but this was definitely a long haul. They had seen what Arch-Mac and PARC had been able to do with 1970s technology. What would be possible with next-century technology? They suspected that this question, and their responses, could lead them to become some of the most influential technology designers of the twenty-first century. Atari Research Laboratory's strategy, however, depended on the continuing financial health of their parent company.

The video game boom didn't last forever, which led to the dissolution of Atari Research. I maintained contact with some of the people I had met during that era, and lost track of others for a while. Not long after that group dissolved, while working on a different book, I looked up a fun-loving young eccentric by the name of Jaron Lanier, who had made some money with a video game he had created and was working at home on a visual programming language. He wasn't there the night of the Atari Research improvisation, but had accumulated the money to explore his arcane pursuits by programming a successful video game, "Moondust." Lanier, an autodidact who has insisted on living thirty years in the future, probably since the day he was born, dreamed of creating something that went beyond computer programming as people then knew it. Instead of typing in arcane commands and designing data structures, Jaron's language-under-construction would enable people to create software by interfacing with musical cartoons. And the means of communication between human and computer included a strange kind of glove with sensors running up the back of the hand, connected to the least expensive personal computer that could be purchased at the time, a Commodore 64. I remember trying the glove, making the basic hand-wave that changed a low-resolution graphic icon on the screen, then writing it up for my book and more or less forgetting about the glove itself until VR hit me. The glove's inventor, Thomas Zimmerman, had also been at Atari Research. I didn't see them again for a few more years, but Lanier and Zimmerman reentered the scene later in a big way, when the VR industry got rolling.

Toward the end of 1988, I saw Scott Fisher again. This time, the technology wasn't theoretical, and it wasn't just a prototype. He had something to show me. Fisher was working at NASA/Ames Research Center, with a full-blown VR test-bed. He and VR ended up at NASA through several convergences of a different sort. Before Fisher had arrived, researchers in physiological optics—the nuts and bolts of human vision systems—at the University of California, working with the Space and Aeronautical Human Factors Laboratory at Ames, had been concentrating on the perceptual and technological aspects of visual instruments. A visual *display* is the CRT or LCD, screen or goggles, on which information is displayed, but a visual *instrument* is a display designed specifically to enhance human judgment in a specific task. They were thinking about designing computer displays to help astronauts perform specific jobs. The design of future space-exploration vehicles was increasingly an exercise in human-interface design.

The way humans perceive and process the kind of information present on CRTs and other visual displays was the focal interest of Stephen Ellis, a University of California, Berkeley, researcher who was involved in the NASA/Ames effort. He is a physiological psychologist who knows his way around computer graphics algorithms, but because he has always been interested in finding out how human beings use their visual sense to know about the world, he told me he considered himself an "experimental epistemologist," the first time I visited his Berkeley office. He's an amiable fellow who definitely falls on the scientist end of the VR social spectrum, and who has a hint of an impish smile at times. Without looking at my notes, the one thing I remember about the appearance of Ellis's office when I was there, more than a year ago, was certain diagrams on the wall. I had read about research in which eye-tracking devices enabled scientists to show various paintings and photographs to people and then map the areas where they focused their eyes. By superimposing a map of the movements of the subject's gaze-points over the experimental image, it was possible to gain information about how our visual systems make sense of the world. In fact, you don't have to be a specialist to conclude, when looking at one of these diagrams, that people move their eyes differently when they are looking at the image of a face, a landscape, or a nude body. Nobody actually scans their gaze methodically over each square inch of a photograph of a face, for example. We "construct" a face with a series of darting glances between one eye and the other, between the eyes and the nose and the lips, with a quick trip over to the lips and the chin, then back to the eyes again. Although we always feel that we are seeing the whole picture, what we do is closer to a process of extracting key features, consulting mental models, and simulating the picture!

When I asked Ellis where his involvement in VR started, he recalled that it started with a graduate student's project. As part of a cooperative agreement between the UC Berkeley physiological optics laboratory and NASA, Ellis and his colleague Professor Lawrence Stark sent one of their graduate students to Ames to examine the possible usefulness of three-dimensional displays to study the characteristics of visual aeronautical instruments.

Ellis's graduate student, Michael McGreevy, was interested in both the psychological and technological aspects of future displays. McGreevy was pursuing an interdisciplinary Ph.D. in the new field of "cognitive engineering." Like Fisher, whom he had not yet met, McGreevy had a long-standing interest in "immersive displays," although he had not yet begun to pursue the issues of virtual exploration, navigation, manipulation, and interactivity. Like Fisher, McGreevy knew about the work of J. J. Gibson, who argued that visual perception evolved in the context of the perceptual and motor systems, which constantly work to keep us upright, orient us in space, enable us to navigate and handle the world. In his more recent work on VR as a form of scientific visualization for planetary exploration, McGreevy has elegantly linked the Gibsonian idea that the environment must "afford" exploration in order for people to make sense of it to the idea that we can begin to learn something important from the data we have retrieved from planetary exploration by flying our points of view through the images themselves. In the early 1980s, Ellis encouraged McGreevy to look into the idea of using 3D displays as part of a new laboratory in "spatial perception and advanced displays."

On one of my visits to NASA, I asked McGreevy about his background. Part of McGreevy's educational background was visible: he was wearing his Phi Beta Kappa key as a tie tack. Like Scott Fisher, he was rather more formal in his dress than most computer jockeys. Over years of interviewing computer scientists and programmers, I learned that it is not necessary for me to wear a three-piece suit when visiting a laboratory. Indeed, I had found that a tie-dyed shirt and a pair of old blue jeans will gain more entrée at places like Apple or PARC, since they mark me as somebody from the idea side, rather than the marketing side of the info industry. Scott Fisher, then and now, looked like one of those young fellows who has been wearing a suit and tie and carrying a briefcase since prep school. I don't think I've ever seen Scott Fisher in anything but a suit, although I have seen him remove his tie for parties.

Michael McGreevy is a bit on the formal side, himself. He has a crisp, riveting–eye-contact delivery. Scott Fisher tends to sit and think about his replies for a moment, and stare into space, almost as if he is putting himself into some kind of mental scene. When McGreevy sat on his desk to chat with me, it had the look of a studied gesture on the part of somebody who was making an effort to seem less formal. Robert Jacobson, who had just joined the Human Interface Laboratory, a new VR research site in Seattle, had scheduled a visit with McGreevy, so I tagged along. McGreevy cited the July 1, 1966, issue of *Life* magazine as an early influence that began to surface when he began looking into the matter of 3D displays. That issue of *Life* included an article on the Surveyor, which had made the first unmanned landing on the moon, swiveled a camera, and transmitted the images to earth. When the scientists on the ground wanted to see what the Surveyor saw, they pasted the thousands of component images onto the inside surface of a large sphere. NASA scientists could then get a Surveyor's-eye view, a kind of primitive immersion experience, by poking their heads up through a hole into the interior of the sphere.

McGreevy also reminded me that besides Robert Heinlein forty years ago and William Gibson four years ago, another science fiction writer, Ray Bradbury, had raised some of the same issues that are now surfacing around VR technology. In his short story "The Veldt," first published in 1950, Bradbury wrote about a sentient, hyperrealistic simulation, in the form of a room that had started out as a psychological experiment and ended up as a product that read the thoughts from the minds of children and created a lifelike illusion around them as a kind of entertainment: "It was forty feet across by forty feet long and thirty feet high . . . the walls began to purr and recede into crystalline distance, it seemed, and presently an African veldt appeared, in three dimensions; on all sides, in colors reproduced to the final pebble and bit of straw." "The Veldt" had been one of the scenarios that kept cropping up during the Atari Research improvisation exercise I had observed. Another Bradbury story McGreevy cited, also from 1950, was called "The Happiness Machine," about a technology that can reproduce the experience of any fantasy and ends up making people unhappy when they have to return to their drab nonaugmented lives. By the time he got to NASA, like many who ended up there, McGreevy was primed by both science and science fiction to take on the challenge of creating something that never existed before. He finished his doctorate and went to work for NASA/Ames.

At NASA/Ames Aerospace Human Factors Research Division, two things attracted McGreevy's attention. In 1984, McGreevy's boss, David Nagel, invited Scott Fisher to give a talk at NASA. He talked about stereoscopic head-mounted displays, about the special optics required to display wide-angle views, and the potential for exploring virtual environments. Stephen Ellis and Michael McGreevy were in the front row. It was a natural technology for the problems in visual perception they wanted to explore. McGreevy started to look at the head-mounted display systems that the US Air Force had developed under the direction of Thomas Furness at Wright-Patterson Air Force Base, near Dayton, Ohio. The VCASS ("VEE cass") system from Wright-Patterson looked a lot like Darth Vader's helmet and contained a considerably more sophisticated and far more expensive system than the ones developed by Ivan Sutherland or Arch-Mac. It could serve as the basis of a superb scientific instrument for studying the human factors side of VR. The Air Force used custom-made high-resolution miniature CRTs and fiberoptics and gobs of computing power; they didn't spare the funding for an application that might raise the survivability of their best pilots flying half-billion-dollar planes. The legendary story, which comes in slightly different versions depending on who tells you, is that McGreevy asked Furness how much it would cost to obtain a helmet for some human factors work at Ames. Furness is alleged to have quoted a price of one million dollars. McGreevy told him that he already had a Polhemus tracker and an Evans and Sutherland display system; Furness replied that the helmet itself was the part that cost a million dollars. McGreevy resolved to build his own.

McGreevy enlisted James Humphries, a hardware contractor with a company called Sterling General, on his quest for an inexpensive head-mounted display. It turned out that liquid crystal displays (LCDs) from portable televisions could substitute for one of the most expensive components. They went to a nearby Radio Shack, bought two inexpensive LCD televisions, and pried out the LCDs when they got them back to the lab. LCDs differ from CRTs in a fundamental way. CRTs generate an image by shooting a beam of electrons at a phosphorescent screen and activating particles on the screen to emit photons. LCDs are made of a material that changes its appearance when activated by a weak electrical current. You can make the pixels out of tiny discrete cells of this material and control it with an ultraminiature circuit. It's cheaper to build an electronic switching matrix—ultimately, that's what integrated circuits are—than tiny vacuum tubes. The resolution of the first LCDs they used at NASA was only 100 picture elements by 100 picture elements (10,000 pixels, compared with the millions of pixels on the custom-made, high-resolution CRTs used by the US Air Force), but the price was right. And that counted for a lot. The "goggles" part of VR was becoming affordable (although it was still more like a helmet than an eyepiece). All that needed to happen was the "gloves" part.

The fact that you could make an affordable if crude HMD with off-the-shelf LCDs and special optics that were manufactured by a company outside Boston constituted a kind of watershed in the emergence of VR research. The next step for the NASA project was to put together a system that would address both NASA's mission and the research goals of Ames's Human Factors laboratory. The Virtual Environment Display System program was under way by 1985, but it had a long way to go before it could do anything new. In 1985, NASA hired Scott Fisher, and Michael McGreevy went to Washington, DC, on a two-year training assignment. Fisher bought an off-the-shelf voice recognition package and began looking for contractors who could build in 3D audio and start experimenting with tactile output devices. Fisher's idea was to build a proper test-bed for exploring all aspects of virtual workstations, telerobotics, and even such applications as surgical simulators and visualization tools for aerodynamic research. In 1985, Fisher started negotiating with a contractor called VPL Research, to add a glove-based input device to the system; in 1986, Fisher brought in another programmer, Warren Robinett.

The NASA VR project was a team effort, and it becomes increasingly difficult and futile to try to sort out who invented what. One could generalize by saying that McGreevy initiated the project and found the first of a series of brilliant contractors; McGreevy and Humphries put together the basic system of LCD displays, wide-angle optics, and Polhemus Navigation Systems' sensor; Fisher brought a broad conceptual framework based on his work at Arch-Mac and Atari Research, a set of clearly defined objectives for building and evaluating the elements of a VR test-bed, and his own network of specialists who made things like 3D audio systems and computer-input gloves. McGreevy advocated the project to NASA funding brass while Scott Fisher pursued his goal of putting the bits and pieces together into a system and initiating application-oriented research that could support NASA's missions. Fisher began to apply everything he knew about stereographics and virtual exploration to the NASA project.

Robinett, who had developed a reputation in the world of personal computer software for his moneymaking Atari *Adventure* and his award-winning educational software, *Rocky's Boots,* started writing application software—programming code that would use all the hardware to put people inside models that the rest of the team had created, including a computer graphic model of a hemoglobin molecule, a model of the space shuttle, an architectural model of the laboratory itself, showing all the furnishings and instruments in place. Douglas Kerr, a software contractor, later joined the team and was responsible for the system software; Kerr remained and took over the software part of the project after Robinett left in 1988. Eric Howlett's firm in Waltham, Massachusetts, Pop Optix, developed special lenses to give the visual display a wide field of view without sacrificing too much resolution. Howlett, a white-haired epitome of the garage inventor, saw the need to create special optics to offer a field of view that approximates the wide human visual window without distortion. The NASA test-bed was a system that needed to be carefully integrated, an assemblage of devices from different contractors. The effort was not so much a one-man show or even much of a formal hierarchy; rather, the virtual workstation research at NASA/Ames had become a web of researchers and contractors. Steven Bryson started out as one of the programmers for VPL; now he is at Sterling, still one of NASA's prime contractors for the VR project.

When McGreevy returned from Washington, there did not appear to be a great deal of coordination between Fisher's and McGreevy's goals for the project. There were disagreements about what should be implemented, who should implement it, and how it should be implemented. Warren Robinett left in February, 1988, to sail the South Pacific. In 1990, Fisher left NASA (to found, together with Brenda Laurel, a commercial VR venture, Telepresence Research). I met David Nagel at an Apple-sponsored function in early 1990. Now at Apple, Nagel had been both Fisher's and McGreevy's boss, so I asked him which of their stories was the most accurate, and he grinned and said, "They're both right."

Scott Fisher and the people who worked with him extended the day-to-day development of VR research when McGreevy left. The first HMD was a crude "proof of concept" prototype. The second-generation system was the one that started sending out the first solid message that VR research was affordable, useful, and legitimate; one of the things NASA has proved best at doing has been to legitimize certain ideas from the fringe. People started finding their way to Mountain View from the time the setup at the Human Factors laboratory was described in the October, 1987, issue of *Scientific American*. I remember that issue because I saw the disembodied computer graphic rendition of a glove on the cover and immediately thought of that early VPL glove I had seen in Jaron Lanier's cottage.

The Virtual Environment Display (VIVED) system used a redesigned HMD, with special wide-angle optics from Pop Optix, a Polhemus position sensor, a pair of fifteen-year-old Evans and Sutherland computer workstations, and an equally elderly DEC host computer. And it used something that no VR system had ever used—a glove. The story of the origin of the glove as a VR input device is another many-faceted saga that will be explored in slightly more detail in the next chapter. The NASA/VPL contract to develop VPL's glove to NASA's specifications was the inflection point. VPL had been selling gloves for a while, but NASA was the first site that had the resources and vision to combine a glove and an HMD. Scott Fisher initiated a contract with VPL to deliver custom gloves and software to the VIVED project in 1985. Among the VPL people who played a role in the contract were Lanier, Zimmerman, Chuck Blanchard, Steve Bryson, and Jean-Jacques Grimaud. A crucial threshold was crossed into VR as we are coming to know it when Scott Fisher and his colleagues put their hands into cyberspace for the first time.

The importance of the glove and the full-body input devices that followed is fundamental. It goes beyond the convenience of being able to manipulate virtual objects by reaching out and picking them up. Fisher was well versed in the work of J. J. Gibson, the visual perception researcher who claimed that it is our navigation of the three-dimensional world and our handling of objects within it that teach us to see as we do. And he knew of the power of gestural input. A glove that controlled a virtual object would also be what Gibson called an "affordance," a means of literally grabbing on to a virtual world and making it part of our experience. By sticking your hand out into space and seeing the hand's representation move in virtual space, then moving the virtual hand close to a virtual object, you are mapping the dimensions of the virtual world into your internal perception-structuring system. Which leads to fascinating digressions about philosophy and science, but tends to diverge from the pragmatic concerns of a high-risk, high-tech, poorly funded organization like NASA. Except for one thing: VR interfaces to semiautonomous robots might be the only possible way to build a space station.

NASA was interested in telerobotics for space repair applications; dextrous end-effectors (mechanical hands that mimic human motion) had been developed at Stanford and the Jet Propulsion Laboratory, as well as MIT and the University of Utah, and a glove would be a perfect dextrous control device. The idea of combining a glove with a stereoscopic display and position sensing worked for both virtual worlds and teleoperation. Fisher carefully described his vision of how such a device should work in conjunction with the VIVED. The technology developed and patented by VPL enabled NASA to couple hand movements to the virtual world. In 1986, cybernauts launched themselves into cyberspace by sticking their hands in it along with their heads.

Fisher remembers that Warren Robinett stayed up all night after the first working version of the VPL glove was delivered, in 1986. Robinett recalls that he had programmed a crude 3D block-version of the virtual hand, and showed it to the others when they arrived at the laboratory in the morning. "We took turns putting on the HMD and the glove and moved our hands around in a virtual world. We were speechless at how cool it was," Fisher recalled when I asked him about the first time he put his hand into cyberspace.

The glove and the virtual hand on the screen turned out to be useful for many things. First, it indeed seems to be essential in a Gibsonian sense—wiggling my fingers and seeing the fingers of the hand wiggle on the screen in slightly delayed synchrony are a key component of the feeling of presence I experienced the first time I tried out the NASA system. When I talked to Robinett later, at the University of North Carolina, he recalled that VPL had delivered some software that created computer-readable output from the glove. The problem from the point of view of the person charged with creating the software that will make a virtual world feel virtual is the information-juggling. In order to keep a world updated from one or more peripheral sensors, the computer has to do some fancy dancing lest the whole progress bog down.

The way in which information from gloves and position sensors and computer models is passed around and processed and transmuted into photons on a display screen can make a critical difference when it comes to systems that are supposed to react in real time, which for humans means thousandths of a second. If you want to make sure you are in the real world, move your head very quickly to one side or another. If the rest of the world doesn't move along with your head for a couple hundred milliseconds, you are in VR-land. Time lag is always a problem in building VR systems, and dealing with complicated computations very rapidly has always been a part of that problem. The illusion of VR, like the illusion of film, comes from making small but global changes on the visible "frame" of information, then presenting those frames to the human viewer quickly enough for them to fuse into the illusion of motion or reality. Unlike film, a reality engine has to recompute the coordinates and light values of the entire depicted world every time the VR equivalent of a frame is changed (every 1/30 second, when the right-eye view is switched to the left-eye view, or vice versa). NASA was hardly the last to come up against the "systems lag." A year after I talked to Robinett, I ran into a small company in the north of England that believed an entirely new kind of computer known as a "transputer" could be capable of dealing with the ever-proliferating data VR systems are called upon to juggle. If VR proves to be useful, and computer throughput—the amount of bits you can process per second and not stumble over your output— proves to be a bottleneck, developments at the fringes of experimental computer architectures, such as transputers, may become more important.

The data Robinett had to handle from the VPL glove gave moment-to-moment information about fifteen different joint angles in the glove. The data from the angle sensors that determined the positions of the fingers every 1/60 of a second was part of a massive data stream constantly coming in to be processed and sent back out again in terms of changes in the graphical world model—in crude mimicry of the hand-eye-finger-arm loop system we all use in the physical world with our biological arms. Creating a hardware and software system that can match the most precise specifications of our existing hand-finger system will continue to be a formidable test for VR engineers for the foreseeable future. The difficulty of mechanically mimicking human hand-eye-brain capabilities to discriminate very fine discrepancies in space or time is one of the reasons that the true potential of VR systems may take five or ten years or more to approach the point of reasonable verisimilitude. The Polhemus sensor contributed another stream of information about the glove's position and orientation with respect to the room, and added its own constant to the delay.

The first thing Robinett did was build his own 3D model of the hand that would be linked to a representation on the screen and could be modified by changes in the data stream. That was the prototype that he worked on all night and showed the others in the morning. The glove-hand-world model itself was crude at first, just enough to get the system working. It took months for VPL and NASA programmers to tweak it. But when I put on the EyePhones and a DataGlove myself in late 1988, the act of moving my hand in the glove and watching representation of my hand and fingers move in cyberspace were like hooks, handles—*affordances*—that linked "in here" to "out there" and dragged my sense of being in a physical space from the physical room that held my body to the space defined by the 3D computer model. The hand that floated in the virtual world was more than a hand. It was me.

The hand was also useful for two other important aspects of VR environments—manipulating the virtual environment itself and navigating through it. Whether the input device is a glove, a joystick, a trackball, or a video silhouette of a moving hand, some kind of direct and smooth linkage between the operator's hands and the objects in a virtual world is one of the fundamental requirements for VR systems. A world that you can navigate but not manipulate is, in the VR sense, not as "real" as one where you can reach out a virtual hand, take the top off a virtual teapot, lean over, and look inside. "Surrogate travel" is the phrase they use at Arch-Mac to describe environments you could look around in from angles of your own choosing, but which you couldn't touch or change.

Douglas Kerr took over the software of the NASA VR project when Robinett left, and Steven Bryson and Richard Jacoby took over when Kerr left. The first system was running on minimal hardware and positively prehistoric computing gear. Kerr's responsibility was creating a new software system for an HP 9000 computer, one powerful enough to provide real-time shaded graphics rather than wire-frame. The smoothness, shading, naturalness of the lightning are obvious indicators of the amount of computing power—and hence the size of the budget—that goes into a computer-generated graphic. The next step from wire-frame is to cover the frame with a surface, smooth the boundaries so contours look more like contours and less like angles on a straight-lined skeleton, and alter the light values of different parts of the depiction to indicate a uniform light source. Again, the reason for the emphasis on shading and smoothing is derived from the way we pick out a wealth of telltale indicators from the sensory stream and check them against our world models. Shading is one of the several important cues that tell us whether the perspective, the lighting, the proportions of the world or a picture of it are "correct." Moving the software system to a hardware base that could offer some decent computing power was a necessary milestone. Once you get over the euphoria of proving that something works and might be useful, you begin thinking about making it reliable enough to use as a scientific instrument.

When I talked to Douglas Kerr, more than a year after he left the project, he recalled that the main problem in keeping the VR system working was the fact that they were trying to perform rather difficult feats with the least expensive possible hardware. "We were at the leading edge of cheapo technology," he recalled. "We were pushing the envelope in every direction—the optics, the sensors, the graphics, the computation. Something in a system like that is always broken. Calibration is a problem when you want to use the system for serious human factors research."

The VIVED evolved into the Virtual Interface Environment Workstation (VIEW), which is the version I experienced in December, 1988. The head-coupled stereoscopic display was streamlined down from a helmet to a boxlike device about the size of a scuba mask. It said NASA in big red letters on the front, which did bestow a certain sense of adventure the first time I donned the HMD. The glove was of light fabric, with cables running up the fingers, down the back of the hand, and then through an umbilical cable to a computer. The laboratory itself was so crowded that it would be entirely fair to use the word "threaded" to describe the way people made their way through it. But that didn't matter much in terms of the virtual explorations, because the amount of physical movement the operator could carry out was limited to a five-foot hemisphere, centered on the sensor attached to the head mount. The display resolution was considerably higher than the first model, with the equivalent of about 300 television scan lines. Along with the glove and the HMD was a small microphone for connected speech recognition—the ability to summon menus by saying "open menu" and dispose of them by saying "close menu." That part of the system wasn't working the first time I flew through cyberspace because it took some preparation time to "train" the system to recognize the operator's voice. And the 3D auditory display had not yet been installed with that system configuration the first time I came. But being able to stick my head and hands in a world, move around it for a pace or two in any direction, and reach out and change it was sufficient to impress me the first time. It wasn't the smoothness of the ride that captured my imagination, but the vastness of the new space this vehicle had opened for exploration.

The system I saw was a proof-of-concept demonstration for the idea of using VR systems in actual NASA missions, with a view toward possible civilian spin-off applications in medicine and scientific visualization. VIEW was part of a system that Fisher and his colleagues presented to their sponsors as a highly effective telerobotic control interface. Marvin Minsky's notion of telepresence (which will be discussed in a later chapter) had been floating around for years, and telerobotic technologies had advanced slowly on a few fronts from undersea vehicles to prosthetic arms, but the VIEW system was presented as an ideal interface for amplifying the power of existing operations. Besides the virtual world display, Fisher's team built a stereo video camera on a remote, gimbaled platform, capable of providing stereo imagery in real time; the plan was to work toward the day when they could blend stereo imagery from robots in the physical world with computer-generated imagery. Space station maintenance, one of NASA's main concerns, would require many more technicians in orbit than NASA could afford to send; the only practical solution would be to build robotic repair vehicles that a single operator could control, inside or outside the vessel, through telepresence. VIEW was a testbed for building such systems and for measuring the human factors aspects involved in operating them. Another mission requirement took VIEW into the virtual world as well as the teleoperator realm. Another problem faced by astronauts was the complexity and sheer number of instrument displays required to operate the most advanced models of spacecraft. Combining Knowlton's idea of a virtual display with the Dataland work from Arch-Mac, the data space part of the VIEW project involved multiple virtual control stations, menus in three-space operated by voice and gesture, 3D sound cues, and navigation through data.

Think of yourself as one of a small crew of human operators in a space station, monitoring and sometimes using your hands, legs, eyes, and attention to guide the operations of a larger crew of remote-controlled robots. You want to be prepared for the contingency that would require you to don a space suit and get out there yourself. And you want to be able to repair robots inside and outside the spaceship. Once you are in that space suit in the near-vacuum of space, the controls you see on your suit are your literal lifelines; the information you have readily available to your eyes and ears can make the difference between life and death for you and the rest of the human crew. VIEW made its entry into the realm of virtual data spaces in its role as a portable information environment for astronauts working outside their vehicles. Not only could you see (or hear) the state of your oxygen supply by saying the words "monitor oxygen" and watching a display pop up in space, but you could summon detailed repair diagrams of complex machinery, grab the virtual diagram, and position it right next to or overlayed on the device you are trying to fix. You might even want to have a voice remind you that you have five minutes to get yourself back into the airlock, and you might not mind an acoustical beacon that guides you to the source of a trouble spot on the vehicle's surface or the location of an emergency entry, or two acoustical sources that change according to their proximity with one another and thus assist visual guidance during precise docking maneuvers.

Mountain View was only an hour's drive for me, so I returned to NASA/Ames several times in 1989. The first time I came back was to *hear* my way through cyberspace. The fact that we live in a 3D world applies to auditory as well as visible signals, and part of Fisher's original vision for VIEW included some kind of 3D auditory display. When Elizabeth Wenzel and Scott Foster had their "Convolvotron" working, I visited their laboratory down the hall from Michael McGreevy's office, put on a pair of binaural earphones, and directly experienced what they meant by "virtual acoustic spaces." Fisher had heard about perceptual psychologist Wenzel's work and asked her to look into the possibility of adding 3D sound to the VIEW system; Dr. Wenzel involved Professor Frederick Wightman of the University of Wisconsin and Scott Foster, president of Crystal River Engineering. They came up with a means of simulating the location of up to four different audio channels within an imaginary sphere surrounding the listener. The device, called a "Convolvotron," was expensive at first, but it involved two microprocessor boards that could plug right into an IBM personal computer. Most important, it worked.

Wenzel, a research pyschologist at the Aerospace Human Factors Research Division, is convinced that 3D audio will prove to be useful for space applications and scientific visualization in ways that 3D video might be limited. Their system works by taking into account some of the ways the human auditory apparatus, from our internal signal-processing circuits to the convolutions of our outer ears, creates an invisible environment around us—a world of sounds to which we are closely tuned, even when our eyes are closed. Again, the bilateral symmetry of our biological sensors plays a key role. Humans have two ears; we can swivel them by moving our head, and the differences in the signals detected from those auditory sensors play a key role in our ability to locate sounds in space. We have the kind of eyes we do— great for pinpointing potential meals—because predators like wolves and humans have eyes that point forward. Prey animals like deer have eyes on either side of their head, to detect potential predators. The external ears, long ignored by psychophysiologists probing the neural bases of audition, turn out to be part of the highly precise human auditory system. The Convolvotron works by taking into account the specialized auditory signal processing that "personalizes" what each person hears and the way the external ear is shaped and folded.

Before I put on the earphones in Wenzel's office, she asked me to select an ordinary commercial music tape. I chose a reggae classic. I entered acoustic cyberspace to the accompaniment of Bob Marley and the Wailers. I closed my eyes. There was the band, where bands are usually located—somewhere in front of me, and occasionally directly to my right or left. Then Wenzel put a Polhemus sensor in my hand and told me to move the cube around. I moved it to arm's length, above and to the right of my head. And that's where the Wailers moved, just a tick short of instantaneously. The Convolvotron made it possible to create an auditory point of view, a specific position in acoustic space that matches the operator's position in visual space, and to locate and relocate auditory objects. It made it possible to track the progress of a sound through space, to make assumptions about the size of the room or the distance from a sound source. A squeaking door twenty-five feet behind me, a trigger cocking forty feet to the right at the periphery of my vision, the tumblers in a lock clicking into place six inches from my ear—suddenly become possible.

Other methods for creating controllable spaces are being investigated by people who design concert halls and consumer audio systems. Like visual 3D technologies, there is more than one way to create a perceptual illusion. The Convolvotron's 3D audio technology works because it models a specific human signal-processing function and attempts to simulate it in silicon. The key to the Convolvotron is something called the "Head-Related Transfer Function (HRTF)," a set of mathematically modelable responses that our ears impose on the signals they sweep from the air.

Sounds are propagated as spheres of disturbance in a medium. The medium through which we hear most sounds is that mixture of gases we call air. The spheres of disturbance, what people commonly call "sound waves," are mixtures of frequencies, travel at a fixed speed, and are absorbed by some surfaces or bounce off others. The "acoustic properties" of a space are determined by the way soundwaves bounce and absorb and intersect with one another. It's the kind of system that you can only monitor by computer, and which requires a minimum amount of computing power because of the complexity of the task. Like our vision system, our auditory system involves an active and integrated relationship between our brain and our sensors. The distance, size, direction of motion of a mosquito or a jet plane are tracked and localized by the way our brain processes the stream of perceptions that come in from the ears. We test that stream of perceptions in ways similar to the tests we apply to visual perceptions—we swivel our heads, in extreme moments we cup our ears with our hands, and we all do some kind of automatic internal filtering that enables us to pick up specific sounds in noisy environments (the "cocktail party effect" that enables people to track specific dialogues in a babble of conversation). Millions of years in the land of mammals have shaped up a superb mobile acoustic warning system; it's simply a matter of finding the handles, and binding them to the most appropriate aspect of a virtual world.

The part of our survival system that has a self-model, that knows how far apart our ears are and how far off the ground they ought to be, is constantly monitoring and testing, filtering, amplifying, comparing, and equalizing the signals swept in by the ears and transduced by the eardrum and innermost parts of the ear. Even the convolutions of the ear play a part in turning what our ears detect into an acoustic model of the world, by creating a kind of transformation of all the signals that come in through the outer ear. That's where the HRTF comes in. Wenzel calls it "the listener-specific, direction-dependent acoustic effects imposed on an incoming signal by the outer ears." Starting with the human part of the system rather than the existing technology, Wenzel and colleagues created a computer model of a specific human HRTF. They sat real people in an anechoic chamber, surrounded them with 144 different speakers, and measured the effects of hearing precisely modulated signals from every direction, by means of tiny probe microphones placed near the eardrums of the listener. The way the microphones in that listener's ears distort sound from all directions is a specific model of the way that person's ears map the environment and a general model of the way most human ears impose a complex signal on the incoming sound waves, in order to encode it spatially. Each "location" filter detected by sounding one of the 144 speakers is mapped against this function. The map of these signal filters is converted to numbers, downloaded from an IBM AT desktop computer into the device designed by Scott Foster. The device performs in silicon 300 million "multiply-accumulate" operations per second and uses the numerical model based on the HRTF to reconfigure any sound source so that it appears to be coming from a specific point or number of different points within an acoustic sphere.

Part of the utility of 3D acoustic feedback lies in its enhancement of the visual and tactile cues in a VR system; there's nothing like the sound of footsteps behind you to help convince you that you are in a dark alley late at night in a bad part of town—sounds have the ability to raise the hairs on the back of our neck. And another way in which auditory displays could be useful is in the transmission and recognition of complex and rapidly changing information in situations where vision is occluded, or to make physical effects audibly perceptible the way computer graphic simulations can make scientific phenomena visible. Some of the applications Wenzel and others have cited for continuing work in "auditory visualization"—a phrase that will either live on in the annals of oxymorons, or be replaced by a more precise description of auditory intuition-amplifiers—are for use by highly trained human pattern seekers. In 1989, Wenzel, Foster, Wightman, and Doris Kisler presented the technology and the underlying theory, noting: "Applications of a three-dimensional auditory display involve any context in which the user's spatial awareness is important, particularly when visual cues are limited or absent. Examples include air traffic control displays for the tower or cockpit, advanced teleconferencing environments, monitoring telerobotic activities in hazardous situations, and scientific 'visualization' of multi-dimensional data."

The "scenarios" that Scott Fisher used to focus the design goals of a VR application project constituted a key component of the VIEW project. One such scenario was organized around the premise that VIEW systems could be the key to developing a surgical simulator for medical students and surgical specialists such as plastic surgeons, which "could be used much as a flight simulator is used to train jet pilots. Where the pilot can literally explore situations that could be dangerous to encounter in the real world, surgeons can use a simulated 'electronic cadaver' to do preoperation planning and patient analysis. The system is also set up in such a way that surgical students can look through the eyes of a senior surgeon and see a first-person view of the way he or she is doing a particular procedure. . . . the surgeon can be surrounded with the kinds of information windows that are typically seen in an operating room in the form of monitors displaying life-support status information and X-rays."

The possibility of using cyberspace as a communication medium in some way that we can barely understand now, was raised by Krueger and Negroponte in the 1970s. In 1989, at a technology convention showcasing the wares of Pacific Bell, I watched Jaron Lanier in the midst of a mob of telecommunication managers that was trying to force its way into the demonstration of "Reality Built for Two," VPL's half-million-dollar demonstration of a shared virtual space. And later I saw how serious Japanese telecommunication research managers have become about the idea of cyberspace as a collaborative or communication tool. In 1990, Fisher wrote an update of another scenario that had been developing since the "transmission of presence" experiments at Arch-Mac, the possibility of "telecollaboration through virtual presence." The origins of this idea are shared by several sources, Lanier among them, and the future development of the technology will require a network of researchers in a variety of fields. Fisher's 1990 description is a concise statement of the vision:

A major near-term goal for the Virtual Environment Workstation Project is to connect at least two of the current prototype interface systems to a common virtual environment database. The two users will participate and interact in a shared virtual environment but each will view it from their relative, spatially disparate viewpoint. The objective is to provide a collaborative workspace in which remotely located participants can virtually interact with some of the nuances of face-to-face meetings while also having access to their personal dataspace facility. This could enable valuable interaction between scientists collaborating from different locations across the country . . . it will also be possible for each user to be represented in this space by a life-size virtual representation of themselves in whatever form they choose—a kind of electronic persona. For interactive theater or interactive fantasy applications, these virtual forms might range from fantasy figures to inanimate objects, or different figures to different people. Eventually, telecommunication networks will develop that will be configured with virtual environment servers for users to dial into remotely in order to interact with other virtually present users.

Fisher's vision was not to materialize at NASA during his tenure, however. He left in 1990 to form a VR research and development business with Brenda Laurel, Telepresence Research. As of this writing, they have been working together on some kind of plan for a VR entertainment and education project; they've both been traveling to Japan, and I think it is fair to say that something new will be coming from his branch of the VR world, probably having something to do with the intersection of telecollaboration and first-person 3D theatrical interactions. The fortunes of the VIEW project and NASA itself have not been propitious for VR in Mountain View. The VIEW project was "zeroed," as they say in research bureaucratese. However, the people at Sterling General—VR veterans Steven Bryson and James Humphries among them—were still working on enabling technologies. Wenzel et al. were still exploring 3D acoustic space, Stephen Ellis is still confident that eventually he will get his hands on a lab-instrument-grade VR system, and Michael McGreevy has been forcefully advocating a new VR development mission for NASA: virtual planetary exploration of the voluminous visual data transmitted by space probes. In McGreevy's words, "the planets are being digitized"—and therefore armchair explorable if the data serve as the basis of a navigable model. To make his point concrete, McGreevy showed me a video of a simulated flight over a digital model of a planetary surface, based on real telemetered data. "Mars the Movie," by Kevin Hussey and others at Jet Propulsion Laboratory in Pasadena, put the viewer's perspective right there on the surface of Mars. "LA the Movie" takes a digital flight, zooming in over Long Beach, and heading for the virtual Hollywood Hills.

Old NASA hands like Stephen Ellis and Michael McGreevy have their ways of responding to temporary funding setbacks. I wouldn't be surprised to see that VR survives at NASA in several forms, and continues to contribute valuable data about the human factors aspects of cyberspace. But it is no longer a VR mecca where R & D pilgrims come for their dose of conversion experience. Nor is it a center of a focused or broad-scale research effort. The contractors who learned how to build components of VR systems to NASA's specifications represent the possible beginning of a new wave in VR development—the first commercial enterprises devoted to supplying the cyberspace exploration trade. The research will continue in universities and the laboratories of established technology companies. Considering the past history of Silicon Valley, if VR is going to break out of the laboratories and into homes and offices, it will take an unknown company, a startup, that has a vision about where the technology might lead. I know of about a dozen VR companies right now: Autodesk, Sense8, Telepresence Research, Fake Space Labs, Pop Optix, Polhemus Navigation Systems, RPI, Enter, Division, Simmgraphics. I think most of them would agree that the prime mover of commercial VR has been VPL Research, and its inimitable CEO, Jaron Lanier.

Chapter Seven

THE BIRTH OF THE VR BUSINESS

VR is shared and objectively present like the physical world, composable like a work of art, and as unlimited and harmless as a dream. When VR becomes widely available, around the turn of the century, it will not be seen as a medium used within physical reality, but rather as an additional reality. VR opens up a new continent of ideas and possibilities. At Texpo '89 we set foot on the shore of this continent for the first time.

VPL RESEARCH, INC.,
Virtual Reality at Texpo '89, 1989

Inside Jaron Lanier (who turned thirty last year) is a precocious eight-year-old who got together with some friends and built a spaceship. Now, he wants us all to take a ride in it. You have to understand that about Jaron before you can deal with the other layers of identities he seems to radiate on an hourly basis. He's the CEO and founder of VPL Research, Inc., the company that developed and markets the DataGloves and EyePhones most VR researchers use; he's a millionaire on paper if VPL stays on course. He's been designing a way for people to program computers with magical gestures. He's a self-taught drop-out. And he's a Character, with a capital C—several Characters, in fact. There's the childlike idealist who seems to believe that VR is going to be an unalloyed force for good in the world, "a new plane of reality that will thrill everybody." There's the shrewd businessman who found venture capital and put a company together, and who has already survived several rounds of the kind of lawsuits that spring up around promising new technologies. There are the computer scientist, the musican, the philosopher. And there's the guy with the amazing head of hair, outlandish biography, and inability to resist whimsical speculation about the surrealistic era his company is trying to usher in.

Lanier's personae have already generated much mass media attention, some of it concentrated on his unorthodox appearance and history as well as the technology. A weird-looking guy for a weird-sounding idea. He's a writer's delight, and many of us have taken a crack at him by now, from *The New York Times* to the *National Enquirer.* Why hunt for a story hook when you can start with Jaron Lanier himself? A detailed line-drawing of his martian-leonine, anachronistically hirsute head has graced the front page of the *Wall Street Journal,* captioned, rather unfairly, "Electronic LSD?" (Unfairly because Jaron doesn't indulge in recreational drugs. He doesn't even drink. He was in kindergarten in a remote corner of New Mexico during the hippie era.) Photographers take one look at him and set up their most bizarre lighting angles.

Collecting other journalists' descriptions of Lanier has become something of a hobby for me. Here are a few of my favorites:

"In Palo Alto, just off El Camino Real, a dead-end dirt road leads to some old, cottage-style homes enveloped by overgrown weeds. The third house on the left, with every window shade pulled firmly shut, is where Jaron Lanier lives. It is late morning and Lanier answers the door, his bloodshot eyes blinking from the unaccustomed sunlight. He is a large, shaggy bear of a man, with his brown hair falling naturally around his bearded face like Jamaican-style dreadlocks. A long, robe-like Indian shirt that is ripped at the pocket flows halfway down his black pants." That was Sherry Posnick-Goodman, reporting in *Pen-*

insula, quite accurately. I've been to the cottage, and the window shades are indeed shut—it's a hard habit to break if you grow up in the desert. I've seen that shirt, or one remarkably like it. And he does tend to be less than completely alert if you come calling too early in the day. Steve Ditlea, writing for *New York* magazine, called him "the wizard of odd." And John Barlow wrote, in *Mondo 2000*, "Amiable, round and dread-locked, he looked like a Rastafarian hobbit." True, true, and true. He is big, he is odd, and he is amiable. And according to Marvin Minsky, "He is one of the few computer scientists who looks at a larger picture." Whatever you think of him, there is no denying that a conversation with Jaron Lanier often involves sudden flights to unexpected places.

Jaron Lanier is a magnet for attention, but that's just one side effect of the raw energy he throws off in the form of theories, visions, proj-ects, tunes, and epigrams ("information is alienated experience," "VR is the first medium that doesn't narrow the human spirit"). In the minds of many who have read about VR in the popular press, he has become identified personally with the technology. Journalists who don't do their homework tend to use phrases like "Jaron Lanier's virtual reality," which can irk those who have been working with head-mounted displays and immersive experiences for years. Lanier brought Zimmerman and his hardware together with a top-notch soft-ware crew, worked closely with the programmers and NASA for years, and secured funding for his enterprises. VPL today is a crew of jolly enthusiasts, their offices a twenty-four-hour hive of programmers, children, and journalists, and their products are found in every VR laboratory in the world.

Lanier does things differently from the way other people do things, mostly because that is what he has always done as far back as he can remember. As a small child, he moved with his father to a remote corner of New Mexico, where they built a huge house consisting of four geodesic domes. From what he says about it today, Lanier's non-conformity didn't fare him well in his early environment. Like other visionaries, he had good reasons to devote his attention to his fantasies rather than the realities around him.

"I was very strange as a kid and was socially rejected by most people," he admits today, with a resigned, positively cheerful expression. Then as now, he didn't attempt to impose order on his anarchic hair nor did he attempt to conform sartorially. He dropped out of high school to become a musician. At the time, he wasn't in the least interested in computers. That was when he picked up his notion that "information is alienated experience." He objected to the way that life had to be chopped up into binary fragments to be modeled by computers. How-ever, he was attracted to the idea that computers could be used as musical instruments, because he has always considered himself to be primarily a musician. He did learn enough about computers to find some programming work at the nearest university, including work on educational simulations that showed him some of the potential of the technology. And he liked the idea of computer graphics. In a 1988 interview, he said that the first computers he encountered started him thinking about how to model worlds with a computer: "I felt that the images on the screen were made up of little realities that could be changed."

Lanier had dropped out of high school at fifteen, talked his way into mathematics classes at New Mexico State University, tried to make it as a composer of music in New York City, returned to New Mexico. In 1981, he headed west in a car that had been abandoned by drug smugglers: "It had no floor, started with a screwdriver, and had bullet holes in the side," he is fond of remembering. He began to learn about computers in earnest when he hit Silicon Valley. It was the height of the personal computer and video game boom, and he made more money creating the sound effects for computer games than he made as a musician. He learned the programming craft quickly and started creating his own video games for Atari. One of them, "Moondust," was a hit in 1983, bringing him enough royalties to quit his job and start pursuing his own venture. Lanier had discovered that he was one of those young people who had the ability to create the symbolic incantations that made computers sit up and do their tricks on CRT screens and enabled companies like Atari to rake in more than a billion dollars a year. Precocious, even among the software prodigies at Atari in the early 1980s, Lanier got the idea that he could create an entirely new computer language, one that nonexperts could use, one that used beautiful images and sounds instead of dry alphanumeric codes.

Jaron Lanier first came to my attention in 1984, when *Scientific American* put a mock-up of his visual programming language on its cover. Instead of the usual incomprehensible lines of programming code (such as "if X > 0, then goto 20"), a musical staff with equally incomprehensible but far more intriguing graphics of kangaroos, melting ice cubes, and chirping birds was portrayed as a computer program of the future, in the language-in-process that Lanier was calling "Mandala" at that time. That cover story led to the origin myth of VPL Research, Inc.: When the magazine's editors called Lanier and asked the name of his company, he didn't feel like telling them it didn't have a name, so he replied: "VPL Research, Inc." The company is now housed in several floors of an office park wedged between the Redwood City yacht basin on San Francisco Bay and Highway 101, the main artery of Silicon Valley. It's a typical Silicon Valley startup operation, with a few exceptional sights. Besides Jaron himself, there is the wall of gloves where twenty or thirty now-obsolete prototype gloves are displayed. VPL had grown to about thirty-five employees, last time I counted. At the moment, they have not quite outgrown the modest set of suites that they will look down upon nostalgically from the heights of the VPL building ten years from now, if things go according to plan. Why not? This is a place where things like that happen: You can see the ultramodern headquarters of NeXT, Steve Jobs's post-Apple company, from VPL's offices.

In 1985, when I paid my first visit, VPL was one corner of Lanier's phantasmagoric cottage in Palo Alto. He still lives there, although he could afford to buy Palo Alto real estate, if he wished. The main furnishings are still the same collection of 300-odd musical instruments that filled the place five years ago. Funky houses like Lanier's are not easy to find in Palo Alto, but he managed to find one of the few unpaved streets left in this tree-shaded, lush-lawned, terribly upper middle class enclave. I returned there recently to ask him about VPL's current projects. The computer equipment is still scattered about, and the walls, floors, and at times parts of the ceiling still are covered with exotic musical instruments from around the world. Bagpipes from Turkey. Drums from Gabon. Shakuhachis and sitars. In no particular order. In the middle of it is Jaron Lanier, in full regalia.

I paid that first visit to find out about "Mandala," the programming language that might enable nonprogrammers to command computers by creating little diagrams with ice cubes and chirping birds and other visual icons. I didn't know until I got there that the input device to this new language was a glove developed by Lanier and a colleague. At that time, I was writing a book about the future of programming languages, and Lanier's words, as quoted in the book, seem just about halfway prophetic now: "I believe that current programming languages are actually the larval forms of something far more interesting that will mature in the next ten years," he told me five years ago—"—a new form of communication on the same level as speaking and writing." It seems to me now that we have more than five years to go before his still-unfinished programming language (now called "Embrace") will become as important as symbolic communication. But if you had asked me five years ago if I would nominate Jaron Lanier's company as the advance wave of a new industry in the 1990s, I might have conceded it was possible, but I doubt if I would have bet on it.

Jaron Lanier has an abundance of dreams for the technological future of the human enterprise, but his oldest dream is that computers will enable people to exchange simulations—images and sounds and dynamic models—just as we exchange spoken and written words today. He calls this future computer-augmented metalanguage "postsymbolic communication" now, and although he was already onto the glove, Lanier had not yet begun working with head-mounted displays. "When you make a program and send it to somebody else," he explained five years ago, "especially if that program is an interactive simulation, it is as if you are making a new world, a fusion of the symbolic and natural realms. Instead of communicating symbols like letters, numbers, pictures, or musical notes, you are creating miniature universes that have their own internal states and mysteries to be discovered." He didn't use the term "virtual world," but that is what it seems he was describing.

"Here, let me show you," he said, the first time I met him, years before either of us uttered the word "virtual" to describe anything a computer was likely to do. We made our way, with a bit of fancy stepping over and around particularly voluminous objects of unknown function, through more rooms in which musical instruments outnumbered the furniture, and seated ourselves in front of a computer. It was literally the cheapest personal computer that money could buy at that time, a Commodore 64, the kind people bought for their kids during the first personal computer boom of the early 1980s. He explained that the whole idea of a gesture-controlled, pictorial programming language grew out of his desire to create new forms of music by orchestrating simulated instruments. Midway into this music-composition project, he realized that his ideas could be applied to a generalized programming language that used purely iconic notation.

This is how I described it then: "Lanier based Mandala on the idea of dynamic representations. The language itself is a graphic simulation of the inside of the computer. The programmer can watch the computer's operations, at an appropriately high level of depiction, while programs are running, so the user is in continuous interaction with the internal state of the machine. The programmer interacts with the Mandala representations through the use of a unique input device—a glove." I didn't connect what I saw in Lanier's cottage workshop with what I had seen Scott Fisher pantomime at Atari Research a couple years previously, but that cheap screen and crude glove were half of a virtual reality. The other half of a working VR was the part that convinces users they are immersed inside the environment they are manipulating. It would take Scott Fisher to bring along the idea of surrogate travel, the HMD, and head-tracking technologies and contract for a version of the glove that could work as an input device for a 3D virtual world. Before the NASA/Ames breakthrough, however, the use of a glove as a computer input device had grown out of another convergence—Lanier and his visual programming language met the inventor of the glove, Thomas Zimmerman.

Jaron Lanier met Thomas Zimmerman in 1983, at a meeting of

people who use personal computers in conjunction with electronic music synthesizers. Zimmerman had invented the glove for reasons that had little to do with Jaron Lanier's aims—he wanted to play true "air guitar," and he wanted to find a way to use the human body as a musical instrument. He wasn't at Lanier's house the first time I visited, in 1985. Journalist Steve Ditlea turned up at my door with Zimmerman himself not long after I started writing this book, so I got his part of the story directly from him. Tall and lanky, a few years older than Lanier, with distinctly mixed feelings about the VR bandwagon, he is Lanier's opposite in many ways. In 1981, around the time his future partner was driving a bullet-riddled clunker to California, Zimmerman was on the other side of the continent, performing his first experiments with body-gesture feedback. "I remember typing a program on an early Atari computer, in my parents' home in Queens, New York," he told me, a small smile on his face. "I was sitting there in front of the computer with my pants pulled down halfway to my ankles so I could attach sensors to my knee with an Ace bandage, and there were wires from my body to a box that was connected to the computer."

It occurred to Zimmerman that the most interesting part of the body as a gestural input device to a music synthesizer would be the hand. That's what "air guitar" is—teenage boys and overage teenage boys often like to pretend they are playing great guitar riffs with their fingers while they are listening to rock music. If you could put on a glove with the right kind of sensors for determining the position of each finger and the position of the hand in the room and tracking the changes in position in real time, plug the glove into a music-synthesizing device, and create some kind of software for translating motions into music, you could eliminate the physical guitar entirely. His undergraduate degree was from MIT and he was an inveterate tinkerer. Zimmerman went out and bought an old work glove and less than ten dollars' worth of parts.

The key was to keep it lightweight. Exoskeleton hands that clamped around the human hand had been built before, as early as the 1950s, but they were too clunky for Zimmerman's purposes. He used thin, pliant, hollow plastic tubes that conducted light—not the same as fiberoptics, but the result is similar in that you can shine a light down one end and it will come out the other end, even if the tube is bent around corners or over knuckles. One tube ran over each finger and down the back of the hand. At one end of each light-conducting flexible tube was an inexpensive electronic light source; at the other end was an inexpensive electronic photosensor. When the tubes are bent, the amount of light that gets through to the photosensor is diminished by a measurable amount. It isn't the most precise approach, but it can work to give a continuous signal from the fingers and knuckles, roughly mapping changes in finger position. In 1982 Zimmerman applied for a patent for his optical-flex-sensing glove and was granted US Patent No. 4,542,291. The scope of the patent was not as broad as it might have been, because a researcher by the name of Gary Grimes, who worked at Bell Laboratories, had patented a glove-based computer interface device in 1981; Grimes's glove used small switches at each finger joint. Bell didn't pursue Grimes's work, however.

While he was waiting for the patent process to follow its course, Zimmerman went to work at an exciting new place—Atari's Sunnyvale Research Laboratory. There he met Scott Fisher, Jaron Lanier, and the other infonauts. Jaron Lanier's visual programming language and Thomas Zimmerman's optical glove converged when Zimmerman attended a meeting of computerized musicians at which Lanier gave a talk. Lanier and Zimmerman made a deal. Zimmerman became one of the founders of, and assigned the patent to, VPL Research, Inc., and started working on the next generation of glove. Lanier came up with the idea of attaching absolute position sensors to the glove. Research is the part of the process when you find out *how* to do something, and development is when you begin looking at the *best,* most cost-effective ways of doing it. The improved model put thinner, lighter, more accurate fiberoptics (invented by Young Harvill) in place of the light guides. The coating on each fiber bundle is inscribed (scratched) at each knuckle. The amount of flex from each knuckle directly determines how much light leaks out through the precisely calibrated scratches before getting to the photosensor at the end of the optic bundle.

But hardware is only half the story. Somebody or, more likely, several somebodies have to painstakingly code and debug a suite of computer programs that turn the hardware into a reality engine. Charles Blanchard and, later, Young Harvill and Steven Bryson, who now is with Sterling General, another NASA/Ames VR contractor, worked with Zimmerman and Lanier for two years to perfect the software that connected the signals from the glove to the world model in the computer. The software requirements for a VR system include several different systems that must work closely with one another. First, you need a means of converting the analog electrical signals from the flex sensors to a stream of digital information. Then you need a computer program that models the virtual world, and it must include or communicate with the rendering software—programs that orchestrate visual, auditory, and other forms of output to depict the virtual world to the operator. Programmers must design the world model, rendering, and input systems in such a way that they don't waste precious response time sending data back and forth. All of this gives you a featureless, blank world. In order to create a virtual world itself, you need a tool for building 3D computer graphic models.

One of Harvill's programs turned into *RB2 Swivel,* the currently available modeling tool for creating the sculptural design of worlds with VPL systems. Mandala, Lanier's visual programming language, changed its name to "Grasp." The visual programming language, now eight years in the making, has changed its name to "Embrace," and Lanier is guiding it to be a VR tool. Several parts of the original Mandala project, along with several entirely new parts that Lanier's programmers invented, evolved into the software system that VPL sells today. The dynamics (movement) program is a VPL product designed by Chuck Blanchard called "The Body Electric." The part that renders the virtual world in real time is called "Isaac." Only the most primitive part of the software infrastructure for VIVED had been completed when VPL provided the glove as specified to NASA/ Ames; it was up to Warren Robinett and then Douglas Kerr to put together their own system for NASA's computers. The time and effort that go into programming VR systems are factors in the maturation of the technology that can't be boosted by orders of magnitude when a new microchip becomes available. It took several years for both VPL and NASA/Ames to perfect their software and upgrade it to the more powerful generations of computers that have come along in the past five years.

I've been to the VPL office-laboratory a half-dozen times, and Lanier and I have met in a series of coffee shops, pastry parlors, and ethnic restaurants over the past two years, as I've tracked VPL for this book. I recall that we were both having donuts and I was having coffee at a Palo Alto cafe when he described the evolution of the company from what I had seen in 1985 to what it is today. One night in 1984, the young VPL crew finished the software interface and connected the glove to the toy version of the visual programming language. "I re- member a friend of mine walked in," Lanier recalled, "and he said, 'This is the most exciting room on earth right now.'

"SRI offered to hire us as a laboratory," Lanier added, "but they estimated it would take twenty-five years for the technology to reach the public. So we decided to keep doing it ourselves. Marvin and Margaret Minsky were my first investors after I spent all the money I made from Moondust." (Both Minskys will reappear in later chap- ters.)

Jean-Jacques Grimaud came to VPL from the French computer industry, and so did one of VPL's biggest investors, Thompson Avion- ics. Grimaud became VPL's president, Lanier remained CEO. Their first DataGlove sale, according to Lanier, was to Thinking Machines Corporation, the Cambridge, Massachusetts, company founded by another computer prodigy, Daniel Hillis. Hillis was and is a pioneer in the field of massively parallel computer architectures, in which many individual processors are linked into a kind of computing network. "It was the first purchase for Thinking Machines and the first sale for VPL," Lanier told me. Then came the NASA contract. It was Margaret Minsky who told Lanier and company to get together with Scott Fisher at NASA/Ames. The NASA contract forced them to refine their technology, and it put VPL on the map when a second *Scientific American* cover story, in October, 1987, featured the DataGlove as an illustration for an article about the NASA/Ames VR lab.

VPL's early ventures reflect Lanier's eclectic range of interests, the innately general-purpose nature of VR technology, and VPL's strategy of working to develop their market in stages and on several parallel fronts. In 1987, a company by the name of Abrams-Gentile Entertainment (AGE) offered VPL access to toy companies. VPL and AGE then developed a version of the glove suitable for selling as an input device for the popular Nintendo video games, a market that is measured in the tens of millions. The "PowerGlove" eventually involved Mattel engineers as well. In the process, the DataGlove technology changed into something else. A lawsuit over who invented what has been quietly settled, and VPL shares royalties from Mattel with AGE. Approximately a million PowerGloves have been sold, and sales projections target two million by the end of 1991. The VPL DataGlove sells for $8800. The Mattel PowerGlove sells for under $100. The key to the price difference is the PowerGlove's use of an electrically conductive ink printed onto a strip of flexible plastic that follows each finger, in place of the far more expensive (and more accurate) fiberoptic bundles used in the DataGlove. Changes in finger flexion change the electrical conduction properties of each strip.

The absolute position of the PowerGlove—where the hand is in the room—is also determined far less expensively, if less accurately, by an ultrasonic technology. Ivan Sutherland had experimented with ultrasound, and some of VPL's earliest prototypes used it, as well. Inaudible pulses of ultrahigh-frequency sound are emitted by three sources around the frame of the television screen that acts as the Nintendo system's display device. A sensor on the glove, and calculations based on the characteristics of the sound, yield an absolute position for the glove. Sound tends to bounce off objects in a room in strange ways, and if anything obstructs linear transmission between the glove and the screen, the readings become inaccurate. But for less than $100 for a workable version of a glove that can be stamped out in printed circuits and vacuum-formed plastic, an ultrasonic and electroconductive system works just fine. I wouldn't be surprised to see a full-body version for around the same price range in a year or two. Technically, there is no reason why the performance of electroconductive ink and ultrasonic position-sensing technologies can't be improved and manufactured even less expensively. If even better technologies come along to dramatically improve performance, there will be an infrastructure for manufacturing large numbers of gloves.

I tried out a prerelease version of the PowerGlove and associated games at AGE's New York office in the summer of 1989. The Gentile brothers (the "G" of AGE) seemed as stereotypically hard-driving and pragmatically New York as Jaron Lanier seems stereotypically California laid-back and impractical. The AGE office was a supermodern loft in the same building Studio 54 used to occupy. I can see how disagreements might naturally arise from two such disparate operations. But I was there to get my hands on a PowerGlove, which wasn't yet available on the open market. As a toy, it was a great-looking item that strapped onto the back of the hand, with finger guides, straps that went around the wrist and extended up the forearm, buttons, and other controls. As VR, it was minimal. As a game, it was very attractive. I found it to be a more than mildly amusing way to box with a video-game heavyweight. I was downright fascinated when I put on an inexpensive version of electronic shutter glasses and tried to grab purple leering gremlins from midair and stuff them in a treasure chest that floated in front of my face.

The PowerGlove would not be my technology of choice if I wanted to dock virtual molecules. But there are probably a hundred thousand teenagers with change in their pockets for every pharmaceutical chemist with a need for haptic visualization tools. And that is where the leverage of the PowerGlove lies. The sheer size, financial clout, and economics of scale that the toy industry brings to the VR game are a wild card in the future evolution of the VR industry. If the glove improves in accuracy and drops in price, if toy versions of full-body input suits become available in conjunction with toy 3D glasses, then the huge potential revenues could drive further research and development. I was to see, months later, at Fujitsu's research and development laboratories outside Tokyo, that other major players in the experience industry are interested in the consumer entertainment potential. And I've recently learned that VPL has contracted with one of the world's largest entertainment conglomerates to develop VR as an entertainment medium. From World War II flight simulators to technological toys, VR continues to be driven by unexpected forces, particularly when Jaron Lanier and VPL enter the picture. As long as

VPL stays in business, I expect to see Jaron Lanier at most of the odd intersections.

VPL dominates the glove-input market, small as it is, but it isn't alone. Others have developed transducers for hand and finger input, using different technologies. It isn't easy to see now how the industry based on hand-input devices will shake out. VPL might end up dominating; there might be parallel developments in research, industry, and entertainment; or a company that nobody knows about yet might spring up. Elizabeth Marcus developed a lightweight exoskeletal device while at Arthur D. Little, then left to form her own company, Exos, Inc., to market it. I met her in Santa Barbara at a VR conference and had a chance to put my hand in it. It is definitely more time-consuming to put one's hand into an Exos device, since each finger must be wriggled through three ringlike mechanisms. At each joint, a sensor measures the angular deflection by means of a technology based on "the Hall effect." Her claim is that while it is more bulky and less convenient than a glove, the Exos device yields the kind of precise position-detection needed for certain kinds of tasks. Yet another kind of glove was developed by James Kramer, a student in the department of electrical engineering at Stanford. Together with Larry Leifer, director of rehabilitation R & D at Stanford University, Kramer developed a glove that used strain gauges to convert finger-spelled words into speech, for use by nonvocal deaf and deaf-blind persons. VPL sued Kramer and Stanford; the suit was settled out of court. I also tested Kramer's device in Santa Barbara; reading from a chart of finger-spelling positions, I was able to use my fingers to slowly spell out words, letter by letter, through a voice-synthesis system. Kramer's system is also usable for VR. When and if a competitive climate in the glove-input-device market develops, the price of high-performance input gloves will drop.

Lanier and VPL didn't limit themselves to gloves. After the NASA glove and the initial marketing successes of the DataGlove, VPL started working on a commercial version of the head-mounted display. They started with the special wide-angle, low-distortion optics manufactured by another wild-and-woolly-haired inventor, Eric Howlett, who had provided the optics for McGreevy's and Fisher's systems at NASA. The optics and color LCDs were housed in a scuba-masklike headpiece. One of Polhemus's position-trackers was mounted on the headpiece, as well. Between 1988 and 1990, VPL was the "gloves and goggles" vendor to the research world—the first vendor of "off-the-shelf" VR systems. Laboratories no longer had to reinvent the virtual wheel and build their own HMDs and input devices. It wasn't an inexpensive proposition to get involved with the ground floor of VR research. As of October 1990, the Polhemus Navigation Systems Isotrak position sensor cost $2500 if you bought it from VPL. The DataGlove cost an additional $6300, the EyePhone was $9400, and the software package cost $7200. That comes to $25,400 without the expensive computer needed to make it all work. As of late 1990, the graphics workstations of choice for the VPL systems were those marketed by Silicon Graphics, Inc., for $95,000, $75,000, or $250,000. For the RB2 system that enables two people to share a virtual world, two of the $250,000 computers are required.

How much reality do you get for all that money? After my first experiences with the NASA wire-frame graphics demonstrations in December, 1988, the next VR experiences I had were in spring, 1989, with the first system that was set up in the "Cyberia" laboratory at Autodesk, Inc., a company that is located conveniently within bicycle distance of my house. The first Autodesk virtual world was an office plan—bland, but solid. Instead of wire-frame outlines, there were solid, shaded surfaces. It was verging on cartoon-quality. Less than six months after my NASA initiation, the Autodesk world, crude as it was, seemed like a major advance. The rate of improvement in the reality level of virtual worlds was so rapid that it seemed like watching a time-lapse history of cinema, from the Edison days to "talkies" to Cine-mascope. My first encounter with VPL virtual worlds, in the summer of 1989, seemed to be yet another frame in that accelerating time-lapse film. The first VPL world I entered, the day after "Virtual Reality Day," was the "Day Care World," which had advanced all the way to cartoon-quality, and had the additional advantage of having another person in it. Inhabiting cyberspace with another sentient being is a distinct experience in its own right.

When he found out that VPL would be publicly demonstrating their RB2 ("Reality Built for Two") system in San Francisco on June 7, 1989, the same day that Autodesk planned to unveil their own VR research at a CAD convention in Anaheim, Jaron Lanier thought it would be a neat idea to make June 7 an annual "Virtual Reality Day." VPL and Autodesk were discussing the possibility of setting up some kind of live video hookup, maybe even a shared VR demonstration between Anaheim and San Francisco. Lanier often makes more plans than he ends up carrying through. There was no hookup between the two cities. June 7, 1990, came and went with no follow-up event. But the first VR day was a signal event. I decided to go to Anaheim first because the Autodesk presentation was scheduled to take place only on June 7. I came back that night and showed up the next day at the San Francisco Civic Auditorium where VPL's demonstration was still happening. The event was called "Texpo," an exposition sponsored by

Pacific Bell. One of Lanier's talents is making deals with unlikely partners, and somehow he managed to convince somebody at Pac Bell that
VR represented the communication system of the future.

Televirtuality is not as much of a mental stretch for the people who
know where the telecommunication technology is heading in the future; nevertheless, finding the right person to work with at a notoriously conservative company like Pac Bell required finesse. The result
was the RB2 demo. There were two sets of EyePhones and DataGloves,
each linked to one of the more expensive Silicon Graphics engines,
connected by a high-speed communications cable. Two people demonstrated the system; the two-dimensional depictions of the virtual
world each operator saw were visible on large display screens. One of
the things they did that went over well with audiences was to play tag
in virtual space. One operator discovered that it was possible to hide
the representation of his body inside the head of his companion!

Because it was a small demonstration in a larger exposition, the VPL
setup was hidden behind partitions. Selected groups of VIPs were
ushered behind the partitions to see the two VPL demonstrators sitting
on their high chairs, faces swallowed by their EyePhones, DataGloved
hands waving in the air, pretending to be designers or simply playing
games in their virtual world. When I got there, Lanier gleefully con-
fided that so many people heard about the demonstration and tried
to crash the orderly procession of guided groups that there was a near-
riot the previous day. He took me behind the partition, and I watched
the groups of visitors, almost none of whom had ever heard of virtual
reality before, watch the demonstration in utter amazement. After the
convention, I drove down to VPL's Redwood City headquarters and
put myself in the world that had been demonstrated at Pacific Bell.
The formal name of the demonstration was "Day Care World," and
it represented a scenario of a hypothetical virtual communication of
the future. The idea was to show how two people in different geographical locations could meet each other in a shared virtual space
and use their body sense and imagination to accomplish work. In this
scenario, the two people were meant to be designers, and their task
was to try out ideas for a day care center.

The day care center itself was a small room in a building with windows and a door. The "grain" of the world was much finer than the
NASA or Autodesk versions. Objects looked rounder, more "solid."
One of the ways that computer graphics are created is to break down
complex objects into a series of small elements known as "polygons,"
because that is what they are—pixels grouped together to make very
small objects with multiple straight sides that are arranged, in turn,
into mosaics that depict the environment and objects within it. "Gour-

aud-shaded polygons" are the currently *de rigueur* technique for rendering objects in the kind of depiction known as "2 1/2 D." The more polygons a computer can display onscreen at one time, the finer the grain and the more lifelike the depiction grows. The number of polygons per second that the reality engine is able to present at any one time, and the number that it can manipulate from second to second, measure in one way, the power of a rendering system. The expensive Silicon Graphics workstations were able to provide the raw computing power to put up a couple of thousand polygons per second, thus making a rather "smooth"-looking world. Nevertheless, despite the price, it still seemed a long way from reality. Alvy Ray Smith, one of the gurus of contemporary computer graphics technology, claims that "reality is 80 million polygons per second."

So the day care world was smoother than previous virtual realities I had experienced, and not terribly complicated or avant-garde. On one wall was a depiction of the Pacific Bell corporate logo, another one of Jaron Lanier's politically savvy touches. I found that it was possible to exit through the door and travel some distance across a featureless plain and see the building from afar. It was possible to fly around the room or manipulate the objects in it. If the other person was inhabiting the world at the same time, it was possible to see him or her represented as an abstract, cartoony solid. Only the hand of each inhabitant moved in direct synchrony with the operator's movements. On one wall of the room was a switch. I reached over and turned it on. A ceiling fan started to rotate, causing a mobile hanging from the ceiling to move. The mobile, a cluster of small triangles, was also grabbable—and flew off unexpectedly when I released my grip. There was a small table with blocks on it. I could pick up the blocks and stack them. There were a water fountain and chairs, all movable by reaching and grabbing. I was able to shrink or expand the world and thus see what it would look like from a child's or an adult's point of view (the command for making the room larger was to point down with my little finger, and I could make the room smaller by pointing up with the same finger).

The next time I tried out a VPL world was several months later, when Kevin Kelly, editor of the *Whole Earth Review*, persuaded Lanier to build a world for him from scratch. Lanier sat down at a Macintosh computer and in a couple of hours was able to use VPL's world-building tool to create an environment he called "Ritual World." I drove down to VPL to check it out myself. Lanier voices many ideas and talks about many plans, and many of these notions never seem to materialize. However, there is no doubt about his talent as a world builder. Ritual World contained groups of vertical objects that looked like very thin pillars, arranged in a circle. When I entered the world, they stretched far overhead. Inside the circle were ribbonlike green things that Lanier called "ferns," but which resembled a real fern no more than Mickey Mouse resembles a real rodent. The stretch Lanier was asking of pepole when he called his graphic a fern is along the lines of his theory that people tend to fill in the blank spots in VR. Underneath my feet I saw one of the features Kevin Kelly had described to me: Several planes of linked green, brown, and maroon polygons that looked rather like kaleidoscope patterns rotated in clockwise and counterclockwise circles.

"That's the oriental carpet," Lanier explained, when he saw me staring at the ground. "We call it that," he added, "because it keeps reorienting itself."

On top of the carpet were the pillars, atop each pillar was a jewel, and from each jewel, a flame shot upward. It was as wild and imaginative as the Pacific Bell demonstration was austere and practical. Lanier pointed out one unexpected feature of Ritual World that his girlfriend had discovered the night it was built: the pillars are hollow, and it is possible to enter them and look up into the jewel. It seems that even the creators of the most primitive virtual worlds have surprises in store when they enter them and explore.

I got down on my hands and knees to get a closer look. You tend to crawl around on the floor when exploring virtual worlds. Like sex, exploring virtual reality seems to require bodily positions that look amusing to others. By doing a pushup very slowly, I was able to stick my head between two of the counterrotating layers of the carpet and saw one of the little surprises Lanier had built into the world—a jewel that I could grab and carry around with me.

The pace of activity at VPL seemed to be accelerating throughout 1989. With all the business to be done and conventions to address and journalists to chat with and photographers to pose for and lawyers to deal with, it became increasingly difficult for Lanier to find time to chat. And it became more difficult for him to find time to build worlds. Like many of the other demo-driven outfits in the computer field, VPL's motivation for creating a new demonstration was usually the pressure of another big convention where people were expecting to see something new. I remember seeing Lanier at SIGGRAPH 1989 in Boston, speaking animatedly with an elder stateman of the personal computer industry—thirty-five-year-old Steve Jobs. For the SIGGRAPH world, Lanier created some new furnishings for virtual rooms. That same afternoon, on another part of the vast SIGGRAPH exhibit floor, I watched Eric Gullichsen demonstrating Autodesk's virtual racquetball to an endless line of enthusiasts. *Ki, baraka, charisma,* whatever you call it, Jaron Lanier has it—and many were convinced that summer in Boston that VR has it, as well. I look back on SIGGRAPH 1989 as the time when that unmistakable sense of a cultural revolution in the making began to build up around VR, in both the circles of technical experts and the popular media, who could sense something beginning to happen, even if they weren't quite sure what it was.

In 1987 VPL introduced a full-body version of the DataGlove, a DataSuit that resembles (and is based upon) a diver's wetsuit. Ann Lasco-Harvill and others joined the team to create a neck-to-ankles companion to the glove. A network of sensors feed data on fifty different degrees of freedom—knees, neck, ankles, wrists. It is expensive—in the $50,000 range—and it took a while to develop a reliable model (with that many sensors, there are lots of ways for the suit to malfunction or miscalibrate). I danced with a virtual lobster at VPL's introduction of the suit. And when I visited Tokyo the following March, I saw an exhibit in which a model wearing a DataSuit demonstrated "interactive television of the future." The introduction of the entire body into a virtual world was another idea that had earlier roots—Stewart Brand's book, *The Media Lab,* has a photograph in it of Nicholas Negroponte wearing an early version of a position-sensing wetsuit. The use of the entire body as an input device continues to be actively investigated, from Myron Krueger's laboratory in Connecticut to a Japanese consortium outside Kyoto.

By the summer of 1990, VPL was moving to the next phase of its plan to develop the market. Most of their revenue had been from sales of equipment to research and development centers, and royalties from the PowerGlove. The time had come to seek out partners to develop new applications. Lanier and I arranged to meet again in the fall of 1990, to talk about the latest developments. As usual, we ended up at a restaurant not far from his office. Trying to maintain a conversation in any place where people knew how to get hold of Lanier had grown difficult. At the office, there were too many phone calls to field.

"What's new?" I asked, knowing that I would get a two-hour flight of enthused plans, some of which would prove to be partially baked, and some of which would surprise me eventually with their solidity.

"Mattel is going to bring out Super Glove Ball for the PowerGlove, and we're working on bringing out a low-cost glove as an input device for personal computers, but right now I'm excited about the virtual kitchen we're building with Matsushita," he replied. It isn't hard to tell what frame of mind Lanier is in at any one moment. He conveys a lot of it in his nonverbal communications—his gestures, his facial expression. He has a hearty laugh, and his brow tends to furrow when he is upset. He had that kid-in-a-toyshop look when he described the virtual kitchen.

VPL and Matsushita Electric Works, Lanier explained, had decided to explore a joint venture. Matsushita has a data base of CAD models of thirty thousand different kitchen appliances that are sold, largely, through major Japanese department stores. With the Matsushita data base and VPL's system, it will be possible for customers to come to department stores and enter the dimensions of their kitchen into a computer, which will then create a virtual model of the kitchen. The customer can then place different appliances from the catalog in their 3D-modeled kitchen and see how they might fit, look, interact with each other. The plates on shelves will be movable, the refrigerator doors will open—and customers will even be able to see heat escape from the open refrigerator door into the room.

"We're planning to include a thermodynamics simulation," Lanier added, "to show how hot and cold air moves through the model."

VPL's scope of strategic partnerships, albeit in the early stages when I spoke with Lanier about them, sounds impressive. Another VPL project in progress is an interior design simulator for a major American automobile manufacturer. Another major client, one of the world's largest entertainment companies, has contracted VPL to plan public access virtual entertainment parlors. (The entertainment arcade business heated up in the summer of 1990 when the "Battletech Center" opened in Chicago. For $7, each participant is enclosed in a small motion-platform, sound-and-visual simulation of a space cruiser, and groups fight virtual battles with one another.)

Surgical simulation is a more serious project that VPL is also developing. The original NASA/Ames project had involved a collaboration with Dr. Joseph Rosen at Stanford Medical Center, to develop "the virtual cadaver." Although human cadavers and living patients are necessary to train surgeons, there is a big advantage in having the ability to simulate, in three dimensions, specific anatomical problems that would necessitate surgery—particularly if the simulations could be based on 3D medical imaging of actual patients. For diagnosis, imaging systems such as those under development at Stanford and UNC are likely within the next ten years. Something as delicate and tactile as surgery might require more advanced technology to make the simulation anything more realistic than a 3D schematic. Ultimately, VR surgical planning could train doctors and enable specialists to "rehearse" difficult operations.

Another VPL project involves working with designers at a major aircraft company—"nonmilitary aircraft designers," Lanier hastened to add, "because we don't take any money from the Defense Department and we don't work on military systems." One of this company's newest wind tunnels was designed with the assistance of a VPL system.

VPL also was dabbling in the new and potentially lucrative field of financial systems visualization. The stock market, after all, is no longer just a group of people shouting on a trading floor. Most of the financial transactions in the world now consist of a flow of data through computerized trading systems worldwide. If there is a way to discern patterns in those massive financial flows by making them visible and manipulable in three dimensions, the payoffs could be enormous. VPL was working with an actuarial company which wanted to represent data bases as forests and individual items in the data base as trees as an experiment in total immersion in a financial data base. Could one get a better "feeling for the numbers" by flying or strolling through them? Predictions or arbitrages might be executed by noticing which tree sticks up out of a forest, for example, and grabbing it with the DataGlove.

In the world of entertainment, VPL was embarking on a collaboration with director Alex Singer, who had directed the film *A Cold Wind in August* and television shows *Hill Street Blues* and *Cagney and Lacey*. I had long suspected that Hollywood was getting in on the VR act when I met Singer at Texpo. The movie deal Singer is trying to put together, tentatively titled *A Man and a Woman and a Woman*, is about two characters who fall in love inside a VR demonstration. Considering Singer's track record, it is within the realm of possibility that a popular movie will put VR on the map the way *War Games* brought the phenomenon of teenage computer "hackers" to public awareness. The RB2 system showed up on American television in 1990 in the form of a music video created by Lanier and guitarist Stanley Jordan, using VR as a real-time improvisatory tool. Jordan is portrayed playing a virtual guitar the size of skyscraper. Another video, by the Grateful Dead, featured a skeletal hand animation that was driven by a DataGlove. (Grateful Dead lyricist John Barlow is a friend of Lanier's. I've seen them both at recent Grateful Dead concerts, where Lanier nodded at the colorful crowd around him and mentioned to me that "a Grateful Dead show is one of the few places I don't stand out in the crowd.")

The music videos and animations point to another viable application for body-based input devices. Because DataGloves and DataSuits provide real-time information on natural body motions, they can be used as control devices for television animation. The kinds of motions that are desired for animated characters can be mimed by a person in a DataGlove or DataSuit, and the output used to drive the animation. In Tokyo, I saw technicians at Fuji-TV, a major production studio, using the DataSuit for exactly that purpose. The technician dons a suit, runs across the floor, picks up a ball, and throws it into a goal. Then the representation of the running body can be transformed into a dragon, a cat, or whatever form the animator desires.

"Reality Net" is what Lanier calls one of his grandest schemes. Only a year had passed since the RB2 demo at Texpo and Lanier had already begun arranging for a national virtual telecommunication network test-bed. The first link was to be arranged sometime in 1991, betwen the UNC VR group and VPL. Seattle's Human Interface Technology Laboratory and the Media Laboratory had also agreed to join the experiment. VR researchers I had met in Kyoto were also eager to join the new experimental network when it finally goes on line. The idea is to create a VR telecommunication channel for sharing information and performing joint explorations of the nature of VR as a telecommunication medium. If worlds could be shared by plugging a helmet into the telephone network, Gibson's cyberspace might happen sooner than we think. At the beginning, such a network could be organized by sending packets of computer data through the existing telephone network, the same way computers exchange information via modems. Eventually, the high-computation-intensive VR systems could be linked to the high-bandwidth optical fiber networks of the future.

Considering how ARPAnet started as an experiment and spawned the enormous interlinked network of public and private computers known as "the Matrix" or "Worldnet," Reality Net might easily expand very quickly beyond the blue-sky stage—as soon as some programmer somewhere comes up with the code for linking reality engines and telecommunications networks. Throughout 1990, the fundamental work involved crafting a set of protocols for sending information over relatively slow public communication lines. The idea of changing the configuration of a virtual world on a transcontinental basis becomes far more feasible when there is a means for keeping the computer data that constitute the world models separately at each site, so that the only information that needs to be transmitted from one site to another is a stream of information about the changes that take place in the world model from the point of view of the operators. That is, it is faster to send the code for "you moved your foot" over a telephone wire, and let your home computer run the software that moves the foot you see, than to send all the instructions for rerendering the foot in the new position.

VPL has licensed the 3D sound technology originally developed for NASA/Ames by Scott Foster's Crystal River Engineering, Inc. For enough extra money, it is now possible to buy VPL's "AudioSphere" to go along with the other equipment or as a stand-alone 3D sound system. At the same time that VPL announced AudioSphere, they also announced "VideoSphere," a technology for allowing operators of the RB2 system to see a panoramic video camera-generated environment beyond the computer-generated objects in a virtual world. An example cited by VPL would be a video background of an actual building site, into which a virtual model of a proposed building could be placed. A special camera configuration called a "Revolvatron" gathers the images. Integrating the output of new electronic imaging systems (CCD cameras), video, digital video, and other devices for gathering and storing imagery from the "real world" into the computer-generated virtual world is another frontier that might spin off into an industry of its own.

Lanier and VPL aren't showing signs of slowing down. Last week, he and Alex Singer were off to Tokyo again. Whether or not VPL continues to dominate the VR industry, there is no doubt that if such an industry does develop, it has been kick-started by a peculiar little company that began in a cottage, presently resides in an industrial park, and plans to change the nature of reality as we know it.

VPL may have been the first, but it is no longer the only company that has taken a stake in the VR industry. Jaron Lanier may be one of the more visible visionaries in the VR field, but he isn't the only one. Indeed, although VPL seems to have its fingers in most of the glamorous possible applications for VR technology, the field itself might be driven in the near future by the entry of a firm that is not well known outside the computer software industry, that was founded less than ten years ago by a group of programmers, and that now grosses more than $100 million annually—Autodesk. John Walker, one of the Autodesk founders, is an intensely private and far less colorful figure than Jaron Lanier. In the long run, he, and the company he founded, may be as important to the VR industry as the company that sells the goggles and gloves.

Chapter Eight

CYBERSPACE AND SERIOUS BUSINESS

Cyberspace is a general purpose technology of interaction with computers— nothing about it is specific to 3D graphical design any more than fifth generation interfaces based on raster graphics screens are useful only for two dimensional drawing. New technologies, however, tend to be initially applied in the most obvious and literal ways. When graphics displays were first developed, they were used for obvious graphics applications such as drawing and image processing. Only later, as graphics display technology became less expensive and graphics displays were widely available, did people come to see that appropriate use of two dimensional graphics could help clarify even exclusively text or number oriented tasks.

So it will be with cyberspace. Cyberspace represents the first three dimensional computer interface worthy of the name. Users struggling to comprehend three dimensional designs from multiple views, shaded pictures, or animation will have no difficulty comprehending or hesitation to adopt a technology that lets them pick up a part and rotate it to understand its shape, fly through a complex design like Superman, or form parts by using tools and see the results immediately.

JOHN WALKER,
Through the Looking Glass, 1988

VPL made off-the-shelf VR research possible for many people in the computer science research community, but many more people outside the esoteric circles of human interface research began to take notice when Autodesk jumped into the cyberspace business. I got there early and grabbed a front-row seat when Autodesk rolled out the first in-house demonstration of their "Cyberia" project in the spring of 1989. The room ended up more than usually tightly packed that afternoon with Autodesk personnel, their guests, and more than a few of us freelance technology-watchers who had sniffed it out and wangled an invitation. I saw Stewart Brand there, publisher of the *Whole Earth Catalog,* as well as Hugh Daniels, the first technical director of Whole Earth's virtual community, the WELL. Among the faces I recognized was another WELLite, futurist Peter Schwartz, formerly strategic planner for SRI International, Royal Dutch/Shell, and the London International Stock Exchange, now president of his own consulting firm, Global Business Network. Theodor Nelson, technoprophet, pamphleteer, Autodesk Fellow, and seasoned performer of the personal computer revolution, stood at the edge of the crowd and smiled that goofy, infectious smile he uses on crowds. Esther Dyson, gadfly of the info industry, made a rare northern California appearance. John Barlow, songwriter for the Grateful Dead, Wyoming Republican, and self-described "technocrank," had also been drawn in via the WELL. I glimpsed Apple people, spotted scouts for the Lucasfilm tribe. The only common feature, I noted, seemed to be an instinct for the edge—the fringes of technology where revolutions can break out.

Most of the people in the audience had taken their first peeks into cyberspace during the preceding week. Considering the technical sophistication of this particular audience, there was no need to explain that what they had seen was just the clunkiest rough first stage, that the hardware base was cheap and inadequate for the time being. People here in Autodesk's large seminar room didn't have to be forced to extrapolate from the demo they experienced to future possibilities. Mention was made of future applications in visualization, games, fitness, sports, education, training, theater, dance, physical therapy, tele-robotics. Everybody's eye was on a broad horizon. The panel presentation was delivered by the demo-dazed Cyberia staff, who had been working night and day for weeks, first to complete the demonstration prototype, then to show it to all the people who wanted to dive in for themselves. It was almost like a baptismal ritual. Sticking your head into the office plan. I recall a deeply emotional speech by William Bricken (in which, according to my notes, he declared: "The overwhelming metaphor for me is that exploring virtual realities is a way of learning how to be a child again.") Then Eric Gullichsen introduced a short film featuring Timothy Leary.

After the panel concluded and received their standing ovation with moist eyes, Esther Dyson was among the first to comment.

"I don't mean to be negative, but this might be very good for some things, and not so good for others," she said. "You can't take away my keyboard, for example." She managed to convey, in one swift, concrete example, the disenchantment one often feels among a group of people who share a kind of religious allegiance or conversion experience that simply isn't one's own.

To which Autodesk cyberspace programmer Randal Walser replied, in that soft-spoken Aikido style of his, that Dyson could get her keyboard back in cyberspace, as soon as something more powerful than a toy system is developed. "You could have a simulated workstation, and that would mean you could have a virtual virtual desktop," he added with a small smile.

Theodor Nelson added his characteristically tangential and provocative observation that cyberspace might enable people to discover, as he has been saying for years, that "computer programming is really a branch of moviemaking."

The context of Nelson's statement was supersaturated, given that we had all tasted just a bit of cyberspace, that most of us had been reading Nelson's amazingly accurate prophecies for at least a decade, and that we were wildly spinning our mental wheels at that moment, trying to get a purchase on what seemed to be emerging. To me, and I could sense, to others, Nelson's brief statement was such a powerful, compact conceptual image it amounted to the equivalent of a spell. When I returned to my office, later that afternoon, I looked through a file I had been keeping somewhere in the garage just-in-case-someday file cabinet. Sure enough, in November, 1980, here is what Theodor Nelson had to say about the deep meaning of virtuality. Keep in mind that he wrote this before personal computers amounted to much, when head-mounted displays were laboratory curiosities as far as the rest of the computer world was concerned:

The central concern of interactive system design is what I call a system's *virtuality*. This is intended as a quite general term, extending into all fields where mind, effects and illusions are proper issues.

By the virtuality of a thing I mean the seeming of it, as distinct from its more concrete "reality," which may not be important.*

An interactive computer system is a series of presentations intended to affect the mind in a certain way, just like a movie. This is not a casual analogy; this is the central issue.

I use the term "virtual" in its traditional sense, an opposite of "real." The *reality* of a movie includes how the scenery was painted and where the actors were repositioned between shots, but who cares? The *virtuality* of the movie is *what seems to be in it.* The *reality* of an interactive system includes its data structure and what language it's programmed in—but again, who cares? The important concern is, *what does it seem to be?*

A "virtuality," then, is a structure of seeming—the conceptual feel of what is created. What conceptual environment are you in? It is this environment, and its response qualities and feel, that matter—not the irrelevant "reality" of implementation details. And to create this seeming, as an integrated whole, is the true task of designing and implementing the virtuality. This is as true for a movie as for a word processor.

At some subconscious level, I'd bet that a quarter of the people in the auditorium at Autodesk that afternoon had read that article almost a decade prior, and immediately unpacked the meaning from Nelson's spell.

After Ted Nelson zinged us, another fellow stood up and made a comment. From magic to the mundane: The young man who addressed the panel had the kind of haircut you can still get at a barbershop for $5; he was wearing an inexpensive white shirt with an actual pen-protector full of pens in his breast pocket. As he talked, it began to dawn on me that this speaker from the crowd was John Walker, the normally reclusive president of Autodesk, a man who occupies the very opposite end of the charisma spectrum from Jaron Lanier. He's a programmer of legendary proportions who founded a company a few years ago that makes hundreds of millions of dollars a year now, and who recently took a voluntary indefinite leave from the presidency of his company in order to get back to programming.

*The closest other term I can find is "mental environment." My students have urged me to retain the term "virtuality," even though it causes confusion among users of so-called "virtual systems," meaning real systems configured with huge virtual memory.

In the eyes of Wall Street, VPL isn't even a blip on the horizon yet; Autodesk is a contender for the big-time. And John Walker is Autodesk's inspirational center. The antiheroic image is just the way he is. He likes to program, he doesn't like to shmooze. He will speak up, though, and rely on his powers of persuasion.

Walker doesn't talk to anyone remotely resembling "the press," but I managed to make his acquaintance at the Hackers' Conference a few months later, an invitation-only gathering of the conceptual movers and shakers of the mind-amplifier business. The word "hacker" has come to mean, in the mass media, "antisocial urchins who break into other people's computers." Originally, the term was a more honorable reference to the virtuosity of some programmers in finding ingenious ways to overcome obstacles. Indeed, without the hackers who invented the term, we wouldn't have personal computers or public computer networks today. The Hackers' Conference is a weekend gathering held each fall, at some rustic location near Silicon Valley. The *Whole Earth Review* and a few now well-heeled hackers hosted all the heroes of the personal computer revolution—the NASA graphics programmers, the young architects of the original Macintosh, the teleutopian computer networkers and game-designers, the veterans of the Homebrew Computer Club of the 1970s, where multibillion-dollar companies blossomed out of the fantasies of hobbyist prodigies.

VR was the buzzword du jour of the 1989 Hackers' Conference. I was given the job of hosting a raucous afterdinner plenary session on VR issues. It was convenient to give me the role of moderator, a role I was to fill at several such gatherings because it was a politically expeditious way to finesse the various egos and competing business interests among the VR crowd. About a hundred motley participants, mostly but not entirely male, gathered in the auditorium of the rented summer camp. It turned out to be less of a watershed than I had anticipated, although the potential for a momentous consensus of some kind had been high: The pace of technological revolutions in Silicon Valley is so fast that many of the veterans who had lived through the now-historic infancy of personal computers now are in positions to influence policy during the next revolution; more than a few of the young engineers, programmers, and entrepreneurs who had created the computer breakthroughs of the recent past were there that night. I wondered if this might be the last time that leaders from the emerging VR industries could speak so freely among one another without their attorneys present, especially if today's R & D skunkworks turned into tomorrow's profit center. People from Autodesk were there. People from VPL, NASA, and other VR research sites and businesses were there. If issues of privacy, property, and free speech in cyberspace were ever to be discussed by a group of people with the requisite skills, experience, and power to do something about these concerns, the Hackers' Conference in 1989 was an ideal occasion.

By Saturday night after dinner, however, more people seemed to be in a mood to talk about cybersex than the legal or epistemological implications of virtual reality. I will admit that I mentioned the possibility of sensual titillation in cyberspace when enumerating the issues we could talk about. This wasn't the last audience that seemed to zero in on that one possibility, no matter where you hide it in the list of potential applications for VR systems. Hackers' Conference tradition has the moderator give a brief intro, and then serve mainly as a means of passing the microphone around a hundred notoriously opinionated people. This was not the kind of crowd that any moderator in his right mind would try to handle. That evening was the occasion when Lee Felsenstein, former anarch of the Homebrewers, resurrected Ted Nelson's grand old word, *teledildonics*.

The discussion turned serious, however, when Walker stood up and reminded us that the potential of 3D design is evident all around us. "CAD isn't just about drawing things with computers," he noted. "It's about designing every manufactured object we use." He was talking about the point of leverage where a technology they were well equipped to understand had the potential to move the world.

Walker was talking about the kind of change his own company had wrought, one that is less visible than many aspects of high-tech culture, but which affects everything and everyone in a lot of little ways. The CAD revolution was indeed as ubiquitous as it was invisible to most people. The levers of change aren't always visible—most people don't see designers and architects working with computers instead of tracing paper to design soapdishes and skyscrapers—but everyone sees and lives among the effects of CAD tools. The entry of John Walker's voice into the VR dialogue took it on a serious turn that was at once visionary and pragmatic. Frederick Brooks wants to create intellect amplifiers for scientists, doctors, architects. Jaron Lanier sees an entire new universe of human communication. John Walker points out an untapped market in making it possible for more people to work smarter by immersing themselves in models of their problems.

Autodesk's entry into the cyberspace R & D field marked it as a serious, if nascent, industry. John Walker commands respect among programmers not just because of his net financial worth, but because Autodesk is known as a programmer's company. It started because Walker and others gambled that architects and designers who had been using personal computers for other applications would pay several hundred dollars for a somewhat less powerful version of the multithousand-dollar CAD software that large architectural and construction firms were using on mainframe computers. AutoCAD became one of the personal computer software best-sellers of all time. As computer graphics techniques matured and the power of computer hardware increased, Autodesk's toolkit began to take on depth, evolving from a flat screen version of a mechanical drawing to 2 1/2 D depictions of spaces and objects. By 1990, Autodesk was reporting fiscal 1989 sales of $117 million and a Wall Street valuation of the company at $926 million. Autodesk's rapid success is direct evidence that designers will pay for intelligence-amplifying design tools, even relatively crude ones, if it gives their own expertise the right kind of boost. Walker believes that the same people who made his company worth close to a billion dollars—AutoCAD users to whom envisioning three-dimensional objects is serious business—are ready to move their points of view through the window of the computer screen and into the virtual worlds they are trying to design.

As mundane as CAD software sounds, it is in itself a powerful version of the intellectual augmentation tools Douglas Engelbart and Frederick Brooks have been envisioning for decades. That Autodesk consciously sees itself in the intellectual augmentation business is evident by the projects it is exploring simultaneously in hypertext, on-line communities, multimedia, and virtual reality: Autodesk bought a large interest in the legendary unfinished *Xanadu* hypertext project from Ted Nelson and his colleagues; hypertext is one way of linking up all the world's knowledge into a kind of automated network, in which creation, publication, and payment of royalties for intellectual property could take place in a system accessible to everyone everywhere. Autodesk also hired Randy Farmer, one of the designers of the first graphically depicted on-line communities (*Habitat,* originally a project of Lucasfilm's Games Division, now owned by Fujitsu), for their Amix subsidiary, to help design an information-exchange marketplace of the future. Most recently, Autodesk opened a multimedia division, for exploring the use of technologies such as compact discs (CD-ROM), videodisk, and digital video interactive (DVI). Autodesk's Thursday in-house seminars in their Sausalito office became popular events. Having a friend who worked at Autodesk became fashionable.

In the world of the very best software wizards, there are the lone wolves and the ones who go to work for a company because the other top programmers are there. For this reason, rumors that a large portion of known wizards are showing up at one company or another is evidence that something interesting is afoot, something that extends the frontiers of the technology itself, not just the marketplace. There was a period in the early 1970s when all the software superstars seemed to be gravitating to PARC. In the early 1980s, those who wanted to push the outside of the envelope in new technologies and save the world through whizzy new software ended up at Apple. Toward the end of the 1980s, Autodesk seemed to be gaining a quiet reputation as a haven for the best and brightest software wizards. Unconventional thinking was welcome, Autodesk seemed to be saying in its hiring pattern, just as long as an applicant's programming skills were Olympian.

Running a large group of programmers is an entirely different exercise from creating great software by yourself. The way he communicates his vision to the rest of the company is Walker's hallmark; he is known for his inspirational memos as well as his software design (a collection of them were published as "The Autodesk File"). In September, 1988, Walker propelled the company into the cyberspace business—or at least nudged it the first symbolically significant yard or two toward that goal—by circulating a paper around Autodesk. It bore the title "Through the Looking Glass: Beyond User Interfaces." It wasn't a secret document, although it saturated Autodesk for a few hours before copies began finding their way to electronic desktops around the world, including mine. Reading that manifesto not long after my own first glimpse of cyberspace was one of the acts that propelled me into a couple years of traveling and VR-tasting following the growth of VR research and development worldwide. John Walker had perceived the significance of a moment in history, and urged his company to take advantage of a vast new market they had, for the moment at least, all to themselves. CAD was the foot in the door, but the territory Walker foresaw beyond the barrier of the 2D screen was enormous.

Walker saw that what they were selling was more than a new way to make blueprints. Autodesk's products for three-dimensional visualization and manipulation of objects and structures could be seen as the spearhead of something at least as important as the personal computer industry, which still reached only a small portion of its potential market. Walker made a case that "today's fascination with 'user interfaces' is an artifact of how we currently operate computers—with screens, keyboards, and pointing devices—just as job control languages grew from punched-card batch systems. Near-term technological developments promise to replace contemporary user interfaces with something very different." The "something very different" is the total-immersion display Walker prefers to call a "cyberspace system."

In the memo that grew into a paper that was later published in a book, Walker presented a kind of revolutionary chart of human-computer interfaces and showed how this long-term development trend is leading directly to that interdimensional window Ivan Sutherland proposed in 1965. Walker knew where the silicon was going, and he knew where 3D rendering software was going. He saw 1989 as the year when it would be feasible to put together crude VR systems for less than half a million dollars. He was another one of us who tried out the Kitty Hawk version of VR in 1988 but conceived a mental image of a 747 in nineteen ninety-something. Another believer had been drawn into the VR conversion experience.

Walker noticed a fine distinction that might make a lot of difference in the long run. Most conventional commentaries refer to "generations" of computers in terms of whether they use tubes, transistors, simple integrated circuits, or very-large-scale integrated circuits as their fundamental switching elements. In his paper, Walker proposed a historical taxonomy based on how people and computers *interacted*, rather than how the computer's components are made, with each "generation" of interaction defined by a characteristic mode of operation, rather than by hardware or software. The first generation of electronic digital computers in the late 1940s were operated by plugboards; computations could be reprogrammed by rearranging the plugs like a giant switchboard. The second generation came along in the 1950s, when batch processing enabled punched cards to substitute for the switchboard. The 1960s brought the third generation of time-sharing, which brought programmers into direct interaction with the computer via keyboard and screen. In Walker's taxonomy, the use of menus by which users could select commands, in lieu of remembering arcane command strings, was the fourth generation. The direct manipulation interface, the ARC-PARC-Mac point and click paradigm, is the fifth generation in Walker's schema.

Having established this historical sequence, Walker pointed out that if one examines the transformation from one era to another, the process is one of removing barriers between the computer user and the computer. The front panel of the computer was the barrier that was removed in the transition from plugboards to punchcards; you no longer had to remove it to run a new program. The countertop over which programmer's handed their decks of punched cards to system operators was the barrier that time-sharing eliminated. The terminal, with its rigid requirements for command-strings, was the barrier that was removed by the advent of menus. The menu hierarchy of commands that hides part of the process of commanding the computer from the user was the barrier that was removed when point-and-click operations made the screen the final barrier. If the screen is the current barrier to be removed in order to ascend to the next level of interface evolution, Walker sees the old quest for an artificial world converging at collision velocity with the evolution of the human-computer interface. A human-computer interface and a reality engine are two ways of looking through the same looking glass.

Seen through Walker's taxonomy, something that would have stayed in the background seems to stand out: *to enter something and look around* is a different metaphor for interaction—with different barriers to be removed—than the act of commanding something and perceiving the reply. We interact with computers in the pre-VR era, we don't wander around inside them. The present "fifth-generation" interfaces enable users to manipulate two-dimensional graphics in a "conversational" manner—exchanging commands and replies with the computer in a kind of dialogue. Here, Walker has found the crux of the problem— the barrier to thinking we must remove but can't see because it is right there in front of us. In his original memo, Walker declared:

> I believe that conversation is the wrong model for dealing with a computer—a model which misleads inexperienced users and invites even experienced software designers to build hard-to-use systems. Because the computer has a degree of autonomy and can rapidly perform certain intellectual tasks we find difficult, since inception we've seen computers as possessing attributes of human intelligence ("electronic brains"), and this has led us to impute to them characteristics they don't have, then expend large amounts of effort trying to program them to behave as we imagine they should.
>
> When you're interacting with a computer, you are not conversing with another person. You are exploring another world.

Explore, not *converse,* is the best verb for describing an ideal mode of human-computer interaction. It is a simple change in perspective that makes an enormous difference in the way everything looks. That insight is the reason John Walker knows exactly where the computer revolution is going next, and where the company he founded ought to allocate resources. He declared in "Through the Looking Glass," a memo that reads more like a manifesto, "I believe that cyberspace is the only technology which is a serious contender to define the next generation of user interaction." The entire document is a fascinating instrument. Here's a visionary whose answer to the question, "Is it prudent to build a company on an idea from a science fiction book?" is "Why not? That's what I did."

Walker acknowledged the science fiction origins of the idea of entering a computer world, citing William Gibson recently and Frederick Pohl in the past, noting in an offhanded way the personal significance to him of Pohl's "Heechee" books, "the second of which, *Beyond the*

Blue Event Horizon, inspired the product Autodesk, which idea played a significant part in the formation of this company." Walker argued that " 'artificial reality' and 'virtual reality' are oxymorons" and proposed that *cyberspace* is a better term, with its roots in the Greek word *cyber,* meaning "steersman." (Several months later, when some overzealous legal counsel convinced the Autodesk crew to attempt to trademark the word "cyberspace," and programmer Eric Gullichsen's name was mentioned in the rumor, Autodesk received a letter from William Gibson's attorney—and the rumor floated back, via their mutual friend Timothy Leary, that Gibson had been looking into trademarking the name "Eric Gullichsen.")

If cyberspace technology is the direction of the next computer revolution, and if CAD software users are the vanguard of the cyberspace market, and if Autodesk and its user base are in the perfect position to exploit that market opportunity, then what did Walker suggest they do? First, the cost of a minimal cyberspace system had to be brought down. In late 1988 it still cost hundreds of thousands of dollars to get into cyberspace, if you count the cost of the rendering hardware and world-modeling computers you had to buy to complement the expensive gloves, goggles, and position sensors. However, Walker noted that it would be possible to use the graphics capabilities of a processing chip that had just become commercially available—Intel's i860, sometimes called "a supercomputer on a chip"—to soup-up a far less expensive personal computer. Buying a video add-on board based on these chips, using inexpensive Amiga 500 home computers as rendering engines, building goggles the way McGreevy and Humphries did, and buying the gloves from VPL or using a less expensive six-degree-of-freedom trackball, could bring the cost of everything but the control computer down to the neighborhood of $15,000, Walker estimated.

This "cyberspace in a briefcase" system, as Walker dubbed it, would be a prototype that could demonstrate to CAD users the virtues of zooming around inside their building during the design process. He wasn't suggesting that Autodesk should get into the hardware business. The role that Walker saw as ideal for Autodesk would be as toolmakers, creators of the software construction kits for all the explorers who are going to come along when cyberspace systems reach the right price-performance point. Existing Autodesk products and projects for 3D computer graphics modeling could be used right away, without modification, for building worlds. Future software could be built by modifying existing products. By the time the hardware components of cyberspace tools grow affordable, Autodesk might steal a march on the rest of the software world. VPL was dividing its energies between the hardware business and the software business, and they were still a struggling if promising startup. VPL was developing the market in a dozen different ways, and CAD was only one of them. With dozens of programmers maintaining and extending 3D rendering programs, Autodesk was far ahead of any other software companies that might decide to jump into cyberspace.

The conclusion to Walker's extraordinary memo reached sermon intensity:

"If Autodesk believes, as I do, that this technology not only holds the key to the next generation of user interaction but will first find applications in our central market, three-dimensional design, then Autodesk should apply resources to developing this technology commensurate with its potential," he wrote. "If we undertake this project, we should commit to it explicitly and allocate adequate manpower to get it done. . . . An 'Autodesk Cyberpunk Initiative' which will yield results within four months and products within twelve is affordable, achievable, and appropriate," Walker added. The prototype materialized almost as quickly as Walker predicted. The memo was issued in September, the Polhemus sensor and DataGlove were acquired in October, the HMD was completed and the first prototype was demonstrated in March, 1989. The products have taken a little longer; as of the spring, 1991, marketable products were still under development. The project has had its ups and downs since Walker kick-started it.

The winter of 1988 and spring of 1989 were heady times around Autodesk. A project was assembled, personnel assigned and hired, offices furnished. A suite of offices and one large workshop area were named "Cyberia" by the Autodesk cybernauts. William and Meredith Bricken, a husband and wife team, and another couple, Eric Gullichsen and Patrice Gelband, as well as Randal Walser, Gary Wells, and Christopher Allis constituted the original contingent. I bounced in there one day in early 1989, before they had a demonstration of "cyberspace in a briefcase" ready to show, and spoke to the project leader, William Bricken. We realized then that we had talked once on the telephone seven years previously, when he had been a cognitive scientist at Atari. He was still a mathematician and a philosopher as well as a psychologist, an AI programmer, and yet another full-blown "religious convert" to the cyberspace crusade. His wife, Meredith, was an educator, artist, and psychologist. Although their visions of the future were different from Jaron Lanier's, the Brickens seemed no less deeply convinced that they were on the ground floor of the biggest thing since the alphabet, maybe since spoken language. The first thing they needed to know was whether I realized how big this cyberspace fantasy might become.

Meredith was busy conducting literature research and documenting the research program; William was sketching out the vision of the cyberspace toolkit they were building, and dreaming of the laboratory-cathedrals they could create with their new tools. Gullichsen, Gelband, and Walser were the main programmers. Gullichsen wears his blond curls down to his shoulders, and the weather has to get fairly cold to see him wearing long pants; Eric is the one who brought his friend Timothy Leary in to make the promotional film announcing their cyberspace program to the public. Gelband has a Ph.D. in physics. Christopher Allis, who dresses in black and wears an Autodesk earring, helped with the hardware systems and test piloting. By now, he's probably logged more hours in cyberspace than anyone else, including Jaron Lanier. Together, they made quite a jolly crew, especially in the early days of the enterprise, buoyed by the Walker manifesto and their company's support.

The first demo the Autodesk Cyberians put together when they calibrated their hardware was the office plan. Autodesk's CAD software now comes with a library of different environments that users can use as the basis for their own creations, and the most well-known item in the library is a floor plan for an office space, with cubicles and desks, bookshelves, doorways. They were demonstrating their first cyberspace all day long for weeks before the first public showing. I dropped by their office one afternoon in the spring of 1989 and watched others encounter their first cyberspace while I waited my turn. It's a little jarring to watch somebody else in cyberspace, especially if you can't see what they are seeing—a bit like watching someone hallucinate. It was easier to watch when I walked behind the programmers and looked over their shoulders as they monitored the system performance. Whatever the participant saw in 3D was represented on a display screen. For that reason, I knew what kind of space I was about to enter before I entered it.

After the NASA world of extremely simple monochrome and wire frame, the colored, shaded solids of the Autodesk system were impressive for their verisimilitude; not so impressive was the speed—the perceptible lag between my motions and the changes in the cyberspace display seems to be more irritating the more one grows accustomed to the experience. The i860 boards can generate about 100,000 polygons per second, which isn't as many as it sounds, because that means that even a cartoon-simple scene can be re-rendered only as fast as five or six frames per second. Thirty frames per second appears to be the lower minimum threshold of performance for any kind of extended work. When chips that are in prototype stages in Santa Clara and Yokohoma today hit the market in a couple of years, that threshold should be affordably achievable. This is one of the areas where economy of scale can drive the diffusion of a technology; if a toy or tool or consumer product makes it worthwhile to ramp up to mass-produce new 3D graphics chips, future VR systems will be more powerful and less expensive than is possible today. The first Autodesk system was clunky, but it was indeed a cyberspace in a briefcase, cost just under $25,000, and it ran on the kind of computer that can be purchased off the shelf for a few thousand dollars, instead of a few hundred thousand.

Not that the constraints of the hardware bothered me terribly. The "office plan" did something else that the models of molecules or space shuttles or wire-frame versions of the NASA laboratory had done— it conveyed a sense of place. So the place was a little funky, reality-wise. It was *there*, in vivid, shadowed, fully parallaxed shades of red and blue. There were things to be discovered, aspects of the world to be understood by walking over, reaching out, and grabbing them with my hands and fingers. The representation of my hand looked less boxlike than it did at Mountain View, although it was not what I would call anthropomorphic. It looked like a robot's yellow glove. I discovered that it was possible to reach my hand and the cyberglove over to the bookshelf, grasp one of the books by making a fist, and take it down for closer inspection. There wasn't enough computing power in the "briefcase" system—which was more like the size of a steamer trunk, when you pile all the components together—for me to open the book and read it. But the hardware's ability to display dynamic books in cyberspace is only a question of time.

I noticed that Autodesk had adopted the convention for navigating in cyberspace that had been pioneered at NASA/Ames—the "point your finger and fly paradigm." The environment wasn't entirely as staid as the kind of offices people are accustomed to in physical reality—after the original office plan was implemented, they added a portal to another dimension, because that's what cyberspace is for: one door led out to a virtual swimming pool, where I could stand under a beach umbrella, or even enter the water, where everything took on a blue tint.

For those of us who had our baptismal experiences at NASA when they were still using wire-frame graphics, this was an impressive way-station on the reality quest. For those who tasted cyberspace first at Autodesk, suspension of disbelief seemed to happen more readily, and the loss of control over speculations about the future seemed to be even more prevalent. At about the same time I got my first demonstration of Autodesk's first cyberspace, John Barlow, another writer who was getting sucked into the fascination of the VR industry, described his first immersion in digital territory nicely in *Mondo 2000*, an avant-garde techno-culture magazine:

Suddenly I don't have a body anymore. All that remains of the aging shambles which usually constitutes my corporeal self is a glowing, golden hand floating before me like Macbeth's dagger. I point my finger and drift down its length to the bookshelf on the office wall.

I try to grab a book but my hand passes through it.

"Make a fist inside the book and you'll have it," says my invisible guide.

I do, and when I move my hand again, the book remains embedded in it. I open my hand and withdraw it. The book remains suspended above the shelf.

I look up. Above me I can see the framework of red girders which supports the walls of the office . . . above them the blackness of space. The office has no ceiling, but it hardly needs one. There's never any weather here.

I point up and begin my ascent, passing right through one of the overhead beams on my way up. Several hundred feet above the office, I look down. It sits in the middle of a little island in space. I remember the home asteroid of The Little Prince with its one volcano, its one plant.

How very like the future this place might be: it's a tiny world just big enough to support the cubicle of one Knowledge Worker. I feel a wave of loneliness and head back down. But I'm going too fast. I plunge right on through the office door and into the bottomless indigo below. Suddenly I can't remember how to stop and turn around. Do I point behind myself? Do I have to turn around before I can point? I flip into brain fugue.

The video of the cyberspace tour of the office-plan demo lacks the depth of the goggles and gloves version, but it does include a moment that made the audience at the CAD convention in Anaheim I attended gasp when the camera's point of view flew up and looked down at the familiar CAD plan from a totally new perspective. The cyberspace announcement and discussion session was titled, at William Bricken's request, "Weird Science." In the suite upstairs in the hotel where that convention was held, selected customers were given brief personal tastes of cyberspace. Timothy Leary, who had been in the film for reasons that weren't quite clear to the rather staid audience, was in the suite for a while, sipping a scotch, stepping into the hall to smoke a menthol cigarette.

The next cyberspaces the Autodesk crew created were examples of unexpected ways cyberspace might find its way into people's homes. The "hicycle" demo used a stationary bicycle, the kind people pedal for virtual miles in their homes while they watch television. When I mounted the stationary bicycle and put the face-sucker on my head, I found myself on a path through an endless cartoon countryside. I started to pedal. My point of view started to move down the road as the world streamed by on either side of my central focus. The immediate mundane use as an exercise simulator that would make indoor aerobics by flabby technocrats more palatable is spiced by the only-in-cyberspace feature that enabled me to *fly* over the landscape when I achieved a pedaling velocity of 20 miles per hour. Golf and skiing are two other sports that could be suitably cyberspace-simulated with the right kind of motion platforms. The next Autodesk cyberspace I entered was also a sports world—the solo racquetball game they had up and running in time for SIGGRAPH 1989 in June.

At SIGGRAPH, I stood in line for half an hour at the Autodesk booth, chatting with the computer graphics enthusiasts about their expectations for cyberspace. When the time came for my demonstration, they put a racquetball with a Polhemus sensor in my hand instead of fitting it into a glove. The "ball" was more like a multifaceted solid than a smooth sphere, and it took a moment to compensate for the lag, but I found myself moving the real racquet in the roped-off demonstration space at the Autodesk booth, and watching it make contact with the virtual ball in the cybercourt. It bounced off the walls and ceilings at all the angles I expected it would, although it did travel in conveniently slow motion. I couldn't feel the moment of contact, but as Eric pointed out, it might not be hard to create a low-tech "thump" in the racquet handle at the moment of contact.

Soon after I experienced the bicycle and racquetball demonstrations, I started a series of conversations with Walser, Autodesk's cyberspace architect and manager of their sports-and-entertainment-related applications. A softspoken fellow, a dedicated father, Randal Walser is the kind of person the adjective "thoughtful" well describes. For the past two years, between long programming sessions to get the demonstrations working, he has articulated, in a series of papers, his visions of what cyberspace might become. Like others who have ended up in VR development, Walser was involved with programming games software in the early 1980s. At one point he worked for a subsidiary of Bally/Midway on *Cyber Ridge*, a prototype virtual exploration adventure game using videodisk technology. Walser and his colleagues found an appropriate ridge in Utah, spent six weeks photographing it with a 360-degree panoramic camera, and began to build a battle simulation based on the digital model of that ridge in Utah. After the company building the game folded, he worked on AI and telerobotics projects before he joined Autodesk, where he seems determined to stay for the long run of its cyberspace research effort.

Despite the high optimism of the previous spring, morale at Autodesk took a dive in the fall of 1989. First, William and Meredith Bricken resigned in a dispute with management over direction of the project. Then Eric Gullichsen and Patrice Gelband quit to form their own VR company, Sense8 Corporation. Eric Lyons took over as director of the project. Carl Tollander and Randy Walser continued to work on the software infrastructure they would need to build the grand new spaces they had proposed in their position papers. But Autodesk's cyberspace initiative seemed to lose momentum until the late summer of 1990, when Autodesk started expanding the number of programmers devoted to the project again. Walker, having set the ball rolling, took his leave of absence and was off programming something. Whether Autodesk succeeds in steering its sometimes erratically supported project to full product rollout remains to be seen.

In my conversation with Walser, and the position papers he has written, several key ideas emerged. The first idea is that economics of state-of-the-art systems is going to dictate the appearance of what he calls "Cyberias," places one goes to experience cyberspace, rather than the immediate widespread diffusion of "personal simulators." Next, the aspect of cyberspace in which a sense of bodily presence is joined with a great plasticity of bodily representation is a matter of potentially profound consequence. Further, Walser shares with Brenda Laurel and others the belief that cyberspace is inherently a theatrical medium, in which people participate in events that have dramatic structure and emotions. Finally, he makes a case that surrogate sports are more than a clever entry into a market where a lot of disposable income is available for experiential gadgets: Sports combine the social aspects of "Cyberias," the importance of physical embodiment, and a concrete instance of a first-person dramatic interaction.

It isn't hard to imagine a place something like a contemporary health club, with booths of various sizes and kinds of padding, and equipment such as stationary bicycles and treadmills, where cyberspace apparatus of various kinds is available on a membership or by-the-hour basis. Who would create such spaces? Walser believes that what is needed is a new kind of generalist-specialist: "a new breed of professional, a cyberspace architect who designs and orchestrates the construction of cybernetic spaces and scenarios. The talents of a cyberspace architect will be akin to those of traditional architects, film directors, novelists, generals, coaches, playwrights, video game designers. The job of the cyberspace designer will be to make the experience seem real. The job is as artistic as it is technical, for experience is something manufactured spontaneously in the mind and senses, not something that can be built, packaged, and sold like a car or refrigerator," Gullichsen and Walser wrote in 1989.

Walser's ideas about embodiment evolved over the two years I talked with him and read his formal papers. In 1989, together with Gullichsen, he wrote:

Cyberspace will not merely provide new experiences, like new rides at a carnival. More than any mechanism yet invented, it will change what humans perceive themselves to be, at a very fundamental and personal level. In cyberspace, there is no need to move about in a body like the one you possess in physical reality. You may feel more comfortable, at first, with a body like your "own" but as you conduct more of your life and affairs in cyberspace your conditioned notion of a unique and immutable body will give way to a far more liberated notion of "body" as something quite disposable and, generally, limiting. You will find that some bodies work best in some situations while others work best in others. The ability to radically and compellingly change one's body-image is bound to have a deep psychological effect, calling into question just what you consider yourself to be.

Imagine a costume party at which you adopt not merely a new set of clothes, but a new body, a new voice, and—in a very fundamental and literal sense—a new identity. Now imagine that you do this not only at a party, but every day, as an integral part of your life. Who, then, are you? It may seem, from your present view in physical reality, that you will be centered as you are now, in your physical body. It always comes back to that, right? But does it, even when you spend nearly all your waking life in cyberspace, with any body or personality you care to adopt? Does it, when the consequences of your actions and decisions in your alternative personalities have physical, social, economic, artistic, technical, and ethical consequences every bit as significant as those in your "original" personality? Does an alternative personality, active only in cyberspace, legally constitute a person?

Randal Walser doesn't claim to have answers to all the questions he raises. In Anaheim, we talked about all the uncertainties of the technology. We had time to talk again in May, 1990, in Austin, Texas, at the First Conference on Cyberspace, convened by the University of Texas Department of Architecture. On the final night of the conference, we convened on the lawn of a hotel overlooking the Colorado River that runs through Austin, waiting for that moment of dusk when a million bats emerge from under the bridge downstream. Walser was increasingly convinced that knowledge of theater, sports, dance, and film was going to become as important as programming. "In important ways, cyberspace goes beyond all previous forms of expression," he told me, to the background of high-pitched squeaks and two million leathery wings. "The difference is in the way cyberspace empowers the audience to shape the outcome of the experience," he added.

"What do sports and theater and cyberspace have to do with one another?" I asked.

"They are all refined forms of play," he replied.

Walser had presented a paper in March, 1990, in Anaheim again, to the National Computer Graphics Association, with the title: Elements of a Cyberspace Playhouse." Acknowledging Brenda Laurel's pioneering work, Walser argued strongly that theater and dramatic interaction are indeed the proper metaphor for conceiving the purpose of cyberspace:

> Cyberspace is a medium that gives people the feeling they have been transported, bodily, from the ordinary physical world to worlds purely of imagination. Although artists can use any medium to evoke imaginary worlds, cyberspace carries the worlds themselves. It has a lot in common with film and stage, but is unique in the amount of power it yields to its audience. Film yields little power, as it provides no way for its audience to alter film images. Stage grants more power than film, as stage actors can "play off" audience reactions, but still the course of the action is basically determined by a playwright's script. Cyberspace grants ultimate power, as it enables its audience not merely to observe a reality, but to enter it and experience it as if it were real. No one can know what will happen from one moment to the next in a cyberspace, not even the spacemaker. Every moment gives every participant an opportunity to create the next event. Whereas film is used to show a reality to an audience, cyberspace is used to give a virtual body, and a role, to everyone in the audience. Print and radio tell; stage and film show; cyberspace embodies.

In mid-October, 1990, an intriguing rumor began to circulate about John Walker. He was in the process of finishing the project he had been working on since he retreated from day-to-day operations of Autodesk in 1989; the kernel of a transportable cyberspace operating system. No public announcement has been made at this point, but it looks as if it is true that Walker has indeed been working secretly on the heart of a software architecture for a commercial cyberspace system. Has Walker poised his company on the brink of another revolution? Time will tell. It pays to bear in mind that there is a sizable population of software developers out there—the people who will build, sell, and lease cyberspaces for various purposes—ready to buy a product sight-unseen if it is true that John Walker was instrumental in creating it. That is where the ultimate test of Walker's vision lies. He sees Autodesk as the supreme vendor to the cyberspace prospectors, the toolmaker who sells the software infrastructure for using goggles and gloves to design objects in cyberspace. As Frederick Brooks noted, the success of a toolmaker takes a while to judge, while those who use those tools have a chance to succeed or fail. When and if the early adopters demonstrate a competitive advantage to VR design tools, then the cyberCAD revolution could happen on a large scale, when the mainstream CAD software users come on-line and permit economies of scale that could force down the price of the enabling technologies.

By the end of 1989, VPL and Autodesk were not alone among the major commercial efforts. A quasi-academic, quasi-commercial effort known as "the HIT Lab," in Seattle, run by an ex-US Air force researcher, had started a research and development consortium that included, within its first year, Digital Equipment Corporation, the Port of Seattle, Sun Microsystems, US West Communications as charter members. This new outfit had suddenly appeared on the commercial cyberspace scene, but it didn't appear from nowhere. An all-important element common to most technological revolutions related to computers and electronics had entered the VR story: technology transfer. When military technologies reach a certain age, they sometimes stimulate waves of civilian development. After the crash of the aerospace boom, for example, the availability of many trained electronic engineers helped propel the consumer electronics revolution that came along shortly thereafter. A great deal of the transfer involves more than the declassification of data and the interest of hardware vendors to find new nonmilitary markets for their wares. A lot of technology transfer happens when key members of military projects begin to interest themselves in civilian applications. The founding of the HIT Lab, and the particular person who founded it, signaled the surfacing of a stream of VR research that had been shielded from the attention of civilian researchers for decades.

Chapter Nine

REALITY ON YOUR RETINA

A hush settles over the operating room. A catheter is inserted into a vein in the patient's arm. At the end of the catheter is a micromechanical device containing a scanning TV and pressure sensor. The chief surgeon places a special helmet on his head. The power to the catheter and helmet display is turned on. A green glow reflects into the surgeon's eyes. Instantly, he is transported visually into the body of his patient as his helmet gives him the sights and sounds received from the probe located at the end of the catheter.

As he makes his incredible journey inside this human being, he sees a whole new world from inside the blood vessel. He "pilots" the catheter probe, navigating toward the heart, while hearing the gurgle of the blood around a defective heart valve. As he approaches the heart valve, he reaches out with his hand to remotely control a miniature suturing machine which corrects the valve malfunction. On the other side of the heart, the surgeon continues his journey into the aorta. He fires his laser ray gun to destroy another enemy—fatty tissue blocking the aorta.

THOMAS A. FURNESS III,
"Fantastic Voyage," 1986

"Don't stay fixated on the idea of screens. Screens might become obsolete sooner than you think," Thomas A. Furness III, until recently the director of the US Air Force's VR research program, said to me one rainy, early winter afternoon in Seattle. We were sitting in his office at the headquarters of the Human Interface Technology Laboratory, overlooking a rain-streaked, panoramic view of a brick building on the campus of the University of Washington. The visible part of the laboratory was a humble beginning, but what Furness carries around in his head is an asset unequaled by any of the other VR labs in the world. Frederick Brooks has been concentrating on VR for just as long, but Furness had better funding and thus knows what is possible with VR technology when price is not a priority. Furthermore, Furness has access to the network of contractors he cultivated during his Air Force research career—engineering firms, many of them small shops, have been working on the components of reality engines for years.

I had mentioned to Furness that the problem with portraying reality on head-mounted displays is that no known methods of portraying an image on a CRT or LCD display screen, or even the expensive fiber-optics the US Air Force uses, is close to matching the capacities of the human visual system.

"A laser microscanner will paint realities directly on the retina; it's just a question of when it will happen," he informed me, in that way of his that makes you want to believe him. He added: "The people we have working on it think we can achieve a resolution of 8000 by 6000 scan lines." His tone of voice and his own steady blue eyes had me half-convinced that he wasn't fantasizing, he wasn't speculating, he was simply pronouncing a fact a few years before it would come true.

I spent a week at "the HIT Lab," as they ended up calling their startup institution. I spent a lot of my time in Seattle talking with

Furness, and I've encountered him at various VR conferences around the country for a year since then. I think I have a fairly good feel for his credibility, and that is what makes it hard not to believe some of his incredible goals for the future of VR technology. If it is indeed possible to safely scan laser beams directly onto the retina at a high enough rate, in an extremely controllable manner that is synchronized with the eyeballs' motion in real time, as Furness seems to believe so firmly, then it will indeed be a breakthrough of major dimensions in terms of the verisimilitude of the virtual display.

Two problems with all artificial displays that have been built so far have to do with the power of resolution and the size of the visual field available to artificial (as opposed to biological) vision systems. To the human visual system, the pixels on even the most advanced graphic displays are far more coarse-grained than the physical world; and as long as you can detect an edge or boundary to what you can see, the size of the field is inadequate, as well. Recalling Alvy Ray Smith's epigram about computer graphics and human visual discrimination, it might require the equivalent of eighty million polygons per second, everywhere you can see, to approach human 3D visual modeling capabilities. A stereoscopic laser microscanner, a marvelous but as-yet hypothetical instrument for directly stimulating the light receptors in the human eye, would significantly boost the reality level of cyberspace—at the cost of massive demands on computing power. The basic element of the human-computer interface would be a retinal light receptor cell, either a rod or a cone, rather than a pixel. Perhaps a massively parallel computer architecture of the future could devote an entire computer to each rod and each cone in each eyeball: in principle, this could make possible total control of the visual environment.

Furness came across as sincere and genial during all our conversations. I think he's genuinely glad to see anybody who appreciates the scope of VR's potential. And I also think he is a person who knows how to pay attention to the other person in a conversation, especially if the other person is writing a book about what he is doing. You don't get your way with Air Force generals for decades if you are completely ignorant of the fundamentals of good public relations. I've seen him work the room with the best of them, particularly at the conference in Santa Barbara, in March, 1990, when it became clear that a science of VR research was emerging. I can recall standing in the auditorium of the Sheraton the last evening of that conference, conversing with other participants after one of the sessions, and realizing that the people in the group chatting around the coffee urn about whether MIT Press should start to publish a journal for VR research included

Thomas Furness, Myron Krueger, Jaron Lanier, Nathaniel Durlach, and Frederick Brooks. If you wanted to mount a Manhattan Project-style assault on VR, these are the people you'd start with; individually, they are beginning to get handles on the scientific and engineering problems to be solved, but how they could all work together is a project of another dimension. VR has certainly attracted a collection of very different, very strong personalities.

One thing I can say about Furness from personal experience is that he appears to have a high tolerance for diversity in personalities and lifestyles. He seems to have a knack for enlisting talented people in his projects. He's a fellow who worked around men in US Air Force uniforms for much of his life, until very recently; I walked into his office wearing a tie-dyed shirt and my hand-painted shoes, and if he blinked it was so quick I didn't catch it. He gets along well with Jaron Lanier, who refuses to accept any military funding, and he seems to be on good terms with all the other key figures in the VR field—a social landscape that is as thick with interpersonal complications as any other human endeavor involving highly intelligent colleagues. If anyone could act as a kind of social catalyst, widely respected enough to offer something approaching leadership to such a collection of confirmed mavericks, it would be Thomas Furness.

Until I saw the strategic approach to VR development the Japanese are taking, the question of whether such broad research planning should be mounted for such a fringe technology didn't really seem very important. The academic-commercial-government interconnections were there in that room in Santa Barbara, however, to make a believable scenario of an effective "reality-industrial complex" that could be pulled together if some person or institution in the United States ever marshals the will to pursue this technology. How to do that without a war, computer science's traditional funding mechanism, is a real challenge, because neither the US government nor the largest American industries put much money or credence into long-term basic R & D. Furness, of all people, has a good notion of what is possible, what needs to be done, and the will to move things forward in the directions he feels research ought to go.

Appropriately, Thomas Furness was one of the people I first met virtually, in the text-based cyberspace of the worldwide computer network. I had contacted him because I saw a match between my own interests and the needs of the emerging VR field. The first time I traveled around the United States, in the summer and fall of 1989, I saw that individual researchers were all so busy with their pieces of the puzzle that they weren't able to keep up with all the relevant research that seemed to be popping up in unexpected, previously unrelated fields. It wasn't like a highly specialized field where all the new developments are quickly available to everyone who ought to know about them. A new algorithm, chip-miniaturization technology, motion-tracking sensor, computer architecture, human factors research discovery, successful business application might change the environment for everyone else.

In late 1989, a year into the project of writing this book, with another year to go, I saw a way to bring more research information to my desktop than I could gather in ten years of unassisted effort; at the same time, I saw a means of connecting optical specialists in Utrecht with 3D camera engineers in Tokyo and computer scientists in Chapel Hill with medical specialists in Palo Alto. The same system of computers linked by telephone lines and modems that enabled me to make contact with Furness at the University of Washington by sending him "e-mail" was also a potentially powerful forum for establishing the cross-disciplinary communication necessary to create such tools as laser microscanners. It occurred to me that if I only could affiliate it with the right sponsor, I had at my fingertips a perfect new medium. Together with an appropriate VR research institution, I could start a virtual worlds "newsgroup" on the "Usenet," and let the world's fastest-growing techno-anarchic community do the rest.

Most people don't know about the Usenet because it is an informal communication medium that has piggybacked and bootstrapped itself, through a grass-roots movement, on the formal computer communication network that has grown up around the world. People who work with computers are more likely to know about it than people who don't, even though it uses the same telecommunications network as ordinary telephones. When you pick up your telephone, you simply aren't aware that the same network that carries your voice anywhere you can afford to call also carries computer-to-computer communications. It is a real cyberspace, known only to a subculture, which is growing at such a rate that it has a real chance of evolving into a spontaneous, bottom-up, surprisingly smoothly anarchic mass medium of the future. When the high-bandwidth communication networks of the late 1990s and beyond link billions of homes, the biggest cyberspace of all will be ready for colonization. The ad hoc network that has propagated itself over the past decade might look crude today, with people sending their global pen-pals written messages on computer screens. But geographically distributed computer-mediated communication is a social mechanism of great power, and the possible antecedent of a hybrid medium of the future, where telepresence will find its place along with alphanumeric messages.

The Usenet is a coalition of subnetworks consisting of hundreds of thousands of host computers ranging in size from amateur bulletin board systems in high-school students' bedrooms to the Internet itself (the modern descendant of ARPAnet). Usenet reaches approximately five million potential readers/contributors worldwide on a continuous basis, and more sites join the network every day. The individual participants, students and scholars and engineers from Helsinki to Austin, are connected via their own desktop computers to host computers that serve as local nodes. Local nodes are where people routinely work remotely via modem—using a larger computer's programs or data to write scientific papers and engineering documents, conduct research, compose business memos, or create computer programs. The sites include many tens of thousands of university research computers, computers at private companies, government computers, or home computers that are connected to one of the other kinds of nodes.

My local node, the Whole Earth Lectronic Link (the WELL), consists of several thousand members who can access the net by calling from their home computer and modem through one of the WELL's modems (there were about 64, last time I checked). The WELL's modem also automatically connects with the modem at the much larger computer system at Apple Computer, 40 miles away, four times an hour; at that time, if there is e-mail coming in for anybody whose home electronic address is the WELL, it is delivered as bursts of data, or if the WELL has outgoing mail, that mail is sucked into the Apple system, which is connected in a different way to hundreds of other Internet sites.

The Internet nodes exchange communication with one another worldwide through very high-speed telecommunication lines dedicated to transmitting data rather than voices, and the nodes that aren't on the Internet are rarely more than a few hops from a node that does communicate with the Internet. It might take a half hour before the WELL's upstream newsfeed calls, but once a message gets to the nearest Internet-linked newsfeed, it is everywhere else in the Internet world it is supposed to go, more or less instantly.

All over the world, people are communicating with one another by typing words into computers and sending them out as data through the telephone network. The conglomeration of networks that includes Usenet is sometimes known as "the Worldnet." Readers of William Gibson often consider the Worldnet to be an embryonic version of "the Matrix," as depicted in his "cyberpunk" science fiction books. ("The Matrix" is also the name of a quite serious nonfiction directory of the world's computer networks.) Usenet itself is simply a forum that rides along on the network-of-networks structure; it is a fully functioning cyberspace universe on its own account, shipping about 12 megabytes (approximately 12 million words) of information, conversation, programming code, kibitzing, scientific reports and journals, love letters and hate mail, around the world every 24 hours. The number of people and locations involved and the number of messages exchanged are growing daily. Someday, when VR "front-ends" become possible, the ancient Worldnet of today's text-based prehistoric era will furnish the infrastructure for true cyberspace communications. Those of us who have been conversing via our keyboards might someday dive right into a place where we can dance in three-space.

The Usenet is in anarchy because the mechanism for routing messages is built into the way the messages are encoded. Every node communicates and shares data with one or many other local nodes, but since all the local nodes worldwide are connected by some pathway or the other, there are always many ways for information to get around. It is almost impossible to shut the whole network down. If nodes are removed, the information will simply flow through a different path from origin to destination. Every computer that joins the network obtains and uses a certain standard kind of software to send and receive electronic messages in a certain format. There are loose rules of re-creating and propagating newsgroups, and many decisions are made by vote. The way Usenet creates a global asynchronous correspondence that amounts to a vast electronic conversation is by passing along public forums for discussion known as "newsgroups" when computers pass private electronic mail via modem from one computer to another.

People today use interfaces known as "newsreaders" to filter out a manageable gush from the daily tidal wave of info. When I connect my home computer via the telephone line to my local node, the WELL, I use one of the WELL's newsreaders to accept a specified subset of all the electronic messages passed through Usenet to the WELL and make them available for my perusal, indexed according to subject. That way, any messages from anyone in the world addressed to a newsgroup dedicated to discussing a certain rock and roll band, or Pakistani or Scottish culture, or the human genetic code, or one of hundreds of other subjects will automatically be displayed to me if and only if I decide I want to see messages about those topics. If I were to "subscribe" to a newsgroup about virtual reality research, I could, on command, see all the new messages received about that subject. If I wanted to reply in public to any message, or reply through private electronic mail to the person who posted any particular message, I could issue a simple command and join the conversation myself.

This new communication medium, the subject of its own intense social and technical commentary, has particularly useful applicability in that great formal conversation known as science. Researchers can squirt electronic versions of their latest preprint, months before it appears in scientific journals, to the two dozen or three thousand desktops of their closest colleagues. It has been happening, more or less unreported, for years.

It occurred to me that we could create a scientific newsgroup for discussing research and development in virtual reality on a less formal, more frequent, more highly distributed basis than the welter of professional organizations, conventions, conferences, and journals that served as communication media for the individual disciplines that were converging on VR. At every research site I had visited in my VR quest, I had seen at least one person in the laboratory using a desktop workstation to "read news," as accessing the Usenet is known among its participants. It could be an instant cross-disciplinary communication channel.

If an institution of suitable prestige could be persuaded to host such a newsgroup, I could "moderate" it—cull incoming messages before sending them out, to prevent the potential overwhelming of useful discourse by irrelevant kibitzing that such a wide-open system makes possible—and thus initiate and set the tone for a self-propagating worldwide search for research material for my book. At the same time, I might succeed in opening a useful scientific communication channel for everyone else involved in the field. Thomas Furness, whom I knew only by reputation at that point, seemed like the ideal sponsor for such a venture. Robert Jacobson, a friend I had met on the WELL, had recently joined the HIT Lab. He furnished Furness's electronic mail address and I sent Furness my proposal. The next morning, I received my reply. Furness was in favor of the idea. Indeed, one of his goals for the HIT Lab was to make research information as widely available as possible to VR researchers. Sponsoring a worldwide newsgroup would be a first step. An on-line bibliographic data base that could be searched by interested parties through the network could be a future improvement.

Furness suggested we talk on the telephone, which we did. We set up the newsgroup. "Sci.virtual-worlds" was the name I picked. It took a while to get off the ground, but began to flourish as more participants worldwide began to join the conversation. I passed along the job as moderator to a HIT Lab employee when the time came for me to stop gathering information and start writing the book. I still monitor it, which gives me a glimpse at what people are doing or fantasizing about doing in VR all around the world—on almost an hourly basis.

Furness had made the move from military-sponsored research to academia himself a few months before we met. He had been twenty-three, during the height of the Vietnam era, when he first joined the Air Force. Although he left the Air Force in 1971, he continued to work in the same US Air Force laboratory for almost two more decades. "The only thing I changed was my uniform," he jokes today. Now he is forty-six, a professor, and a civilian technology entrepreneur. When I visited Seattle, early in 1990, we had become acquainted through e-mail and telephone via the newsgroup project. We conducted our face-to-face conversations over a period of days, most of the time in a small clearing amid cardboard towers of what was undoubtedly the world's richest collection of data about how to build head-mounted displays, still in the boxes, filling his new office. The return address was Wright-Patterson Air Force Base, Dayton, Ohio.

Furness was another true believer in VR as the new breakthrough technology, capable of changing the world as thoroughly as had the light bulb and the transistor. If anyone has reason to believe in what you can do at the high-fidelity, spare-no-expense end of VR, it would be Furness. He's been creating head-mounted realities for nearly a quarter century by now, with funding support orders of magnitude greater than NASA's or UNC's. Everyone on the VR scene had known about Furness forever, but because the best of his work as founder, director, and chief of the Visual Systems Branch of Armstrong Aerospace Medical Research Lab at Wright-Patterson AFB was classified, his research was part of mostly a separate, parallel research field. "Visionics" is what the US Air Force and their contractors now call the field Furness was instrumental in founding. Visionics is big enough now to hold its own conferences for vendors and subcontractors.

Consider the article about the microsurgeon's tiny journey, quoted at the beginning of this section, written by Furness himself, as a sincere vision of what the technology he was building might enable people to do someday. Then contrast it with the capabilities demonstrated by this scenario published by another author, in *Air & Space* magazine, around the same time:

When he climbed into his F-16C, the young fighter jock of 1998 simply plugged in his helmet and flipped down his visor to activate his Super Cockpit system. The virtual world he saw exactly mimicked the world outside. Salient terrain features were outlined and rendered in three dimensions by the two tiny cathode ray tubes focused at his personal viewing distance. . . .

Once he was airborne, solid cloud cover obscured everything outside the canopy. But inside the helmet, the pilot "saw" the horizon and terrain clearly, as if it were a clear day. His compass heading was displayed as a large band of numbers on the horizon line, his projected flight path a shimmering highway leading out toward infinity.

A faint whine above and behind him to the left told the pilot . . . that his "enemy" . . . was closing in. . . .

The pilot glanced at the weapon system he wanted and raised his left hand. Tiny devices sewn into the fingertips of his flame-retardant gloves gave off signals that were tracked by the associate. When he "pushed" on a phantom button on the virtual display, a confirming click and slight pressure to his fingertip was fed back to the pilot to verify selection. To anyone watching, he appeared to be poking at thin air.

The "Super Cockpit" program isn't a scenario; it's the most recent name of a long-standing research program sponsored by the US Air Force. It wasn't designed to assist surgeons in microtelerobotic adventures, nor to augment the theorizing of pharmaceutical chemists, but to increase the survival rate of combat pilots flying high-performance aircraft. It pays to remember that aircraft training simulation was one of the driving forces behind the emergence of VR technology, and it has continued to drive the development of computer-graphic-based simulation. It isn't "pure" research when a human life, a hundred-million-dollar piece of equipment, and the outcome of a war might be at stake. The Super Cockpit idea evolved over two decades of Air Force sponsorship simply because a top-notch virtual display is the most useful picture of the world to a human who is steering a large quantity of high explosives at supersonic speeds, making life-and-death judgments in fractions of seconds, based on reports from radar sensors, paying attention to a couple of dozen equally vital instrument displays, listening to system status reports and command communications, all at the same time.

FROM FLIGHT SIMULATORS TO PERSONAL SIMULATORS

How can pilots be trained for such complex tasks without killing them in the process? The preferred solution, introduced during World War II, is the flight simulator, in which the pilot can practice many flying tasks while sitting safely on the ground. He or she can even practice responses to unlikely events, particularly those that might lead to disaster.

. . . Losses in combat are concentrated almost exclusively among pilots with five or fewer combat missions; if the fifth mission is survived, the probability of surviving the remaining ones is more than 95 percent, regardless of how many additional missions are flown. These figures suggest that if all pilots could be given the equivalent of their five combat missions before they face the enemy, losses could be minimized drastically, perhaps altering the outcome of the engagement or war.

RALPH NORMAN HABER,
"Flight Simulation," 1986

World War II era pilots remember the Link. They walked into a hangar, and there would be an amputated airplane—just a cockpit, mounted on a movable platform. Inside was the first flight simulator—one of the key historical antecedents of VR. The pilot climbed in, pushed the ignition button, grabbed the stick, and felt the cockpit tilt and roll, even vibrate, in response to the pilot's actions. In a sense, cyberspace is the intersection of stereoscopy and simulation. Insofar as virtual realities are walk-through simulations of virtual worlds, the Link trainer must be listed in the ancestry of cyberspace systems. While one converging technology was moving in the direction of putting people inside computer-generated worlds, a different technology was developing in order to create those simulations.

The Link trainer used during World War II consisted of a full-scale replica of the cockpit of a propeller-driven fighter plane, mounted on a small motion platform. A fellow named Link patented his original design in 1929. Originally a designer of pipe organs and air-driven player pianos, Link used bellowslike devices to pitch, roll, yaw, and turn the simulator when the pilot moved the controls. (In London, sixty years later, a fellow by the name of Jim Hennequin picked up the pneumatic gauntlet again, with his prototypes of air-driven motion platforms and pneumatic tactile-feedback gloves.) Tilting the cockpit simulated banks, turns, climbs, and dives, or even turbulence. Springs and other mechanisms emulated the way it feels to move the stick.

Student pilots would climb into the cockpit and learn how to use the controls to maneuver the airplane. Moving the control stick activated the controls of the motion platform in synchrony with the primitive visual display that showed, in the earliest models, a simple horizon line that shifted when the simulated aircraft rolled or yawed. Engine sounds were played through speakers. Link trainers gave novice pilots a feeling for the way airplanes handle without the risk and expense of logging hours in the sky. More complex training simulators began to use multiple film projectors back-projected on multiple screens where the windshield ought to be. The war-driven technologies of aviation and aeronautical design provided the impetus for creating lifelike simulations that could be "piloted" by an operator.

After the war, the aircraft industry as well as the US Air Force continued to develop, refine, and expand the technology that grew from the Link simulator. Military aircraft were growing faster, more complex, and with thermonuclear armaments, the stakes of air combat increased steeply. At the same time, the commercial aircraft industry was experiencing rapid growth. In both cases, simulation remained so much cheaper and less risky than in-the-air training that simulator technology advanced with other aspects of aviation. Motion platforms and servocontrols were state-of-the-art technologies immediately after the war, but upgrading the visual element of the simulation—jazzing up the view from the cockpit—had to wait for the advent of video cameras. Video, unlike film, doesn't have to be developed; you can view what the camera views when it views it. The dawn of the television era coincided with the early years of flight simulation. When the first video cameras became available, they were coupled to the controls of a flight simulator, and the view through the windshield was provided by a high-quality projection of the video image. Using the kind of miniature sets pioneered by Hollywood, it became possible to steer the cameras in tandem with the pilot's commands, over a miniature landscape. Elaborate physical models of aircraft carriers, landing strips, and various kinds of terrain were constructed. A model that measured thirty by sixty feet could be scaled to represent about six square miles. Larger geographic areas could be modeled, but only at a high-altitude level of detail.

In combination with the motion platform, the projected television view of the modeled landscape gave student pilots an increased sense of realism, and it became clear that increasing the realism of the visual display could markedly increase the effectiveness of flight simulation training. By stacking screens in front and to the sides of the pilot's visual field, it became possible to convey the illusion of immersion in the simulated world without using a head-mounted display. Evans and Sutherland came along in 1968, and computer graphic simulations began to replace the camera-and-model visual displays in flight trainers. Although camera-and-model displays are still used, computer-generated graphics have been the mainstream of simulation technology for decades.

Graphics programmers who built simulators for the increasingly sophisticated jet fighters of the past several decades grappled over and over again with the necessary trade-off between dynamism and detail required by computational constraints. Ever since Sutherland found it necessary to create the clipping divider and matrix multiplier, the intensity of calculations required for creating detailed, fast-changing graphics displays has posed a problem: to present a detailed view of a world or object, the number of calculations required from the computer is very high, and if the displayed imagery moves very quickly, the calculation load is similarly strained; you can see leaves on the trees only when you fly low and slow. The state of the art of computer-generated simulations seems to be destined to continue juggling detail against dynamism for the foreseeable future, no matter how rapidly the enabling technologies evolve, because as soon as the hardware grows powerful enough to make more detailed displays, people also want to make the displays change more rapidly, and vice versa.

Furness is trying to midwife the invention of display technology that doesn't even exist yet and sounds pretty scary to a nonengineer like me and millions of other people who aren't eager to have lasers shined directly into our eyeballs. But I might just try it out if Thomas Furness does it first.

"Look at this," he said, when we talked in his Seattle office. He reached carefully into one of the few boxes that he had unpacked. A very well-padded container held something that looked like a silvered test-tube. He handed the contents to me—a genuine demonstration of trust, I realized, when I saw what it was. The glass cylinder was several inches long, a bit larger than dime-sized in diameter. You don't see handmade CRTs every day.

"Your normal television screen scans a few hundred lines. This miniature CRT scans two *thousand* lines. You only get one color, but you'd be surprised at the level of detail," he explained, while I hefted the delicate, expensive gadget in my hand.

"And that might be as good as we can get with miniature CRTs," he added.

Furness started working on visual displays for the military in 1966, when the Air Force first realized that the major problem in pilot training had to do with the instrumentation that went along with modern avionics. All those gauges and dials and indicators relating not only to airspeed and position and oil pressure, but to enemy radar, electronic countermeasures, weapons systems were becoming too complex for the humans who were operating them. An F15 pilot has to be so agile with his fingertip mastery of the nine buttons on his control stick that this task is known as "playing the piccolo." Now aviators were not only fighting gravity and the enemy; they were fighting the complexity of their own aircraft, as well. Furness began to understand that the pilots were already operating in an artificial reality, even though the older pilots who were responsible for research funding seemed to be inordinately attached to the old-fashioned "steam gauges" they were using to inform the pilots' perceptions and judgments.

Like others who had been captured by Ivan Sutherland's dream, Furness started steering his human factors research toward a new technology that would couple the pilots' perceptual systems far more effectively with the information displays from the aircraft. In 1982, Furness and his colleagues, notably Dean Kocian, produced the first working model of the Visually Coupled Airborne Systems Simulator (VCASS). In the slideshow he presented to his virtual worlds class when I sat in with them in 1989, Furness said the words that were in

most of our minds the moment he displayed the first VCASS prototype: "Of course, everybody called this the Darth Vader helmet." Indeed, the huge black eyeless helmet looked exactly like something out
of a George Lucas film.

Inside the helmet, the pilots began experiencing a kind of flight
simulator view of a physical flight. Instead of looking through a physical windshield, the pilots saw a synthetic 3D "map" of the territory
they were navigating, projected onto their field of view via tiny CRTs
and mirrors. The landscape below may have looked cartoonlike, but
the features were real enough, based on a detailed digital model from
Defense Department maps (a "stored-terrain data base") and synchronized with real-time radar. Forward-Looking Infrared (FLIR)
technology gave a readable depiction of the environment, even at
night. Wherever the pilot aimed his head, VCASS position sensors
aimed avionics systems to cut through cloud cover and the dark of
night—aviators' two old foes. The ranges of antiaircraft missile bat-
teries located on the maps, or whose radar lock-ons were detected by
the aircraft's avionics, were depicted as large red bubbles, or "zones
of lethality." From the illustrations Furness still has, VCASS was like
being inside a life-sized video game of a dogfight. Projected flight paths
at current range and speed were visible as tunnels in the sky. All the
grahics were presented stereoscopically in real time, using optics and
head-mounted CRTs.

The whole idea of wearing a three-dimensional instrument panel
rather than gazing at a flat dashboard display was a total paradigm
shift from the days of silk scarves and open cockpits. The sensibilities
of aviators, their visceral reactions to a totally new kind of tool, were
a vital element in developing such a technology. The first thing Furness
and his colleagues did was to calibrate the equipment until they were
sure it worked. The next thing they did was to try it out with some
combat-tested fighter pilots, to see if they loved or hated it. If experienced pilots loved it, the program might find it possible to obtain
further funding, even though the project had a long way to go until
an operational system could be contemplated. If the experienced pilots
hated the way VCASS looked and sounded and felt, the program was
dead.

As Furness recalls, "Those first pilots were pretty skeptical before
they put on the helmet, but when we turned on the terrain map they
would say, 'Hey, this is neat,' and then we would start them up and
let them fly through it, and they would say, 'This is *really* neat.' And
then we would tap them on the shoulder and ask for another quarter."
The pilots loved it. The project went into the next phase—gathering
sufficient human factors data to know how to build a far more effective
system.

"The first thing you discover when you start looking at the way humans operate in the world is that humans are spatial beings," Furness said to me, said to his class, continues to say in his lectures and publications. "We have two eyes for stereoscopic vision, two ears for acoustic location, two visual systems," he points out. With regard to the two visual systems, Furness noted that the center of visual attention, the foveal area, gathers detailed information about shape and pattern by making many rapid eye fixations. The surrounding peripheral retina gathers less detailed information and is sensitive to sudden changes in the environment, often signaling to the central focusing system to change the direction of focal fixations. In an address to the Society for Information Display in 1988, Furness noted: "It is believed that these two visual systems operate in parallel using different processing centers in the brain. An example I use to describe the simultaneous operation of focal and ambient systems is the fact that most of us can walk down the sidewalk while reading a book and still maintain our spatial orientation (i.e., staying on the sidewalk)." If he was trying to find a technical match between a display system and what they had learned about human spatial perception, he couldn't find a more perfect one than the HMD idea that had been floating around computer science circles since the late 1960s. Immersion in a 3D video game might be a more natural control interface than a panel full of gauges.

Simulators also brought psychologists into the act because knowledge of the way human perceptions operate is fundamental to simulator design. What is the relationship between the size of the field of view and the perceived "reality" of the simulation? How high does the resolution of the display have to be to transmit a sense of presence? What is the optimal focal distance to place a simulator display? All the painstaking research and precise measurements required for the "human factors" side of simulation, paid for by the Air Force, NASA, and others who needed it to perform vital tasks, also created a body of knowledge about the ways human senses can be fooled into accepting virtual versions of the physical world. The cognitive scientists who are beginning to tackle the "interior" of cyberspace—the part that resides in our brains and bodies rather than the displays or computing engines—have a rich body of findings to guide their research, thanks to the human factors aspect of flight simulation. Indeed, one of the motivations of NASA's pioneering involvement in VR in the 1980s grew out of the need by human factors researchers to gather more precise information on the experience of *telepresence*—the psychological experience that results when simulator technology works well enough to convince users that they are immersed in virtual worlds.

Head-mounted virtual world displays of the same kind that Sutherland and colleagues had demonstrated at Utah and that Brooks et al. were applying to scientific modeling at UNC were also the perfect answer to the problem of an aviation interface. As early as 1969, Furness supported a small contractor by the name of Polhemus Navigation Systems, which later was purchased by McDonnell-Douglas Aircraft. By using a stereoscopic, gaze-directed visual display and an accurate three-dimensional acoustic display, coupled with fingertip tactile displays, voice-operated computer control, and eye-tracking technology, VCASS and its successors showed how it is possible to use the human's most highly developed skills for navigating three-space to pilot a high-performance combat aircraft. Our biological command, communication, and control systems evolved over millions of years in the direction of greater effectiveness at performing much the same tasks jet fighter pilots face: locating targets, evaluating threats, making ballistic calculations, knowing the right moment to hurl weapons and the direction to aim them. The spatial accuracy of human hearing was developed, over hundreds of thousands of years, because humans have been hunters for a long time. It is no accident that we can detect a tiny fleck of red at the corner of our field of vision, or sleep through a storm but awaken to the faintest creak of a floorboard or snapping of a twig. If Furness was right, pilots would take to virtual control systems naturally, as extensions of their hunters' instincts.

The pilots liked it, all right. They liked it even more when eye tracking and voice command were added. When radar systems verified a target, a symbol would pop up in the pilot's visual field, and his own attentional mechanisms would zero in on it. Simply by looking at a target, evaluating the information encoded in its shape, color, position, velocity, and saying the appropriate word—"bang," for example—it became possible for pilots to pit their nervous systems directly against the enemy, without the intermediation of dial reading or piccolo playing. The pilot, freed from the physical overload of being a one-man band, only semiconscious at times in supersonic combat due to the effect of G-forces on the brain, can devote all his attention to doing what humans do best—detecting patterns, making evaluations, sensing subtle changes in context. The pilots were enthusiastic about the "look-and-shoot" version of VCASS. More funding was provided for the next stages. By 1986, it was called the Super Cockpit program. New helmets, using half-silvered mirrors to create virtual overlays on the physical cockpit, fiberoptics to provide expensive but effective high-resolution visual displays, and three-dimensional acoustic systems were developed. A glove that used piezoelectric vibrotactile actuators—tiny crystals that vibrate when an electrical current stimulates them—was designed. The first prototype of a tactile glove was a trigger, more or less.

Dr. Furness (he received his Ph.D. in engineering and applied science from the University of Southampton, England) never limited his imagination regarding the potential of VR technology to only the military applications. He knew that reality engines were no more confined to military applications than ENIAC, the first electronic digital computer, was confined to the ballistics calculations the US Army created it to perform. Both civilian and military applications required what he called a "virtual world generator," which receives and processes the information to maintain a virtual world and to present it to the human operator. In an aircraft, the sources of input to the virtual world generator are the digital terrain data bases, radar and infrared sensors, the position of the pilot's eyes and head. In a nonmilitary system, eye- and head-tracking are important, and the operator is more mobile than a pilot, who is locked into position in his seat. In both military and civilian systems, the virtual world generator has to create the signals for presenting visual, auditory, and tactile displays to the pilot or operator.

The other general category of VR systems that Furness saw as a common element of military and civilian applications is what he calls "mindware"—the systems that make VR useful for a fighter pilot or a paraplegic, a pharmaceutical chemist or an architect. Like Frederick Brooks, Furness has always been a believer in the driving problems, the pragmatic applications that provide a focus for research and a motivation for commercial development.

"In 1986 and 1987, I started getting telephone calls from people who had read about my research and wanted to know if they could apply virtual environments to problems in their own fields," Furness recalled in 1990. "People working with cerebral palsied patients wondered whether tongue-steered VR could free trapped minds from dysfunctional bodies. Firefighters wanted to know if an HMD could steer a robot firefighter through a burning building. Anesthesiologists wanted a better way of displaying vital signs from all the instruments they had to monitor—the surgical theater is getting like a jet cockpit that way."

In the summer of 1989, Furness decided that the importance of the technology was too great to confine to military applications. He started packing his documents before I had a chance to visit him at Wright-Patterson. The rumors were rife that he was going into the VR business in some way. When he finally turned up somewhere, he had moved to Seattle. The state of Washington, together with the University of Washington, had started the Washington Technology Center, a hybrid institution for encouraging collaborations between academic research and commercial development—a Pacific Northwest version of the kind of partnerships that had made Silicon Valley and Boston's Route 128 so successful. In November, 1989, the Washington Technology Center announced the founding of a new member institution, the Human Interface Technology Laboratory. Dr. Thomas A. Furness III was the founding director. He was given a budget, some buildings, and a professorship, and the mandate to start making deals with local industries to develop appropriate mindware. He started hiring some interesting people. William and Meredith Bricken, for example. And Robert Jacobson, whom I had known from the WELL's information policy discussions.

When Furness first told me about his agenda for HITL, I could see why my proposal to start a Usenet newsgroup was received so well.

"We want to establish a national knowledge base regarding the ergonomics, technology and application of virtual interfaces," he wrote in his first memo to his newly hired laboratory personnel. Once he unpacked his boxes, Furness would have one of the world's most comprehensive libraries on the human factors aspects of virtual systems. Part of his Air Force work had to do with establishing a data base on human factors. Furness had in mind for HITL a bibliography, a central library for paper-based research reports he had collected, and a computerized data base, all of which would be publicly available. The newsgroup proposal fit right into Furness's ambitions to make HITL a central information clearinghouse for virtual worlds research and to stimulate the emergence of a real science of VR research. To that end, his next goal is to build the tools for measuring that all-important but thus far unquantifiable component of VR, the verisimilitude of worlds. How does one go about measuring reality?

The idea was to get the computer programmers started on creating an operating system for plugging diverse operations into a virtual worlds generator, and to design this master software in such a way that it could be used with many different kinds of computer hardware. If this painstakingly created software could be made widely available for free, it would serve to stimulate both the science and the marketplace for VR. At the same time, HITL would begin development of better position-measurement technology, better and cheaper eye-tracking technology, create a laser microscanner, create mindware to augment disabled people. In the Air Force and in civilian research planning, Furness has a habit of thinking big.

"We have a lot of hard problems to solve," he admits. "We don't understand the human factors dimensions of virtual space. We don't know how to measure how real a virtual world seems. Except for the more expensive military versions, virtual displays lack sufficient resolution for wide-field-of-view presentations. Position-sensing technology needs improvement. Current graphics engines are inadequate.

The headgear is too heavy. There are too many wires. There are no accepted software architectures and toolkits for supporting future development and application of virtual worlds. The whole system is far too expensive. We lack interdisciplinary forms for discussing virtual worlds research." He enumerated the problems so cheerfully, considering the fact that his human interface laboratory had yet to obtain computing equipment. Indeed, the week I was there I saw a disappointing setback when the representatives from NeXT, Steve Jobs's post-Apple computer company, informed Furness and colleagues that they could not donate a half dozen of their newest workstations simply because VR was the hottest frontier in the computer universe. I spent part of the week I was in Seattle sitting in the corner while the HITL people met and grappled with the fact that they had a startup budget, great prospects, and so far no concrete commitments to long-term research funding.

The idea was to build a consortium, consisting of companies that would contribute money and computing resources and in return would share the earliest results of the research. Certain consortium members would have specific mindware projects that would be created by HITL associates. As he passionately declared at the Santa Barbara conference, Furness is committed to "beating his swords into plowshares," so one of his first steps was to hire Suzanne Weghorst, a biomedical systems analyst to work on virtual prosthesis prototypes. The project was initially called "Advanced Adaptive Aids," and the first products were planned to be virtual classrooms for both the physically and learning disabled. In fact, Furness spent some time lobbying for support in Washington, and managed to get then-Secretary of Education Lauro Cavazos excited about the plan. "He's really turned on by the possibilities," Furness told newspaper columnist Denise Caruso. But the funding would have to come from other sources.

Another one of HITL's first projects had a more commercial end, but it was also an intriguing attempt at augmenting cross-cultural communication. "You should talk to Cecil Patterson at the Port of Seattle," Furness told me when I had looked around and noted that his plans were pretty terrific but I was still searching for something they were actually doing on the scale the Washington Technology Center was supporting them to do. So I drove down Capitol Hill early one morning and parked at the edge of Puget Sound. Here were the planners for one of the world's major seaports, one that is growing more important as a gateway to and from the Pacific Rim. "Yes," Patterson assured me when I called him, "we really do want to use virtual reality to plan future changes in the Port of Seattle."

The information systems office of the Port of Seattle is a busy place. We met in an office that looked out on other offices. There were computer workstations on people's desks, but other than that it looked just like any office. The heavy computing machinery has its own area. Patterson, information systems director for the port, confirmed Furness's claim that he was eager to try out a VR system for port planning. Patterson offered two reasons for spending the port's money to pursue VR models. First, he sees VR model-building as a kind of "what-if?" machine for CAD, the way spreadsheets are "what-if?" machines for financial thinking. This process is also called simulation-dynamic modeling, the art of creating a simulation with certain internal rules, then running the simulation through different situations to see how the whole system reacts. Aircraft simulators, designed to help train pilots, converged on cyberspace from one direction. Aerodynamic modeling, which would later become part of the interdisciplinary field of scientific visualization, was a computer tool used by the people who designed the aircraft those pilots flew.

The common element of simulation and modeling is the act of creating a model of a real system in the form of a computer program. Whether you use it as a simulator or a model depends on the effect you want to gain from manipulating the computer model, whether you want to *experience* it from the inside or *observe* it from the outside. Simulation as it exists in flight simulators and cyberspaces involves using that computer-based model to convince you that you are elsewhere—a perceptual effect. The kind of modeling and simulation that creates a mathematical model of an aircraft wing or a cardiovascular system, then subjects it to various tests to see how it behaves, is a cognitive tool for helping people comprehend complex systems. Flight simulation is geared to reproducing the experience of flying an airplane; simulating an airplane wing is a way of observing how it behaves under different conditions. One application is a tool for learning how to operate something; the other is a tool for understanding how something operates. And in the cyberspace education systems of tomorrow, both tools will be capabilities of the same microworlds.

"Designers understand renderings of designs on a computer screen far better than their clients," Patterson told me. He agreed with Frederick Brooks and John Walker that the best way to find out how you feel about a three-dimensional design, whether you are a designer or the person who uses the designed structure, is to walk around in it and handle it. When engineers, facilities planners, architects, contractors, and major clients start talking about long-term plans, they are talking about multibillion-dollar projects. Any device that can help facilitate communication between the people planning such a project is worth what it costs to build. On the scale that the port planners think, the price of a VR system is a bargain. Six months later, when I lunched at Shell Centre in London with one of Shell's strategic planners, he evinced a great deal of interest in VR models of refineries, precisely because Shell already spends large amounts of money on physical models. A good model is a thinking tool and a communication device that gains value when it is used by a group of people who are organizing some kind of 3D spatial structure like a building or a port.

Patterson sees a group of planners entering a model of a new dock in cyberspace. If somebody in a virtual dock model wants to know what the structure might look like if the dock was moved twenty-five feet or half a mile, it is possible to reach over, grab it, lift it up, and move it. Now, if all the planners could be in there together, via some kind of telecommunicated telepresence, they could save on communication and transportation costs as well as making their VR investment back manyfold by eliminating expensive misunderstandings.

Patterson's second reason for wanting to explore the group planning and decision-making properties of VR CAD is, to my mind, even more far-reaching than his intention to use it as a planning tool. Most of the port's biggest clients, the ones who participate in the planning stages of multihundred-million-dollar new facilities, are Japanese, Chinese, Korean, and others for whom English is not a native language. Patterson hopes that misunderstandings, delays, and costs that are caused by the spoken language barrier might be mitigated if the engineers, planners, and clients on both sides of the Pacific could walk through VR versions of the proposed construction during every stage of the planning process. That way, even though the spoken language barrier may remain, the pictorial mental models of what they are planning will be much more in accord. When the subject of the conversation is a three-dimensional object, the similarities and differences among their mental models may vary widely among a group. When the members of the same group walk around the same physical model while they talk about it, their mental models, Patterson suspects, are likely to be much more highly synchronized.

Through 1990, Furness and Jacobson continued to find other partners for the consortium and get the computing engines they would need for their R & D. And William Bricken started the specs for the programming opus he was eager to begin, as soon as the hardware arrived—the design of a virtual worlds operating system, a master program that could turn any computer of sufficient power into a reality engine. A year after I visited, it looked like the HIT Lab was finally beginning to move into gear; they had signed up US West, one of the regional Bell Operating Companies, the Port of Seattle, Alias, Digital Equipment Company (DEC), and VPL Research. Bricken was finished with the protocols for VEOS, the Virtual Environment Operating System the HIT Lab was planning to donate to the cause of VR research.

A proposal to develop the laser microscanner was submitted to the National Institute of Standards and Technology, while Jacobson claimed that "we found two geniuses on campus, in the fields of compound semiconductors and microsensors, who have helped us put together a plan for a prototype."

The HITL seems to be moving on its commitment to serve as an informational clearinghouse, as well. Sci.virtual-worlds continued to flourish, with people extending theories of cyberspace data structures and exchanging plans for building cheap systems, reporting new algorithms or optical inventions, speculating on how to materialize their fantasies. The last time I checked, the NSF was requesting a proposal from HITL for a conference on the science of virtuality, Furness was getting together with Robert Prior of MIT to produce a VR journal, tentatively titled *Presence,* to be edited by Furness and Durlach of MIT, and DEC was delivering the kind of computing power Bricken wanted, in the form of networked high-performance workstations. Everything was on target. I'll have to check back in a couple of years, because that is how long it is going to take, at the fastest, to build their hardware, software, and mindware systems, then test them, apply them to their consortium clients' problems, and measure their effectiveness. They ought to have some pretty neat demos in a few months, though. One of the great positive aspects to the extremely limited availability of VR equipment is that the people who saw the klunkiest early versions constitute a small population. While the price of access starts to drop enough for more people to experience cyberspace, the reality level is rising by the month.

The world does not lack for forces that will tend to drive the development of a reality industry in the 1990s. At the fringes are the small-scale entrepreneurs, the "homebrewers" who will make their appearance in a later chapter. At the other end of the size scale from homebrewers are giant computer and communication corporations, who tend not to pay a great deal of attention to phenomena at the fringes until they look like they might be of large enough scale to require attention, but who can focus considerable resources on a technological goal when one is agreed upon. I found that brand-new, multimillion-dollar facilities have been working on VR technologies for several years now, amassing a research and development force numbering in the dozens, perhaps hundreds by now. Consortia are forming. The government and the top companies are huddling about the significance of VR research and development for their strategy for the nineties. I have seen with my own eyes the impressive early stages of this concerted VR research and development effort. *Where* I saw it is the surprising part.

Chapter Ten

COOL GADGETS AND INDUSTRIAL POLICIES

In the popular mythology the computer is a mathematics machine: it is designed to do numerical calculations. Yet it is really a language machine: its fundamental power lies in its ability to manipulate linguistic tokens—symbols to which meaning has been assigned.

TERRY WINOGRAD,
Scientific American, 1984

A number of years ago, whenever one of my son's friends called him on the phone, he would ask "Is Mikey here?" Being in command of the facts of geography, adults usually think of communication as the transmission of information from one point to another. Children, on the other hand, believe that if they can talk to someone, they must be in the same place. In other words, our concept of "place" is based upon the ability to communicate. The place created by the act of communication is not necessarily the same as that at either end of the communication link, for there is information at each end that is not transmitted. The "place" is defined by the information that is commonly available to both people.

There is a definite trend towards expanding the sense of being in the same place. We can see this in the development of transmission systems from Morse Code to the telephone, to radio, to black-and-white television, and finally to color television. Each of these broadcast and dissemination systems allows us to perceive events from afar more completely than its predecessor. Systems that are now being developed will allow us not only to perceive distant events, but also to act at a distance.

MYRON KRUEGER,
Artificial Reality, 1983

I looked out a giant window that filled my field of view, at an aerial map of a city. If I wanted to be morbid, I could consider it a bombardier's point of view. I could fly over the landscape or stop and hover. The most interesting thing about the scene was the way it changed according to how I looked at it. The fine details of the scene were crisp wherever I fixed my gaze, then grew concentrically fuzzier out to the periphery of my vision. I could move my head or my eyeballs from one part of the city to another. It was an experimental apparatus for exploring the research agenda Negroponte and Bolt had pioneered in the 1970s—a full-scale attempt to match the capabilities of computing machinery to the perceptual capabilities of humans. Human vision is sharp at the center and fuzzier at the periphery—although cognitive engineers are discovering that it isn't quite that simple. The researchers who were demonstrating this two-dimensional version of what would be a 3D experience told me they were planning to install a speech recognition module into this brand-new prototype in a brand-new laboratory. Then it would be possible to freeze or mark or zoom in or out of details in the virtual world by combining eye movements with spoken commands to the computer.

The ten-thousand-dollar gaze-tracker and million-dollar projector that made such an experience possible weren't the only technogoodies they had to show me. Down the hall were the beginnings of force-feedback apparatus experiments, gloves and goggles telerobotic controllers, and the most awesome collection of raw computing power I'd seen since Chapel Hill. Probably the most interesting thing about this laboratory was their goal of putting entire populations into cyberspace by the end of the first decade of the next century. It reminded me of what it must have been like in the early days of the US space program, ramping up a decade-long effort to put humans on the moon and bring them back alive.

The world's richest VR research laboratory also has one of the grandest visions. The Advanced Telecommunications Research Institute International (ATR) intends to build mass-media cyberspace communication systems for the world of the twenty-first century. This well-funded organization, created in 1986, is focusing the work of hundreds of researchers and engineers in an effort to integrate what they call "communication with realistic sensations" into the telecommunications technology of the 2000–2010 era. This is not a scenario or a hypothetical example. Happenstance put me in ATR's central R & D facility in March, 1990.

Why would people need to put themselves into a simulated reality in order to communicate? Why would anybody invest decent money today in order to merge telecommunications and virtual reality twenty years from now? The answer lies not so much in what today's state-of-the-art VR provides, but in what today's state-of-the-art telecommunications systems do *not* provide—the delicate, complex balance of nonverbal cues such as posture, gesture, facial expression, direction of gaze that characterize our oldest and highest-bandwidth technology, face-to-face communication.

I sensed the answer at a gut level the first time I shared a cyberspace with another person; like spoken language and alphabets, VR *feels* like a medium for communication forms as-yet unborn. Even a few minutes experimentation inside VPL's half-million-dollar (but still crudely car-toonlike) Reality Built for Two suggests a kind of "communication modeling clay" of the future that our descendants might shape into a blend of dance and visual communication, conversation and multi-media sculpture, intellect and performance, that we can barely imagine today. Today, in the crystal-set era of VR, individuals in shared cy-berspaces can be recognized by *the way they move*, even when people represent themselves as androids, teapots, butterflies, or crustaceans. It's a bit of fun, and at the same time a serious linguistic exploration, to waltz with a lobster in cyberspace.

When it gets down to serious questions about the purpose of human beings, communication skills appear to be near the heart of the matter. One of the things we've learned in the pursuit of mechanical intelli-gence is that inventing, using, and transforming languages seems to be what featherless binocular bipeds do with our ecological niche, the way coral colonies create reefs with their niche. We are the creature that communicates, and thinks about communicating: *Homo sapiens fiberopticus*. Our human propensity for embracing new communication tools and using them to remake ourselves is where telecommunications research converges with virtual reality.

If humans are best at encoding and decoding brocades of meaning from patches of light and modulated sound waves, and excellent at hand-eye-brain coordination, while computers are getting better at painting three-dimensional simulations and volleying trillions of sig-nals into living rooms around the world, perhaps global communi-cation in cyberspace has the potential to grow into an ultimately *human* environment, in a silicon-chilly way. If our destiny as organisms is to become wetware symbiotes of our own tools, multimedia communi-cation games constitute one of the more pleasant alternate scenarios for how we would use our time. At the beginning of the twentieth century, novelist Herman Hesse wrote about such a romanticized fu-ture mind-language in *Das Glasperlenspiel—The Glass-Bead Game*. Hesse's was a Mandarin view of an intellectual elite who spent their time stringing concepts like beads on a gameboard. Their predecessors already exist, hungry for bandwidth—the communication freaks and idea hackers of the Worldnet, bouncing their personae around blind cyberspace through the narrow channel of on-line text.

Also in the early twentieth century, E. M. Forster took a more cynical view of a similar possibility in his novella *The Machine Stops*. Forster's world is one where people attend marvelous illustrated lectures by one another all day long, never leaving their cubicles, nor wondering about what would happen when the machine might stop. Aldous Huxley's *Brave New World* included "feelies" as a form of manufactured experience, cleverly substituted for authentic freedom by a technologically advanced dictatorship.

Virtual reality technology, now in its most primitive early stages, will be maturing technologically at the same time the high-volume information conduits now in the prototype stage come on-line worldwide. The institutions that now bring telephones into our homes are betting that we will find uses for unprecedented increases in the information-carrying capacity of existing communication pipelines. Voice and high-fidelity sound, full-motion video, text by the Library-of-Congress-load will be transportable through the optical cables of 2010. Besides saving a trip to the video rental store, in what ways are ordinary people going to use very fast, very high-capacity information conduits? What is the sense of building such high capacity into the network if there is no way for people to use it? Virtual reality as a communication medium constitutes one potential answer to these questions: nothing eats up computing cycles and communication bandwidth like a reality engine. Billions of networked reality engines would create a whole new global economy. An investor with a broad understanding of how information technologies evolve, deep pockets, and long-term patience, might start planning now for that coming convergence of VR and communications, in anticipation of an extraordinarily large payoff later. How much is the attention of the human race worth, at today's rates?

The history of communications-based industries is the main evidence for believing in such a hypothetical high payoff: communications and cultures, by their nature, are locked into a cycle of perpetual corevolution. Technologies that make it easier to create new technologies are self-accelerating. And the ripples of effects are not limited to technologies. Ways of life change, public perceptions change, the limits of what is possible or permitted change, when a new communication medium is adopted by a population. New technologies enable new cultural institutions as well as newer technologies; buggy whips disappear and drive-in movies pop up, to be replaced by video stores. Even a simple, ultimately low-bandwidth communication technology like telegraph made multinational corporations possible (Edison, Carnegie, Sears, and Sarnoff, founders of the American electrical power, steel, retail sales, and radio-television industries, were all telegraph operators as young men). The entire roller-coaster of twentieth-century history has been driven by the raw transformative power of innovation in communication technology. There is no reason to suspect that this continual and highly unpredictable process of technological and cultural change will cease during the next ten to twenty years.

If somebody had tens of millions of dollars to spend, access to top-notch researchers, access to world markets, and an ambitous ten-year plan, televirtual communications might go global within twenty years. If such an apparently unlikely scenario should come to pass, communications technology on a whole new level might redefine "reality" in the twenty-first century, the way telephones and televisions redefined old notions of time, space, and human possibility in the twentieth century. That's where the big stakes lie. Upheavals are good for some, bad for others; fortunes can be made by those who discern the new scenario early. New communication technologies bring new political regimes, new social institutions, new economic opportunities, new mental diseases. Some of the major players in the current worldgame are going to find themselves too rigid to respond to the changes, and many of them will fracture. Many of the minor players, or players who don't yet know they are players, will be drawn to the new communications-based worldgame, find that they are flexible enough to adapt to the transformed worldview of twenty-first century players, and end up winning big or even remaking the rules.

Somebody—actually a group of companies—is making that gamble. The Advanced Telecommunications Research Institute International (ATR) has a budget of about $50 million a year, and two of their four laboratories focus on the fundamental sciences and technologies underlying televirtual communications. According to Daniel Lee of the AI department, the "Communication with Realistic Sensations" project itself has $5.3 million per year guaranteed for ten years, to create the basic technologies for cyberspace communications. The Systems Research Laboratory opened in Kansai in spring, 1989; a year later, when I visited, they were still hauling in supercomputers and installing million-dollar projection systems. About 140 communications and computer companies contribute money and researchers to the consortium that funds ATR. It was a sobering experience to go from my first shared VR experience at VPL (where my disembodied hand floated in one of Lanier's cyberscapes and danced with a partner wearing a DataSuit, who was represented in the virtual world as a purple lobster) to a shiny new research institute full of engineers dedicated to creating VR communication for the masses.

For a year before I found out about ATR, I had been asking research directors all over North America whether they felt the potential of virtual reality was too large, too profound, for one company or even one country to handle alone. Those who understood the question were skeptical of such a notion. How could anything be more important than competitive advantage? ATR was a pleasant surprise in that regard. At a time when the question of global cooperation in developing new information technologies is entangled with notions of economic competition and nationalistic passions, it was refreshing to encounter an institution that wrote a commitment to global research cooperation into its charter. One of ATR's goals is to increase international participation in its research; indeed, one of ATR's four "Basic Principles" is a commitment to make a "contribution to international society." These people not only think big, they think cooperatively—or at least they *say* they do. And that in itself is interesting. To quote a favorite expression of a great American entertainer: "Trust, but verify." ATR sounded so good I had to look for myself.

I saw their new facilities, tried out their prototypes, looked at their collection of high-end reality engines. Although the whole scene is slick and corporate compared with operations like VPL and Autodesk, the ATR research managers I met aren't exactly the "suits" one might suspect them to be. They are competent scientists and engineers, who have a clear idea of what they are trying to do; the senior researchers have been working separately on parts of VR systems for years. The company really does proclaim publicly that "ATR commits itself completely to doing its part to help usher in the coming international society." Their main facility includes researchers from the US, UK, Germany, France, Japan, Sweden, Argentina. One of the two ATR researchers who briefed me was Daniel Lee, from Palo Alto, a researcher for one of ATR's shareholders, Yokogawa-Hewlett-Packard. Whether they succeed or fail in their bold plans, ATR is betting that the products of their sponsoring companies are going to create a new kind of global telecommunity.

When I tell people about ATR, I tell them all this before I mention that most of ATR's shareholders and researchers are Japanese, and their laboratory is the centerpiece of a new "science city" under construction outside Kyoto. Sometimes, I can watch my audience's excitement shift gears. Most Americans in the 1990s, given information about the industrial potential of virtual reality, would see a large-scale Japanese effort at creating televirtual communications as an economic threat. Viewed through the perspective Americans have been using to look at Japanese technological success, this is a legitimate and significant reason to worry. Another, ultimately more interesting cause for concern is whether there are people in power *anywhere* who do realize that the coming alteration of reality might change the whole notion of marketplace, redefine what we mean by wealth, redistribute power in unexpected ways. I've met a few such long-range thinkers on both sides of the Pacific.

As of this writing, various assistants to various policymakers in Washington are thinking about whether to think about virtual reality. I've talked to an assistant director of the US Congress Office of Technology Assessment, who was fascinated, but saw it as distant from the concerns of policymakers. I did get a call from a member of Senator Albert Gore's staff; Gore follows new technology, had been reading about VR, was planning a trip to Silicon Valley to try the gloves and goggles for himself. A few big companies in America have an enthusiast or two tucked into a small laboratory in a massive research center somewhere. But in spring, 1990, in a building just outside Kyoto's suburban ring, I discovered that key Japanese telecommunication and computer companies had been committed for nearly two years to accelerating the development of virtual reality capabilities and integrating them with the highest-level research in future broadband communications, computer architectures, and human interfaces. ATR's major patron is Nippon Telephone and Telegraph (NTT), the largest communication company in Japan. Other large sponsors include Nippon Electronic Company (NEC), Toshiba, and Hitachi. The technological big boys of Japan, Inc., are heavily backing VR. Even if every research funder in America was enthusiastically committed to a televirtual development effort, it would take a good chunk of the 1990s to ramp up to the level of research planning in Japan.

The best minds of the leading Japanese research and development laboratories have carefully examined the trends and possibilities in the realms of computer and communication technologies and decided to focus on a clear vision of what they want to achieve in ten and twenty years. In their offices, I saw ATR's slick-paper renderings of virtual meetings of the future. They were strangely reminiscent of the drawings of VR futures in Jaron Lanier's office in Redwood City. I wouldn't be surprised to see ATR join the Reality Net that Lanier has been instrumental in planning.

INSIDE ATR: BUILDING THE GLOBAL MEDIA ROOM

Japan's economic growth has mainly been based on the development of applied technology. We have blossomed into one of the world's most prosperous nations. Now, we need to promote Basic Research more than ever.

Telecommunications is evolving in the drive toward the information driven society, and research areas are being dramatically and drastically expanded. Thus, it is most important to simulate international cooperation in telecommunications research and development.

It is our goal to promote basic technological research, the exchange of researchers and information between research institutions in Japan and abroad. In short,

ATR commits itself completely to doing its part to help usher in the coming international society.

ADVANCED TELECOMMUNICATIONS
RESEARCH INSTITUTE,
Basic Principles, 1989

ATR is the centerpiece of Kansai Science City. Kansai is the name for the plain where the great cities of ancient Japan ruled for 1000 years before upstart Edo became Tokyo. But Tokyo has reached the limits of centralization; there's no place left to grow, and Japanese economic survival has been staked on maintaining a high rate of technological growth. Kansai Science City is a national project to build a new town in the area bordering Kyoto, Osaka, and Nara, centered around a cluster of research complexes devoted to twenty-first-century technologies. The countryside was pleasant, glimpsed through the window of the high-speed Japanese commuter train. It was the last week of March and unseasonably warm. The cherry blossoms, early this year, were just beginning to appear. Of course, at the speed my train was moving, I had about four and a half seconds to take in each cherry grove. The pink daubs on the moving scene grew more frequent as we moved south and west. The modern suburban single-family dwellings were recognizably Japanese, with blue-tile high-peaked roofs (in the *gassha* style that I later learned goes back to the days when raising silkworms in the attic was Japan's foremost technology).

A window seat on a railroad, even at a brisk pace, brings a certain rhythm to the traveler's perceptions of the scenery, from the modern hearts of ancient cities, through suburbs, past industrial complexes with familiar names like Kawasaki and Sansui, across mostly unchanged farmland, bamboo groves, and mountainsides both logged and pristine, that helps convey a sense of the pace of Japan. The old cliche about "land of contrasts" tends to jump out at you when you see the rice paddies and microchip factories strung out along the railroad route.

I rode the Shinkansen, the "bullet train," from Tokyo to Kyoto. Having spent three days in my blue suit, white shirt, and tie, I was traveling less formally—corduroy pants, orange longsleeve tie-dyed shirt, plastic-paper Tyvek jacket with glow-in-the-dark map of nearby galaxies, and shoes painted with a replica of van Gogh's "Starry Night" in six acrylic colors. As an American, I figured I had already attracted attention to myself, so I might as well relax and dress the way I do in Seattle or London. It's nice to know in advance that nobody is going to make an overtly threatening scene over my sartorial style. I love how people on Japanese trains, particularly in the cities, can act blasé in response to American eccentricities. *Gaijin,* "outside people" like me, are expected to look and act abnormal, and people in major cities are blasé. It wasn't until I left the Shinkansen and boarded a train for the town beyond Kyoto that schoolgirls couldn't restrain their giggling.

The ATR Systems Research Laboratory is slightly outside the perimeter of the present population center at Takanohara. The presence of university-size empty lots occupied by heavy construction equipment signaled my arrival at the outskirts of the new science city, which is mostly virtual at the moment. Biotechnology and robotics laboratories, for the present, exist only as sites in various stages of landscaping or construction. A tiled structure that looks from a distance rather like the governmental gray marble monoliths of Washington, DC, ATR's Systems Research Laboratory is a large building on an even larger lot. Room for expansion seems to be part of the plan. ATR's operation is clearly designed to be the vanguard of the new city, reflecting Japanese planners' beliefs that information technologies will be the vanguard technology of the coming century, for the whole earth as well as their company.

I used my memory of the size of the Xerox Palo Alto Research Center (PARC) in Palo Alto to gauge the size of research centers I visited overseas. PARC is three stories high and about half a city block square, built into Coyote Hill in a series of terraces I called "Silicon Valley Neo-Mayan" style. ATR is about one and a half times the size of PARC, more orthodox in its exterior architecture, and the interior is similar, but more modernistic than PARC's—the stainless steel and blue-gray tile more evident at ATR, the creative individualism of American computer hackerdom (rock and roll emanating from open office doors, bizarre posters on the walls, bicycles in laboratories) less visible. Like PARC, ATR consists of several related laboratories occupying different floors and wings of the same building. There are two international conference rooms with access to English and Japanese simultaneous interpretation.

The different laboratories at ATR are devoted to intelligent communication systems, interpreting telephony, the human sensory perception mechanism, optical and radio communications, advanced telecommunications devices. In other words, they aim to build the foundation for the worldbrain's next generation of technological infrastructure, from microcircuits to global networks. In the VR lab, the goal seems to be mass market "wireless VR"—three-dimensional graphics without the need for head-mounted displays, coupled to remotely sensed gestural input that uses cameras and scene interpretation software rather than gloves and suits to track posture, gesture, and direction of gaze. ATR wants to build a computerized communication terminal that looks at you and responds according to the way you direct your attention. In the words of an ATR brochure, they are putting together the pieces of "reality-enhanced visual communication, which uses 3D computer graphics to create a virtual conference room matched to the content of the problem solving."

The idea of a media room, an intelligent theater, or a responsive environment is not new. ATR seems to be picking up where Myron Krueger and Media Lab left off. Krueger used early machine-vision algorithms, coupled with video, to digitize the outline of the user's body; he also created gestural command languages. Ivan Sutherland, in "The Ultimate Display," wrote in 1965: "The computer can easily sense the positions of almost any of our body muscles. So far only the muscles of the hands and arms have been used for computer control. There is no reason why these should be the only ones, although our dexterity with them is so high that they are a natural choice. Machines to sense and interpret eye motion data can and will be built. It remains to be seen if we can use a language of glances to control a computer. An interesting experiment will be to make the display presentation depend on where we look."

Richard Bolt at MIT and others had experimented with infrared gaze detection as a computer interface technology. At the time of this writing, however, Krueger continues to struggle valiantly, underfunded, at the University of Connecticut, which clearly doesn't understand the potential of the artificial reality project tucked away in their natural history museum. Ivan Sutherland has been off designing walking machines and new software architectures. And the MIT Media Lab doesn't seem to have succeeded in spinning off very many actual products for the public (although the military has been using videodisks and spatial data management systems for decades). It looks like ATR users have deliberately examined the state of the art of the technology available to them from their member companies in light of the main directions of American VR research, then planned research programs that would develop the technologies in the directions described by Sutherland and pioneered by Krueger, Bolt, and others into powerful, manufacturable systems.

Although Japanese research and development is often accused of being imitative, the irony is that they have rediscovered VR research paths that were blazed, and not followed, by American innovators decades ago. One lesson American industries seem to be fated to learn again and again, in economic markets, is that innovation isn't enough in high-tech industries. Prototypes must be optimized before they can be manufactured and distributed. It takes engineering and research planning—and a commitment at the highest levels of research management—as well as individual genius to bring a mind amplifier to market. It was easier to build revolutionary new technologies back when revolutions happened only once a decade instead of semiannually. It takes a lot of science and technology and an effectively managed large-scale organization to make commercially viable breakthroughs beyond today's state-of-the-art systems.

The only demonstration I saw at ATR that was far ahead of other VR researchers was the scope of their plans, but I did see the size of their laboratories, and the areas reserved for their planned prototype media room systems. When I visited them, they were in the fifth year of research in their ten-year plan; basic research had been accomplished, but their plans to go beyond older studies are only beginning to progress to prototype-testing. Only the early and partial prototypes of ATR's experimental reality systems exist now; they are still assembling hardware for their core research, which has already begun to map out the territory beyond the place where Krueger's and Bolt's work left off. ATR's graphics engines include the first Connection Machine in Japan, alongside rooms full of fifty-thousand-dollar graphics workstations, and even more expensive Stellar and Stardent supercomputers, networked in various ways. They were in the process of bringing a million-dollar high-resolution projector on-line when I visited. With ten years guaranteed funding and some of the best minds from Japan's research establishment, I wouldn't be too quick to dismiss ATR's chances of giving Japan, Inc., the technical foundation for global leadership in televirtuality.

It remains to be seen whether ATR succeeds in attaining the goals they are throwing money at. I have talked with several knowledgeable critics who argue that throwing money and researchers at a problem is notoriously ineffective and especially so in Japan, but nobody argues with the fact that ATR, NTT, and other Japanese communication giants have formulated a clear set of reasons why VR is the next stage in the evolution of communication technologies, have soberly assessed the obstacles to developing desired technologies, and have articulated reasonable approaches to overcoming those obstacles.

According to Yukio Kobayashi, head of ATR's Artificial Intelligence Department, the concepts of thinking tools and collaborative environments, reminiscent of Licklider and Engelbart, are again dominating the research strategy of a research consortium (except this time, it is ATR, not ARPA, that spearheads the augmentation revolution, and it is 1990, not 1960):

> We think of the objective of telecommunication as problem solving but also as doing things that are enjoyable. We think this could well be accomplished through the understanding and communication of

intentions between human to human, human to machines, and machines to machines over distance and time.

In modern telecommunication technologies, the capabilities for efficient transmission of signals over space and time are readily becoming available. However, for communicating intentions—the understanding of intentions—it is not enough to have only signal transmission. The deduction of motives, meanings and concepts from the intentions and their communication for correct comprehension becomes very important.

In promoting the creation and understanding of information, systems that deliver germane information promptly and systems that organize thought processes expertly are among the many methods to be considered. In the act of creating information, especially for enhancing human creativity, we think that the spatial environment where the human is present and where problems are being solved is extremely useful.

In other words, in the case where complicated problems are being solved with the cooperation of different partners, a shared space (environment) where common sensations can be shared is important. In such a shared environment, solution creation will become highly effective. At times when physical proximity is not feasible, the experience of a simulated shared environment would be helpful for communication.

For people located far apart, if distance and time can be overcome by such a shared environment, creativity will be enhanced and understanding of intentions will be possible. Communications will indeed be enriched.

In the Artificial Intelligence Department, we are conducting fundamental research mainly in the visual media area of such a shared environment, which we call Communication with Realistic Sensations, where common sensations are integrated for the understanding and communication of intentions.

Akira Tomono, supervisor of Kobayashi's laboratory, briefed me on ATR's general objectives and handed me over to Daniel Lee and Haruo Takemura, whose research was directly related to this area. We looked at a couple of videos, took a walking tour of the laboratories, and talked about the research programs that supported the VR project. The VR orientation of the laboratory was clear in the videos, which showed an artist's rendering of a group of people conducting a meeting in cyberspace. Even the agenda of the laboratories not directly aimed at televirtuality was addressed toward the central questions confronting VR designers.

One of the four ATR laboratories, the Auditory and Perception

Research Laboratory, aims at uncovering basic scientific knowledge about the mechanisms of perception and cognition. While previous communication revolutions were rooted in electrical engineering, this laboratory was founded on the premise that future breakthroughs will be based on deeper understanding of how human perceptions work, as well as the way information-processing machines work. Their goals "are to obtain breakthroughs in the technologies of visual pattern and speech recognition, and to develop efficient and friendly man-machine interfaces."

The Visual Perception Department of the Perception Research Laboratory aims at understanding the ways humans process visual information and applying that knowledge to the creation of automatic pattern recognition and synthesis technologies—computers that can recognize that you are trying to attract their attention when you wave your arm or can recreate the way your face looks when you pronounce your name. Scientific study of the pattern recognition capabilities of the human information-processing system is linked at ATR with co-evolving technologies in machine information-processing systems, such as scene analysis and understanding. By learning more about the ways human cognitive and perceptual functions turn various visible light frequencies into the coherent picture we call a "scene," ATR researchers hope to create a computer that tracks and understands the user, a breakthrough that would revolutionize the entire field of human-computer interfaces, as well as make wireless VR possible.

In the Cognitive Processes Department of the Perception Research Laboratory, the key research projects center on the psychophysics and computation of pattern and spatial vision (how do our senses detect and code signals, and how do our brains decode them?), principles of parallel computing and neural networks (how can effective techniques used by human wetware and computer software be adapted to augment one another?), and learning and motor theories of perception (how do we build our original mind amplifiers, our learned set of thinking and perceiving skills?). The Hearing and Speech Perception Department studies and models human hearing and speech perception and directs findings toward automatic speech recognition, synthesis, and coding (how might computers and humans speak to one another audibly?). At ATR, science and technology seem deliberately coupled, and the research staff includes scientists from universities as well as corporations.

The Auditory and Perception Research Laboratory is a virtual reality engineer's dream. Engineers and programmers know how to build digital systems, but the basic science for building digital systems that interact closely with the human brain requires the kind of knowledge that only a highly focused psychological research laboratory can provide. And because the nature of the technologically enhanced human capabilities and possible mechanical technologies depends upon new knowledge about human nature, we are going to learn more about the way human minds create the world, how we encode and decode communication, how we map and remap physical and cognitive processes. As Nat Durlach told me: "When you look into them, these questions about how far human senses can be extended in virtual environments are also deep questions about what human beings are capable of becoming." Perhaps we will want to change ourselves when we find out how the right kind of machines can bridge a few barriers to optimal functioning. The human experiment still has some major bugs in it, several of them potentially fatal, that must be solved in some way very soon; changing ourselves to suit our machines might not seem so terrible when it is contrasted with suitably grisly alternatives. Perhaps we will want to resist changing ourselves when we learn more about the kind of human beings virtual reality produces. In either case, large-scale virtual reality research has already deeply intermingled the work of cognitive scientists and philosophers with the products of the hardware hackers and software designers. Although it isn't explicitly in anybody's research plan, the entire stainless-steel and silicon apparatus of fundamental VR research seems to be targeted directly at the heart of whatever human nature is, or whatever we decide to make it.

The Communication with Realistic Sensations project is where the overt virtual reality action is to be found. Assembling the elements provided by the other laboratories into an experimental media room is one of the projects in progress. The bulk of the VR effort is focused on the enabling technologies for "virtual meeting spaces" that could link teleconferencing technologies with remote presence technologies. The key elements of this vision include a 3D display (preferably one that doesn't require special headgear), eye-tracking devices that would enable people in virtual meetings to maintain eye contact with one another, gesture-sensing technology (again, preferably one that would not require gloves or other encumbrances), and a large visual field display that can create a sense of total or partial immersion in the virtual meeting room. The reason ATR researchers want to do away with face-obscuring goggles is that they regard the face as an organ of communication. Each of the nonverbal elements of human communication—facial expression, body language, hand gesture, eye contact—is the basis for one or more of the enabling technology projects under way in connection with "communication with realistic sensations."

Telephones and computer-mediated communications do not convey the subtleties of nonverbal context carried by body language and by small, rapid changes in facial expression; and video teleconferencing eats up a large amount of communication bandwidth. One approach is to teach a computer what your face looks like and how it moves, and then send the computer model of your facial dynamics to your communication partners, who will keep it stored in their computers' memory devices. Nicholas Negroponte and Richard Bolt pioneered the notion of computerized work-arounds for transmitting facial presence, but their apparatus required each participant to have a mask of the others present in his or her communication space; images of facial movements would be projected onto the masks, creating an inexpensive and relatively effective three-dimensional simulation of a literal talking head. The ATR vision takes the MIT setup into a virtual dimension: The facial "mask" is created solely within the computer, through the use of shape-acquisition and facial modeling systems. When you talk with your colleagues on the NTT RealityPhone of 2007, the processors in their local workstations might present to them an image of your face that moves artificially, based on the way your own face moves, in synchrony with your speech. The demonstration I saw took a video image, broke it down through stages that look rather like topographic maps, and then created a synthesized image that looked like a recognizable if clearly artificial replica of the model's face.

The possibility of valuable spinoffs is one thing to look for in any research program, especially one that is so ambitious there is a good chance it will not achieve its goal. The notion of a shape-acquisition camera, for example, is an intriguing possibility as a general reality construction tool. Optical scanners can now turn any pattern of graphics and text on a flat page into a digital file suitable for processing by a computer and transmitting over communication lines. I can take a photograph, scan it, and send it via telephone and modem. Someday, it might be possible to scan three-dimensional objects into your 3D data base by aiming a shape-acquisition camera at them. Your home movies will not only enable your captive audience to see what you saw in Rome last summer, your friends will be able to stroll along the Via Veneto with you, on a perpetually frozen hunk of captive space-time. And with highly evolved versions of the techniques developed at ATR for transmitting facial information, it might be possible to capture the way objects move in three-space, as well as their stationary form, a possibility that leads to another intriguing spinoff—voice-commanded, 3D reality-sculpting tools.

The advent of computer graphics in the 1960s and direct manipulation interfaces in the 1970s meant that it was possible for humans to use our highly developed visual senses to look at the computer's operations, and to use what we saw to communicate with the computer. In the 1990s, ATR researchers and others are trying to endow computers with comparable powers: ATR has several prototype systems that look at the user, particularly at where the user is looking, by tracking eye and head positions. I sat in various experimental stations where I was at the focus of video and infrared cameras that were in turn connected to image-extraction software in computer workstations. Both the facial recognition and gestural recognition work under way at ATR is based on a long tradition of attempts to build robots that can navigate visually. ATR is working on human and electronic vision systems simultaneously, in tandem. It isn't that hard to convert light into electronic signals and feed the signals into a computer. The difficult part comes in when you have to devise a computer program that knows the difference between one face and another, or even can discriminate a face from a frying pan.

One thing computers can do is perform large amounts of calculations very quickly. To the degree that cognitive scientists can uncover the human visual processing system's tricks for analyzing scenes, it is possible to narrow down the possible approaches to creating scene-analysis software. Edges, for example, seem to be important in the human visual system. Our light-sensing apparatus seems to be hard-wired for edge detection, a technique that is at least partially replicable in a computer program: calculate what every pixel is doing in relationship to its neighbor, and if linear arrays of pixels are all on or off, test them to see if they constitute an edge. It is possible to analyze digitized scenes in this way by comparing arrays of picture elements according to complex mathematical calculations, thus detecting edges and putting the edges together into forms, then comparing the digitized forms with a data base of previously recognized objects. NTT Human Interface Laboratories, one of the contributors to ATR, has already built a camera capable of recognizing automobile license plates. At certain test sites, robot speed-enforcement prototypes are already in place: If the radar detects a speeding vehicle, a video camera captures the license plate, and scene recognition software feeds the number into the motor vehicle department data base. Someday soon, future-shocked speeders will receive robo-tickets in the mail.

Automated highway police robots are not exactly the research goal at ATR. Researchers here think it is possible to build a vision recognition system that not only will replace the DataGlove, but might help understand the dynamics of human gestural language. Like other technical inquiries underlying VR, this one leads to the core questions of what human communication really is. What, for example, is the place of gesture in conversation? Even if the visual depiction of your conversational partner is a computer-aided cartoon like the ones provided by the shape-acquisition camera or cleverly kludged together in Krueger's Videoplace, or the purple lobster I encountered at VPL, there is something about the way people move in space that adds a lot to the act of communication. For that reason, ATR automatic vision systems are designed as research instruments for the psychologists looking at the fundamentals of human communication, as well as prototypes for gloveless gestural input. Twenty years later, and half a world away, these researchers are concentrating their efforts on the human interface breakthroughs Americans pioneered and forecasted but never convinced anybody in America to commercialize.

Richard Bolt's early efforts at MIT to create prototypes of highly visual, human-centered computer interfaces also included eye-tracking technologies. A computer that could determine precisely where the user is looking would have two distinct advantages in televirtual communication systems. First, eye contact is an important component of the "transmission of intention" that ATR is trying to bring to electronic communication systems. You might not care if your conversational partner looks like a lobster or an android, just as long as it maintains eye contact (or avoids it), in synchrony with the real person behind it. Second, and equally important for human interface design, the precise location where you fix your gaze is an important indicator of where you are directing your attention. An intellingent data base would allow you to expand features of the visual landscape simply by looking at them. To say nothing of the communication advantage an eye-tracking system can bring to a person whose mobility might be limited only to his or her eyeballs. When it comes to VR, a computer that knows where you are looking is more effective in its job of creating the illusion of being inside a simulated world. And when it comes to Japanese, a highly context-dependent language, it is important to know where your conversational partners are looking and to see the expressions on their faces.

In 1984, when Richard Bolt explained why it is important to watch the user's eyes, he pointed out how the nonverbal redirection of attention plays a key role in early learning: "One reason for the system to watch the eyes is to open a new channel through which we can detect where the user's attention is directed. The effect can be compared to what children gain when they discover that where the parent is *looking* is useful to them in comprehending what is transpiring between them and their parent, and, in turn, with the world about them."

Mr. Kobayashi, Mr. Tomono, and Mr. M. Iida had worked on an eye-movement detector that used a pair of spectacles and tiny infrared-light-emitting diodes to detect gaze points. Tiny differences in the refraction of invisible infrared light bounced off the surface of the eye can be translated by a combination of hardware and software gimmicks into a relatively precise indicator of gaze direction, especially in combination with magnetic 3D head-trackers like the Polhemus. Kobayashi et al. had also built an electronic camera system and software to extract feature points from the digital image to track gaze direction and face position. Even a fairly unintelligent pattern-recognition algorithm can track the two precisely invariant circles of your pupils on a video image. In some cases, infrared light reflected off the eyeballs provided the feature extraction points; in other cases, small dots of reflective material adhered to the forehead, adjacent to the eyeballs, provided the feature extraction points. The procedure for "suiting up" for this system is something like sticking a tiny postage stamp on each of your temples; you quickly forget they are there, and they are easily removed when you are finished with your interaction. Changes of angle and absolute position between extraction points correlate with eye movements. The signal from the gaze extractor can be used as input to the computer, which can respond in various ways to changes initiated by the user's eyes.

Again, although none of the scientific principles behind these eye-tracking technologies were originally created by ATR, their use of electronic lensless (CCD) cameras and their concentration on the question of what makes virtual reality seem real are intriguing new directions for VR research. Just as every Japanese laboratory I visited is making good use of VPL's DataGlove, researchers around the world would make progress in numerous other VR-related fields if they could have access to inexpensive, precise, nonintrusive eye-tracking technology. The idea of eye-tracking has been around for decades; affordable and unobtrusive systems are not widely available for VR researchers, however. Any advances in this direction would advance VR research everywhere—particularly in Japan, of course.

Like every other self-respecting VR laboratory, ATR has a glove and head-mounted display prototype. The one I tried out resembled the molecular docking setup I had used in North Carolina. I stood in front of a large screen, about six feet square, wearing a DataGlove and PLZT effect "electronic-shutter" glasses. Like the molecules at UNC, the blocks displayed on the ATR screen did not surround me, but floated in space in front of me. Using the glove, I could reach in, grasp a block, move it, and release it. The ATR display included solid objects in the "real world" that could interact with objects in the virtual world. As a prototype for studying telerobotic operations, this setup made it possible for me to pile virtual blocks on top of solid blocks (although, of course, not the other way around). Within a very few years, a system like this could train specialists in heavily roboticized factories (such as those in Japan) how to oversee the semiautonomous operations of a number of robots on the factory floor.

ATR, like NASA and UNC, is not neglecting the auditory and spoken components of reality. Suppose you are using a hyperrealistic VR system of the future and you are too lazy to wave your hand to change an urban hive to a desert panorama. Suppose you just want to say: "strike the city, create a desert." Devices capable of translating spoken words into computer commands constitute another technology that has taken a long time to develop, but which appears to be converging on the other elements of VR computer interfaces. The problem has never been one of imagination, but of engineering, when it comes to the field known as "connected speech recognition." If you have the imagination to envision a virtual space, it isn't that far to leap to a virtual space that changes itself according to voice commands. The integration of voice recognition, eye-tracking, and gestural communications is another dream that goes back to the Architecture Machine Group at MIT in the 1970s; expensive but high-quality speech recognition capabilities were built into the NASA VR system in the late 1980s. The technology available to ATR in the 1990s, however, is quite a bit more capable than the systems Chris Schmandt and Eric Hulteen used to create "Put That There" at MIT's Architecture Machine Group in the early 1970s, or even the devices NASA used a few years ago.

The voice-gesture paradigm is captured in the title of the original Arch-Mac demo: You point at an object on the screen and say "put that," then you point to a location and say "there." ATR researchers are working on human-computer interfaces that integrate voice input with gestures, and their colleagues at NTT's Visual Perception Laboratory outside Yokohama are working on visual-verbal languages that will enable people to describe shapes to the computer by saying the right words to control computer graphics images: "Round, expand, stop, squeeze, dent, stretch, red, stop" or some similar incantation might cause your reality engine to sculpt the image you hold in your mind's eye. While it might take years to develop an affordable speech recognition technology that could parse conversation, it is possible right now to couple a graphical construction tool to a relatively inexpensive, highly accurate (although not perfect) device for recognizing a limited vocabulary of commands. The art is not in the hardware here, but in the design of the visual-verbal command language.

When I heard about ATR's intention to expand the capabilities of a "Put That There" system, in conjunction with a 3D graphical construction spoken-command language, I recalled the night, seven years prior, in Silicon Valley, when I watched Brenda Laurel, Scott Fisher, Mike Naimark, and other young infonauts of Atari's Systems Research Group create a scenario for such a system. I took notes when Brenda took her turn imagining herself at the helm of a voice-commanded, total-immersion personal simulator. This is what I wrote at the time:

"Give me an April morning on a meadow," she said, and the gray was replaced by morning sunshine. Patches of cerulean sky were visible between the redwood branches. Birds chirped. Brooks babbled.

"Uhmmm . . . scratch the redwood forest," Brenda continued: "Put the meadow atop a cliff overlooking a small emerald bay. Greener. Whitecaps."

Brenda was reclining in the middle of the media room. "The background sounds nice," she added: "Where did you get it?"

"The birds are indigenous to the northern California coast," replied a well-modulated but disembodied female voice: "The babbling brook is from the acoustic library. It's digitally identical with a rill in Scotland."

"There's a wooded island in the bay," continued Brenda, looking down upon the island that instantly appeared below her where only green water had been a moment before.

The fine details of Brenda Laurel's fantasy are still challenges to future technologists, but the enabling technologies are healthy and growing fast at ATR. In 1983, among a group of competent prodigies during the height of the personal computer revolution, it was fun, and not terribly threatening, to imagine a media room of the distant future that could respond to voice commands and create three-dimensional illusions at the whim of the user. In 1990, in Kansai, looking at the laboratory space in ATR where the components of such a prototype media room are being wheeled into place, speculations about a world that becomes whatever you tell it to be seem less like science fiction—and more like yet another mixed blessing to contemplate from the electronic experience industry.

THE SCIENCE OF PRESENCE AT ATR

Effectiveness of visual communication and remote operation can best be achieved if the observer becomes totally involved in the displayed image through a feeling of "being there." This sensation of virtual existence or enhanced reality is activated by the capability of a precise pixel expression and a large visual field display.

However, these two requirements are contradictory because view angle per pixel deteriorates as the size of the visual field increases. To solve this problem, a new large visual field display system, based on the fact that human vision is most acute over a narrow area centered on the point viewed, is proposed and its construction is shown.

YAMAGUCHI ET AL.,
"Proposal for a Large Visual Field Display
Employing Eye Movement Tracking,"
1989

From their research into the perceptual elements of VR—the characteristics of 3D simulations that evoke a sense of presence—ATR researchers have been concentrating some of their efforts on creating enormous "intelligent" electronic projection screens. Like the Omnimax and Imax theaters, and the early prototypes of Krueger and Heilig, this approach goes the opposite direction from the paradigm of the head-mounted display. Instead of putting tiny displays right in front of the eyes, why not put the user in front of an enormous screen? What are the characteristics of VR displays that make a difference? Stereoscopic three-dimensional graphics are possible with electronic-shutter glasses and head-tracking devices; 3D graphics without special glasses are available with lenticular screen systems such as those developed at NTT. It is possible to create stereo 3D without an HMD, as long as the user is seated or otherwise fairly immobile. The other purpose of head mounts, and the special wide-angle optical systems that are used in these displays, is to present a wide field of view. When you look at an average television screen from an average viewing distance, the image covers (or subtends, as the specialists say) about 6 degrees of your field of view. A head mount can give the user around 120 degrees, and that difference seems to be crucial in creating the sense of immersion in a virtual world.

When you use optics to expand a stereoscopic video display to fill a user's field of view, the individual light elements on the screen, the ubiquitous pixels, do not change their size relative to one another. The "grain" of the image looks coarser. The only remedy is to create display devices that accurately represent images with many more picture elements per inch. High-resolution displays are at the forefront of another convergent technological battle over high-definition television. Sooner or later, affordable, high-resolution, miniature screens will be available. Right now, if you want to project an image of sufficient definition and still fill the user's field of view, a large screen is a more feasible alternative. At ATR, the strategy is to use a very large screen, about five feet in front of a user, and the trade-off between field of view and high acuity is compensated by projecting a sharply focused image only at the user's gaze point, with concentric circles of less sharply depicted image expanding to the periphery of the field. We are accustomed to ignoring the differences, but that is the way we see the world—sharp in the middle and progressively fuzzier around the edges.

The projection system ATR was putting together to create this effect cost $1 million. The projector, eye-tracker, head-tracker, and other components had been installed and tested individually when I arrived, but had not yet been hooked together. I sat in the viewer's seat in front of the large screen, and they faked the eye-tracking by moving the projector manually according to the gross movements of my head. It looked like an aerial view of Kyoto, with a sharp image in the center of wherever I looked and a fuzzier image everywhere else. If ATR companies can ever work out a way to manufacture affordable remote, multiple-user eye-tracking apparatus, people like Krueger and Heilig could build the responsive theaters they have dreamed about for years.

On the table behind the large-field screen, I saw an apparatus that signaled their interest in the multimodal aspects of the sense of presence. Whenever you see a joystick or a handgrip device connected to a lot of tiny motors and joints, you can be reasonably certain an experiment in force-reflectance feedback (artificial touch) is under way. The box here differed from electromechanical versions used elsewhere in that the ultrasonic motors did not generate magnetic fields, which made it possible to couple them with magnetic hand-position sensors. The ATR force-reflecting joystick had three degrees of freedom. Their next planned addition was to put a force sensor on the end of the teleoperated robot arm, to enable a person to feel the edges of a remote object. I asked them if they were experimenting with texture as well as the kind of force-feedback they used at UNC for molecular docking.

"Yes," replied Dan Lee, "and not just texture recognition, but other components of touch, like the feeling of penetrating virtual objects or a feeling of manipulating magnets in magnetic fields."

The folks at ATR have done their homework, they have shopped for the best equipment, and they have invited the best researchers they can entice to join them. Does this mean we will be buying our telepresence services from Japanese telecommunications companies? It might. There is a dissenting view, however. Those who remember the famous "Fifth Generation" project mounted by Japan's Ministry of International Trade and Industry (MITI) in the early 1980s will recall that Japanese research plans do not always live up to the visions that guide them or the goals set forth by research managers. Now 1990 has come and gone, and neither the Japanese nor anybody else has a desktop knowledge-processing machine capable of automatically translating Japanese to English. And at least one expert who saw several of the same research facilities I visited is not terribly concerned about Japanese competition with US VR research and development. When I returned from Japan, I reported to Sci.virtual-worlds—the computer network discussion group—about what I had seen at ATR. The next day, I got e-mail from Donald Norman, chair of the department of cognitive science at the University of California, San Diego. Norman, whom I had interviewed for a different project a year ago, believes that I might have tapped into an influential but still minority subculture of Japanese researchers who understand the importance of the human side of VR and computer interface design, Norman's specialty.

"Yes, I can see they are pouring huge amounts of money and people at the project, but I also see that they don't understand the human side of the equation," he said, via electronic mail, adding: "I think it has something to do with the way Japanese culture is tuned to 'getting along.' Individuals adjust to the group. Americans of course, glorify individuals. If you are predisposed to adjusting yourself to the machine, it isn't easy to design a human-computer interface in which the machines cater to the idiosyncrasies of the common user. The idea of studying psychology along with electronics wasn't easy to get across. The fact that Japanese research laboratories tour my own laboratory and send postdoctoral students and invite me back to critique their work is a sign that at least a small subset of Japanese researchers is trying to change things. ATR and NTT in general are part of that subset. They are a powerful force in the research world, but not the only ones."

A conversation with Donald Norman, especially about interface design, is bound to be sharp and salty. Most design is bad or nonexistent, and he has a habit of pointing out exactly where that is true, from impossible doorknobs to inhuman computer interfaces. Norman has consulted for many of the modern interface designers, from IBM to Apple, and when he was invited to speak to Japanese human interface designers, he told them in his usual direct manner that they would likely remain in the dark ages as long as they design systems to be used by humans without hiring anybody who understands people— psychologists. Now, it looks like some Japanese research institutions are following his advice. I queried him again via e-mail: "What then? What if they throw money and people at VR and hire psychologists, too?"

"The Japanese researchers have big plans," Norman replied, "but they are no different from researchers everywhere: dreams and words outstrip reality. Many Japanese labs remind me of MIT's Media Lab: lots of flash, but I wonder how much substance there is to it. The MIT Media Lab is all demo. If you ask me, they do very little serious thinking about the ideas they demonstrate. The Japanese are more modest, but even so, I have not seen any fundamental advances in our understanding of the experience of remote presence coming from Japan. The fact is that it is very difficult to accomplish these dreams. Sometimes it take ten years to gain the knowledge, and another ten years to engineer it. The Japanese are not superscientists; they can't go any faster than anyone else can. In my opinion, the Japanese scientist or engineer is very much like the American scientist or engineer. No better, no worse. We should not expect any miracles from them, especially creative breakthroughs. However, where projects require dedicated work and effort, they will probably shine."

Norman reminded me of another rule of thumb that is particularly applicable to large-scale VR efforts anywhere, Frederick Brooks's "Mythical Man-Month." As Donald Norman put it: "The right team could do better R & D with fewer workers if the people are charged with being innovative and creative. Large projects lead to huge management and lots of meetings, and special efforts to maintain 'harmony' and 'morale.' But so much time is taken with these efforts, especially in Japan, and so much concern is taken about treating everyone equally, that surprisingly little tends to get done. I would be a lot more concerned if you told me that one dozen world-class researchers were secretly working to develop virtual reality."

He might be right. On the other hand, people have been wrong about Japan's technological efforts before, I reminded Norman.

Norman countered: "I am not underestimating the smart, industrious Japanese engineers and scientists. But we shouldn't overestimate them. Should we be alert to the possibility that the Japanese might seize dominance of some future VR industry? Yes! But not because the Japanese are superpeople. We should take their effort seriously because they are picking the right problems and solving them. With a number of exceptions, American technological development has become more of a matter of talk than action. We lost the automobile game, the video game, the audio game because we didn't even try to compete. We may lose the high-definition TV game, and we will probably lose the VR game—not because they are better, but because of American inactivity and lack of willingness to put the required time, money, and people into the project. Time is one place where the Japanese method is indisputably superior: Americans want results in months or years. In Japan, they are prepared to wait for years and decades."

Jaron Lanier, who travels to Japan regularly and does business with Japanese companies, echoed Don Norman's assessment when Lanier and I discussed Japanese VR capabilities: "Because the USA has experienced some financial difficulties and some types of industrial decline, there is a tendency to overestimate the Japanese. There is unfortunate talk that the Japanese will 'take over' VR. The Japanese believe in planning, and will tend to announce a long-range plan even when they are in fact groping beginners at VR, just like the rest of us. If we ascribe a supernatural quality to the Japanese, we aren't really seeing them, and that's destructive to collaboration."

An objective evaluation of what I saw at ATR took months of reading their research reports and consulting experts like Donald Norman. But I was buzzing when I left the big blue tile laboratory (the laboratory's director sent me back to the railroad station by means of his car and driver). I took the late afternoon train out of Takanohara for my first night in Kyoto, a city that has a soul, like San Francisco or Venice. And like other cities that seem to have a personality, Kyoto has its own inevitable forces of mortality. What earthquakes are to San Francisco and floods are to Venice, the hypermodernization of an ancient culture seems to be to Kyoto, a lovely old swan of a city suddenly invaded and surrounded by concrete and antennae, condos and discos. Will the technological wonders of Kansai Science City be entangled in the dissolution of the glory that was Kyoto? I had dinner with a fellow who was active in environmental preservation efforts in the Kyoto vicinity. He was worried about the future that Japan, Inc., might be cooking up in their science cities, yet his network of Japanese environmental activists kept in contact with each other and other grassroots groups through a computer conferencing system. Black and white moral judgments about the effects of technology, I've been discovering, are very unstable foundations when communication systems change radically.

Chapter Eleven

"VISUAL, INTELLIGENT, AND PERSONAL"

Our company, NTT, has recently declared our future vision of the 21st century technology. The key phrase is "VI&P (Visual, Intelligent, & Personal)."

YOSHINOBU TONOMURA,
Senior Research Engineer, NTT, 1990

We believe human interface technologies will play a crucial role in making communications and computer technologies more easy-to-use and effective in augmenting human abilities to think and share ideas . . . Communication has long been interpreted as electronic tele-communication in a physical sense. We believe we should take a higher view of communication: inter-personal communication (or human-human interaction) that can't be captured by the framework of ISO seven layer communication model.

TAKAYA ENDO AND HIROSHI ISHII,
Senior Research Engineers, NTT, 1989

Nifty prototypes are one thing. Ultimate workstations that arouse personal technolust are another matter. Nippon Telephone and Telegraph had the highest percentage of mind amplifiers that I would like to take home with me since the first time I sat down at an Alto at PARC in the early 1980s. They started me out with something that looks like an intriguing amusement today but might be a key part of the virtual communities of tomorrow, and which I only mildly desired to slip into my suitcase. Yoshinobu Tonomura sat me down in front of a prototype of a form of cyberspace communicator. A small stereo camera built into the unit transmitted a 3D video image of my face and torso to the receiving unit; on my screen, I saw a miniature but surprisingly realistic 3D real-time depiction of the person on the other end of the line. The person running the demo hit a button on a switcher, and I found myself looking at a small version of myself, floating in the air between me and the screen, moving as I moved. The most exciting thing about it was the fact that I didn't need to put on gloves or glasses. It was a small device, but could be a big step toward "wireless VR"—cyberspace where the participants can make eye contact, use body language, convey facial expression. When they start selling these things, and there is a network capable of handling the blasts of information they will exchange, I'll probably be an early adopter.

After the lensless 3D demo, the NTT people showed me something I could actually use today, if I only had a network of colleagues similarly equipped. Hiroshi Ishii proudly showed me his own "TeamWork-Station," one of a small number of prototype multimedia workstations that linked him to other researchers in the laboratory. The specially equipped researchers all were devoted to the discipline of "computer-supported cooperative work," and Ishii spoke confidently of foreseeing the day when the TeamWorkStations converged with the telepresence technologies to create virtual workgroups in cyberspace. It gave me great pleasure to introduce Mr. Ishii via electronic mail to Douglas Engelbart, after I returned home. Video cameras, voice communications, computer graphics, and various combinations thereof, together with a keyboard, a mouse, a penlike drawing device, and a screen considerably larger than today's generations of personal computers gave Ishii a turbo-powered, state-of-the-art version of the kind of collaborative work system demonstrated at ARC in the late 1960s. It wouldn't surprise me to see something like this on my desk in a few years. I stumbled onto an Alto in 1983, and seven years later I have a considerably more powerful Macintosh on my desktop.

I got e-mail from Mr. Ishii this morning, and a video from Mr. Tonomura that captured a conversation I had with Mr. Ishii several months ago. My visit to NTT's Human Interface Laboratory in Yokosuka keeps coming back to me, months later, via electronic media of one kind or another, which isn't so unusual when you consider how I got there in the first place.

I started out with very little idea of what was going on, VR-wise, in Japan. Japan's VR effort wasn't highly visible if you didn't read Japanese. Annual proceedings of the Society for Information Display and other specialized journals had published English language translations of articles from researchers at NTT and ATR. The titles of these reports were suggestive of an ongoing VR research effort in Japan: "Full-Color Stereoscopic Video Pickup and Display Technique without Special Glasses," "3D-TV Projection Display with Head Tracking," "A Virtual Common Space on Broadband ISDN." At the Santa Barbara conference, I heard Dr. Susumu Tachi describe his research into teleoperated robots in the context of Japan's long-term national robotics development effort. Tachi's sponsor, Japan's Ministry of International Trade and Industry, wasn't a lightweight organization. There were other inklings. Jaron Lanier had mentioned that VPL seemed to be selling more and more VR-ware to Japanese researchers. All that remained for me to do was to transport myself to a notoriously expensive and very foreign culture I'd never visited before, sniff out a story based on my leads, then convince the key VR research engineers and research managers to let me stick my head in their realities, even though I don't understand more than a dozen words of their language.

VR synchronicity struck again: Out of the blue, I was invited to deliver a keynote speech in Japan to an assemblage of several hundred of Japan's info-elite. A person I knew from electronic networking circles, an American, called me on the telephone and said she was speaking for somebody in Japan who was organizing a conference about the future of telecommunications. The sponsor, Izumi Aizu of the Institute for Networking Design, had wanted Douglas Engelbart to give the keynote speech, but Engelbart couldn't make it. I was next on the list, apparently because I had drawn their attention to the work of Douglas Engelbart.

"Well, yes, I might be willing to do that," I replied.

The American intermediary called again the next morning and asked me what I might want in return for speaking. I told them I wanted to see what Japan, Inc.'s R & D wizards had to show in the realm of virtual reality.

About three rounds of telephone conversations into the process of accepting the invitation, I began to talk directly with my Japanese host. Izumi Aizu was the person who wanted me to make the speech for the conference he was organizing. He told me in English that is almost better than mine that he would act as my interpreter, and he mentioned that he knew the best person in Tokyo to ask about Japanese VR research. In fact, the person who had recommended me to Izumi Aizu was the same person who knew how to get me into the highest levels of VR research efforts. It turned out to be a perfect arrangement. I made my speech, then caromed around Japan at Tokyo speed, according to an itinerary that was more or less handed to me, almost always through intermediaries. The process became less mysterious when I met the people who drew me into this exploration from the Japanese side; they are now friends, and I can see that their indirect social protocols are more a matter of their contextual coding than any intent on their part to convey mystery. Communication in Japan, and with Japanese people anywhere, involves a lot more than the text of the spoken words.

Mr. Tonomura, a senior research engineer at NTT's Visual Media Laboratory, had been in the audience when I gave my speech about "thinking tools and communities of minds" in the future "hypernetwork." The term *hypernetwork* was proposed by Izumi Aizu as a name for the expected convergence of telecommunications and multimedia technologies that had Japan's informational avant-garde buzzing. This communication superchannel was well into the decade-long process of implementation globally, so images of the kind of world this communication technology might make possible were serious topics for discussion in Japan around several quarters of the info-technology complex, from chip designers to virtual reality researchers. When the vertical and horizontal industrial networks in Japan begin thinking about the same thing, the results can affect the fortunes of millions of people in Detroit or Silicon Valley. Whether or not they always succeed at their technological goals, visions are something the Japanese planners have discovered and created in abundance. It is not so important whether they invented the slogan themselves; it is more important that the new term or slogan sum up a set of decisions and values about the direction of the long-term project. The slogans that drive these efforts might be as mundane-sounding as "Fiber to the Home" or as bold and mysterious as "The Fifth Generation," but they signal a shared mindset among people who have the power to make things happen.

Unlike the more diverse (some would say "chaotic") situation in the United States, there seems to be general agreement among television, telecommunications, computer, and electronics industry leaders in Japan on a national telecommunication strategy. Japanese planners operate under the certainty that most of their own nation will be linked to a vastly more powerful communication infrastructure than the ones we know today, via very-high-bandwidth information channels. Right now, while the major communication systems in America wage techno-religious wars over incompatible technical standards, American policymakers concentrate their energies on worrying about whether these new networks will bring pornography into the home, American newspapers try to keep the television stations from putting them out of business by delivering information through cables, American telephone companies and entertainment companies wonder whether they should marry or go to war, US defense research diverts funds from potential breakthroughs and concentrates on politically savvy but technically trivial projects, and American innovators find themselves mired in litigation and bureaucracy when they try to move the state of their art forward, the Japanese leadership has aligned itself to a specific shared goal it calls "Broadband ISDN." They have a clear vision of the microchips they need to manufacture, the optical fiber cable infrastructure they need to install, the computing architecture they need to design, the virtual reality research they need to pursue.

Although there are serious technical issues underlying the way these

channels are implemented in hardware and software, the central idea
of an Integrated Services Digital Network (ISDN) is simple: all forms
of signals—libraries full of text, orchestras, sitcoms—now can be reduced to digital form and sent as bits at very high speeds over very
wide networks, to be decoded and displayed over increasingly "intelligent" home information-entertainment-communication-education
terminals. The chips inside the standard television sets of the late 1990s
will be able to outcompute the mainframes of the 1980s. Various American companies are experimenting with different flavors of ISDN;
international standards bodies are trying to find a way to mediate
between a global ecology of competing technical standards. Japan's
largest telecommunication vendor has decided to use a combination
of special communication hardware and software that will deliver on
the order of 600 million bits of information per second to each of
millions of nodes (you could send an encyclopedia in about four seconds or a half hour of compressed, full-motion video in the same
interval). With a reality engine at a certain number of nodes, and the
right kind of clothing, it will be possible to start squeezing realities
through the ISDN tube. Hence, NTT's contribution to VR research.

Nippon Telephone and Telegraph, whose stake in creating that
high-speed information highway is substantial, had politely requested
through an intermediary that some of their senior researchers would
be honored if I could tell them what I knew about the state of VR
research worldwide, followed by a short session in which I would help
them pick my brain. And then they would show me their VR goodies.

NTT videotaped the proceedings and sent me a copy, so I now have
the opportunity to see myself wearing a gray charcoal pinstripe suit,
peach shirt, and yellow tie, sitting in a meeting room of a research
laboratory on the outskirts of Yokohama, lecturing a dozen research
engineers about the unpredictable changes the products and services
they were building would unleash on society. Watching the video is
painful and instructive: I use my hands a lot more than I would ever
be willing to admit if I hadn't seen it. I fiddle with fingers and coffee
cups in rhythms that I seem to be relying on to pull the words out of
my mouth and string them into sentences. The figure on the screen,
whom I know to be myself, but who does not behave entirely in the
way I believe I behave, makes tents and spiders with his fingertips,
conducts his audience like a silent orchestra with a pencil between
thumb and forefinger, wielded as a baton.

It came as a surprise to see how I use gestures I never consciously
recalled using; it never came to my attention before I saw the tape
that I often employ a windmill-like interchange of forearms and palms
to indicate systems that go both ways, transactions, feedback loops.

The landscape outside the window, on the eighth floor of the NTT R & D organization, drifted in and out of focus as the morning's misty rain cleared and returned. The sight of me standing there, with the blossom-dotted early spring landscape intermittently visible through the window behind me, was enough to bring it all back.

The lecture-demo trade seemed like a perfectly equitable arrangement to me, and a challenging intellectual opportunity: By a series of circumstances beyond my control, I found myself spending close to two hours discussing the future of VR with the managers and research engineers who spearheaded the Japanese campaign to shape future telecommunications worldwide. What was the most useful thing I could do with an opportunity like that? What did I know that I thought they ought to know? If I could tell them what was important to think about regarding VR and telecommunications, what would that be? Fortunately, I had some help figuring that out, just in time.

I rode to NTT in a limousine provided by *Asahi Shimbun,* Japan's leading newspaper. Riding along with me was Katsura Hattori, *Asahi Shimbun*'s science reporter, and Akihiko Okada, a photographer who often worked with Hattori. As the pieces of the story I had figured out on my own fit together with what he told me, I realized that Hattori was also the man who had reached through virtual space and drawn me to Japan in the first place. The day after I had told my intermediaries what I was seeking in Japan, even before I talked to Izumi, I found e-mail in my network mailbox from Hattori. He introduced himself as an enthusiastic reader of the Japanese translations of my books and told me he was doing a story on VR for his newspaper. We exchanged e-mail for weeks before we met in Tokyo. Serendipitously, my proposed tour fit into his story, so he and I ended up traveling together to various VR research sites in Japan. With the state of traffic in Tokyo environs, we always left early, and found plenty of time to talk in the company limo about What It All Means.

Hattori had spent two years at MIT's Media Lab, was interested in new information technologies, and was an electronic mail enthusiast. As the most broadly technologically savvy person in a huge information company, Hattori is more than just a journalist; as his Media Lab stint indicated, he was also his company's point man in the world of new information technologies. I recalled that he had participated in a discussion among people all over the globe that had taken place on my local computer conferencing system. ("Taken place?" Hmmm.) When my book *Tools for Thought* was published in Japanese, Hattori had seized on it. He learned how many researchers in computer and media-related fields were also ripe for the ideas presented in the book. The heroes of the story were technological visionaries, and a clear guiding vision is exactly what both the Japanese technologists and the wider Japanese population had been seeking in response to the expansion of Japanese economic power that accompanied the success of the technology-driven Japanese industries.

The vision of personal augmentation was attractive to a Japanese audience, as well; a strong vision of a future world of knowledge workers who could be empowered through the creation of new technologies has been an important theme in Japan for some time. The idea of using electronic machines to amplify intelligence is a natural fit with other trends in Japanese culture. "Groupware" that emulates the systems for working in groups pioneered by Engelbart is a hot item in a society where group values are uppermost. Personal computers are intrinsically individual items—until you plug them together with telecommunication systems and groupware.

I wasn't in Japan on business, strickly speaking, but I was still there to consummate a transaction. I wanted information from them. They wanted information from me. That's what science and journalism are about. Although I'm neither a scientist nor a journalist by the strict definition of either term, by this time my travels had made me a courier, a carrier of ideas heard at conferences in Texas or Kyushu, reprints from obscure journals, and e-mail addresses of people half a world away from each other who are converging on the same research goals. The Japanese research managers had an opportunity to transmit a message along with their information, and so did I. They certainly didn't need me to tell them about what was happening at UNC or the conference in Santa Barbara. These people did their homework, and they weren't afraid to travel. What they probably didn't get in their standard review of the research was something I was by then prepared to provide—strong opinions about the profundity of the social transformation televirtuality might trigger, and explicit suggestions about what we might do to anticipate such an upheaval.

The twenty-first century is very much on the minds of people like the ones I met at NTT and ATR. Indeed, in early 1990, they were in the process of fleshing out the vision of a communication-enhanced future they had created to guide their research for the next ten years. This was no idle exercise, since they happened to represent a healthy chunk of the R & D elite of the foremost communication company in Japan, Inc. What the heck I was doing there, lecturing to them about the human side of technology assessment, was slightly beyond my comprehension—another one of a long list of activities I never dreamed my interest in VR would draw me into until it did.

First, I told the NTT researchers everything I could tell them in an hour on the state of the art in VR research worldwide—about what it felt like to grapple with molecules in Chapel Hill, about the tiny, high-resolution CRTs I had seen in Tom Furness's office in Seattle, and the cross-cultural communication experiment proposed by the Port of Seattle, as well as a few obligatory Jaron Lanier stories. Having laid out the territory as I saw it, I pointed at it and asserted that televirtual communication researchers are dealing with something bigger than one company, industry, or nation, and proposed that now is the time to create new global communication channels and mechanisms for cooperation across these antiquated boundaries, before the technological momentum grows beyond anybody's ability to manage. I talked about the VR disciplines' scientific and engineering foundations, starting with Engelbart, Licklider, and McLuhan, and worked my way forward to Fred Brooks, Myron Krueger, and Scott Fisher. There were a dozen engineers and researchers in the room, a couple of senior engineers who managed the several research departments, and Mr. Hiroshi Yasuda, executive manager of the Visual Media Laboratory. Dr. Yasuda and several of the others were fluent in English, so we proceeded without an interpreter.

When I started talking about social systems and global cooperation, Mr. Ishii's face lit up. You can tell a lot from facial expression, when people want you to read them, in Japan. The problem is that people don't often want you to read them, especially in situations like this. I could gauge Ishii's excitement, however, by the wideness of his eyes. Later, when he showed me around his laboratory, we began talking about VR groupware and intercultural communication. After I returned to America, we kept in touch via electronic mail, and eventually I encouraged him to write an article about computer-supported cooperative work and intercultural communication tools for *Whole Earth Review*, a publication that had printed my earliest reports from cyberspace.

"I have a five-year-old daughter," I told the NTT research brass toward the end of my little speech, adding: "I bet some of you have children or grandchildren who will be coming of age when televirtual technologies come on-line. We know from the history of communication technologies like the telephone and television that communication revolutions change the way people live, perceive, believe. Now we are talking about a tool for changing what we mean by reality. What kind of reality, then, do we want our grandchildren to live in? Is it more important that it be an NTT reality or an American reality, or that it be a humane, empowering reality for our children?" They didn't have any answers to my rhetorical questions. And popping up with an answer to a question like that is an un-Japanese thing to do. The continuing eagerness of people like Mr. Ishii and Mr. Tonomura, however, has proved to my satisfaction that they are interested in responding to the social challenges posed by their technologies.

Then they showed me the future communications systems they were planning and building. Which was more than one kind of answer of its own.

VIZTHINK: THE VISUAL MEDIA LABORATORY AT NTT

Visual thinking pervades all human activity, from the abstract and theoretical to the down-to-earth and everyday. An astronomer ponders a mysterious cosmic event; a football coach considers a new strategy; a motorist maneuvers his car along an unfamiliar freeway: all are thinking visually. You are in the midst of a dream; you are planning what to wear today; you are making order out of the disarray on your desk: you are thinking visually.

Surgeons think visually to perform an operation; chemists to construct molecular models; mathematicians to consider space-time relationships; engineers to design circuits, structures, and mechanisms; administrators to coordinate and administer work; architects to coordinate function with beauty; carpenters and mechanics to translate plans into things.

ROBERT H. McKIM,
Thinking Visually, 1980

. . . research into new broadband information services that allow users to quickly and easily access multimedia information from diverse sources has become an important target. . . . An advanced interface between people and databases is essential to realise the "Visual Thinking" environment for creative work using visual materials, such as decision making in offices or laboratories, artistic or commercial visual design in various fields.

NTT VISUAL MEDIA LABORATORY,
Annual Report, 1989

After talking for about two hours, it was a relief to walk into a laboratory, sit down, and get my hands on a demo. Like my first taste of cyberspace at NASA, more than a year before I visited NTT, my first demonstration at NTT's Visual Media Laboratory made me a believer. I sat down at a desk, and there on a 7-inch color screen a young woman reached her hand forward through the screen into space, toward mine. It was a liquid crystal display (LCD) screen, the kind other Japanese manufacturers were busily engaged in miniaturizing for shirt-pocket televisions. I had not put any special glasses over my eyes. At NASA,

I stumbled around in their HMD, saw the state of their primitive prototype world and imagined what it might become with a decade of concerted research and development. At NTT I looked at the efforts they were putting into the display components of wireless VR and realized that even the most lightweight, nonintrusive head-mounted display might be a temporary artifact of the present state of the art, as Myron Krueger had asserted in the 1970s. Just as future generations of the gesture-recognition cameras I saw at ATR, or versions of Krueger's Videodesk, might eliminate the need for gloves or bodysuits someday, the future versions of the stereoscopic visual display system at NTT might eliminate the need for head-mounted displays.

The NTT 3D display system combines one old optical technology—the lenticular lens autostereoscopic system—with the modern enabling technologies of liquid crystal displays and head-tracking devices. The lenticular lens is the optical stereo trick that gives the image its sense depth. *Autostereoscopic* means that the stereo effect is accomplished without requiring viewers to don special eyeglasses. The display console also had a stereo CCD camera (another leading-edge Japanese technology) aimed at me, so I could switch back and forth between their prerecorded demonstration and a 3D image of myself looking at a 3D image of myself. I reached out and touched someone—myself. My hand passed through a tiny ghost-image of itself. The figure popped right out of the screen and floated in space. When I moved my head horizontally, vertical black bands passed through the image, but it retained its three-dimensionality. It certainly didn't surround me. It was more like looking into a small window onto another three-dimensional world.

One might ask why a telephone company is so interested in ways to simulate human visual perception of three-dimensional scenes. The answer lies in another slogan that represents a whole way of conceptualizing future technologies: "visual thinking." Visual communication encompasses everything from calligraphy to gestures, images to facial expressions, visual models to body language, and the amount of exposure we have to images has increased dramatically since the introduction of electronic media. At this moment, billions of human visual systems around the globe are absorbing information from photos, videos, films, symbols, pixels, signs, logos, and pictographs, launched from billboards, magazines, newspapers, movie screens, television sets, graffiti, and computer terminals. Despite the incessant storm of visual symbols, few of us in America's high-viz culture consider ourselves capable of expressing ourselves visually or of using images to communicate.

We go to professionals for that. In light of the evidence that binocular eyesight has been a mainstay of our species' bag of survival traits for millions of years, isn't it strange that visual communication is not a common language outside advertising agencies and art schools? Eyeballs, after all, are where brains meet the world. Which means that something unprecedented must be happening to our minds as a result of this recent upturn in optic input. Some Japanese researchers believe that this is a point of leverage for Japanese technology, where new computer interfaces and communication media can mesh with the special capabilities of the mindset that accompanies fluency in the Japanese written language.

Japanese writing, based on Chinese-originated kanji pictograms that still retain a pictorial as well as a semantic content, is naturally a visually rich language. People and trees, oxen and rivers, natural forces and recognizable human activities seem to jump out from certain kanji figures when they are pointed out. The act of learning how to express oneself in kanji requires a kind of training of visual thinking skills. American children learn 26 letters, usually written with pencils. Japanese children learn to paint hundreds of pictograms. At least one distinguished Japanese researcher has claimed that because of the pictorial thinking stimulated by kanji, Japanese readers use a different part of their brain than readers of alphabetic languages.

In America and the rest of the nonkanji world, new communication technologies and new knowledge about how the human mind functions are changing the traditional attitude that visual expression is reserved only for the "talented." And the pace of change seems to be quickening because of the continuing development of graphics-based personal computers. Most important, the new revolution is not just about new tools. It's about a whole new way of using tools. Like all revolutions, it's about a new way of seeing. An American expert, Rudolph Arnheim, wrote a book about visual thinking, stressing the importance of the visual sense in thought and communication, and pinpointing a continuing deficiency in our educational institutions:

> The arts are neglected because they are based on perception, and perception is disdained because it is not assumed to involve thought.
> In fact, educators and administrators cannot justify giving the arts an important position in the curriculum unless they understand that the arts are the most powerful means of strengthening the perceptual component without which productive thinking is impossible in every field of academic study.
> What is most needed is not more aesthetics or more esoteric manuals of art education but a convincing case made for visual thinking quite in general. Once we understand in theory, we might try to heal

in practice the unwholesome split which cripples the training of reasoning power.

Some Japanese research managers appear to have read Arnheim. When Mr. Tonomura sent me the videotape of my conversation with Mr. Ishii and the research engineers, he wrote a letter, informing me that "Our company, NTT, has recently declared our future vision of the 21st century technology. The key phrase is 'VI&P' (Visual, Intelligent, and Personal). Our division is responsible mainly for the visual part of the technology needed for the vision."

THE VISUAL PERCEPTION LABORATORY AT NTT

As we combine text with verbal and animated graphic presentation, are we really constrained to communicate in the same black-and-white typography that has carried our intellectual freight since Gutenberg? It seems clear that the old alphabet may no longer prove adequate. We may, over a long period of time, evolve new symbol systems that employ color and position and movement in three dimensions to represent ideas. Instead of reading a book by a left-to-right, top-to-bottom scan, as we now do, we may enter its knowledge space and travel through it. This tendency towards an ever richer representation of problems and proposed solutions may lead to the definition of problems of such complexity that they can only be attacked by the total physical as well as intellectual involvement of the problem solver, who may effectively have to live in the represented world. Each day would be spent exploring the problem space, learning about it, and intellectually and physically seeking a solution.

Myron Krueger,
Artificial Reality, 1983

Artificial and high reality telecommunication services are very interesting and attractive applications of 3D image database technologies.

NTT Visual Media Laboratory,
Annual Report, 1988

If you were convinced that wireless televirtual technology of the future holds great promise for enhancing visual thinking and human communication, what basic research would you do to build a prototype? Many of the key researchers at ATR were on loan from NTT, which conducts its own fundamental perception research. And if they listen to Don Norman and my own ranting and raving, the NTT research managers will also be expanding their psychological, sociological, and

anthropological research. The role of the Visual Perception Laboratory, which seems to take the good part of a large floor at the Yokosuka
facility, is to develop computer vision and computer graphics techniques for accomplishing four goals: remote identification of people
and recognition of their intentions, based on their facial expressions;
scene interpretation by computer recognition of objects and comprehension of associated objects; synthesis of realistic images of humans
and their surroundings in real time; invention of new image-process-
ing techniques based on new knowledge about human visual perception.

CCD cameras can capture detailed stereoscopic images and feed
them in digital form to specially designed signal processors and to
computer software designed to extract features and detect patterns.
The shape-acquisition camera prototypes I had seen at ATR were part
of this effort. The NTT researchers were focusing on a pattern-
recognition system that would enable future computers to recognize
changes of facial expression, particularly the eyes and lips, and transmit those changes in real time, to be decoded into a synthetic head
image that moves in synchrony with the remote user's face. An automatic lipreading system that could extract spoken language from
the visual image—both as a human-computer interface technology and
as a form of person-to-person communication—is a parallel project.
The "head reader" prototype was linked to a high-powered graphics
workstation. In the demonstration, I watched the programmer move
his head and speak while a cartoon version of his face on the screen
reproduced his head and lip motions. The cartoon image is the subject
of converging research into various methods of creating and presenting artificial images from stored data bases and camera input. The
camera feeds visual information to a processor that creates a standard
facial model and applies special algorithms for everything from the
way the synthesized hair moves to the likely phonemes corresponding
with different lip configurations. Neural network systems that "learn"
to recognize a specific pattern are being applied to the problem of
creating computers that literally recognize their authorized users. It's
all part of VI&P.

The infrastructure of cyberspace will involve a lot more than high-
speed communication links and powerful reality engines. The objects
in virtual worlds, whether they are created from graphic models stored
in huge graphic data bases, or transformed from digitized camera
inputs, will have to be stored and retrieved in some way. In fact, one
of the early applications of VR interfaces at NTT involves methods
for flying through 3D graphical data bases using voice commands and
VR navigation. MIT's pioneering Spatial Data Management System
has been used by the US Navy, and nobody else who can talk about

it, for over a decade. It looks like NTT is interested in picking up that long-neglected implication of the MIT demos. In conventional text-oriented data bases, a user can obtain information about objects in the data base by entering queries in the form of command strings that say things like "show me all the Joneses" or "calculate the latest total expenditures in the Smith file." Such "symbolic" searches often involve specialized logical languages, although another trend in that field is toward "natural language queries." In an image data base, however, new ways of storing images, of querying the data base, of navigating through information will be required. The need for a browsing facility is particularly important if these image spaces are to be used for creative thought, as the NTT engineers seem to recognize when they write, in the 1989 annual report of the Visual Media Laboratory:

"A new image base navigation architecture and kernel have been proposed. They create an organized image database browsing environment in which we can see and examine a flock of visual objects effectively. While browsing, visually associative objects can be located without using symbolic query procedures, and we can find unexpectedly significant information. It is very important, but difficult, to automatically organize visual objects. A trial approach in which the computer extracts object images that match input key-words has been developed."

I keep coming back to that phrase, "examine a flock of visual objects." It reminded me of the room full of people in Austin, convened for the First Annual Conference on Cyberspace, who gasped when a medical educator showed them how he could reach in and rotate the modeled data points of an epidemiological data base of trauma injury case histories. Highly esoteric information about the statistical relations between the survivability of injuries and other variables can be used to save lives, if only the right informational relationships can be extracted from the data base. It turns out that you can three-dimensionalize the data, then rotate it in such a way that patterns of relationship between certain variables pop out into high visibility in the form of straight lines amidst disorderly clouds of dots. Image base navigation, like VR itself, looks like the kind of tool we can use to make better sense of the information flood we find ourselves riding every day. And the use of VR models to visualize software architectures—Jaron Lanier's dream of a visual programming language, but a programming language for experts, not for the masses—shows signs of capturing the attention of those who are trying to plan the software that will make all this powerful hardware coming down the pipeline work properly when people get their hands on it.

Virtual reality research at NTT is just beginning to roll into high gear. Before the autostereoscopic screens can offer an improved sense of presence, the LCD technology must develop further, and it seems that this enabling technology is indeed progressing; a wall-size display is in the works, using new high-resolution LCD screens and LCD projectors. Larger and smaller and higher resolution LCD displays are inevitable, driven by the huge financial flows associated with electronic entertainment. Hard work into solving serious problems and years of methodical work to create the basic software tools to accomplish the VI&P goals remain to be done. What I saw in Yokosuka looked more like a bold and well-planned assault on well-identified problems. It will take a few years to determine whether the plan succeeds. The lipreading and facial expression software is currently operating at only about 70 percent accuracy. Synthesized images still in the cartoon stage will have to progress to more lifelike simulacra before they can become useful media for transmission of intention. Shape-acquisition cameras and software also need to improve. These are not all inevitable progressions, the way the cost/performance of computer memory chips can be forecast years into the future. It isn't clear whether the problems remaining to be solved are soluble, even when enough money and attention are focused on them. The human perceptual and cognitive system has a long track record for outwitting those who would emulate its functions computationally.

If one assumes a reasonable amount of improvement in each of the enabling technologies over the next two decades, not taking breakthroughs into account, and assumes that there will continue to be a high degree of coordination between cognitive-perceptual research, hardware engineering, and special-purpose software at NTT and related companies, it is not out of the question to wonder whether we will be creating the new languages Myron Krueger dreamed about, by means of our NTT RealityPhones, twenty years from now.

Telepresence as a communication medium is a rich subject for speculation, but it does not exhaust the possibilities of VR as front end for communication systems. The ATR focus on "Communication with Realistic Sensations" presumes that cyberspace is the place where humans communicate with one another. And the "Visual, Intelligent, & Personal" workstations of the future will include new ways for people to communicate with their cybernetic assistants in the form of personal information-seeking programs, as well as with other people. Yet another possibility remains. As Marvin Minsky foresaw a decade ago, and science fiction writers explored in detail decades before that, telepresence could open up whole new avenues of human-*robot* communication. In the United States, "teleoperator" is the term researchers use to refer to human-robot interfaces. In Japan, they call it "tele-existence." In either case, the foundation of the technology is the human experience of seeing out of the eyes of a machine, and using natural gestures to direct machines to manipulate the physical world. One way to remember the distinction of VR and teleoperator research: In VR, you can demolish the building you are in, with no inconvenient side effects; if you are commanding a teleoperator, the building might not be there when you take your goggles off.

Chapter Twelve

OUT OF THE BODY

A person wears a comfortable jacket lined with sensors and muscle-like motors. Each motion of arm, hand, and finger is reproduced at another place by mobile, mechanical hands. Light, dexterous and strong, those hands have their own sensors, through which the operator sees and feels what is happening. Using such an instrument, you can "work" in another room, another city, or another country. Your remote "presence" can have the strength of a giant or the delicacy of a surgeon. Heat or pain is "translated" into informative but tolerable sensation. Dangerous jobs become safe and comfortable.

We can do a little of this today, using the crude machines made for handling radioactive materials. But once we improve those clumsy instruments—the basic point of this proposal—we will gain in many ways.

MARVIN MINSKY,
*Toward a Remotely-Manned Energy and
Production Economy,* 1979

My consciousness suddenly switched locations, for the first time in my life, from the vicinity of my head and body to a point about twenty feet away from where I normally see the world. The world I saw had depth, shadows, lighting, a look of three-dimensionality to it, but it was depicted in black and white. A few slight head motions confirmed that the laws of binocular and motion parallax held true. Hand-eye coordination had to be refocused. After a moment of disorientation and a few seconds of practice, I could pick up a pencil and put it through a hoop a few feet away, although my fingers had to work in ways that seemed alien at first. I moved my neck and shoulders. Twenty feet away from my body, my view of the world changed in response to my physical motions. I began to accept the odd sensation that accompanied the act of transporting my point of view to that of a machine—until I swiveled my head and looked at myself and realized how odd it seems to be in two places in the same time. Using your eyes and ears and hands to control a robot equipped with cameras, microphones, and mechanical manipulators sounds like a bit of fun, but I never thought of it as a thrill. What you don't realize until you do it is that telepresence is a form of out-of-the-body experience. It tasted to me like a little advance sample of the way it feels to be part of a silicon symbiosis. Telepresence is the name of a concept, a tool, an experience. It might also be one of the names for the way VR is changing what it means to be human.

Telepresence is another vector in VR history that was rooted in MIT, always a historic center of computer breakthroughs. The Artificial Intelligence Laboratory, led for the past thirty years by Marvin Minsky, has influenced psychology and cognitive science as well as computer science. Marvin Minsky likes to think big. He and John McCarthy, now at Stanford, are credited with coining the term "artificial intelligence" and creating in the 1950s the bold subspecialty of computer science now known as AI. Minsky's visions have never been about obstacles, but of opportunities. He looked at the relatively puny computers of the 1950s and sat down to design ways to make them emulate human thought, a job that he still pursues with Engelbartian tenacity more than thirty years after he helped spread the notion that it might be possible to build thinking machines.

Minsky was the first to clearly articulate the general usefulness of another potential aspect of interactive simulations, head-mounted displays, and computer graphics—their use as a natural way for humans to control robots. From the beginning, Minsky recognized the dual nature of this possibility—it not only would furnish the human's sophisticated perceptual and cognitive faculties as a control system for a remote robot, but would create a specific state of consciousness in the human user, an experience of being present in a remote location. He also recognized that although this technology was firmly rooted in science fiction, the implications for the world's economy and for the way people live their daily lives were very pragmatic and potentially enormous.

Minsky coined the term "telepresence" in 1979 in a bold funding proposal:

> We can solve many problems of energy, health, productivity and environmental quality by improving the technology of remote control. This will produce nuclear safety and security, advances in mining, increases in productivity, economies in transportation, new industries and market. By creating "mechanical hands" that are versatile and economic enough, we can shape a new world of health, energy and security. . . .

To do this we must improve our "telepresence" instruments—that is what I will call them—so that they feel and work like our own hands! This could take twenty years and cost a billion dollars. It will repay itself a hundredfold. Three Mile Island showed the absurd inflexibility of current nuclear technology. We still wait until men can rush in to look—and absorb a year's dose in a few minutes. Ridiculous! With better telepresence tools we could quickly inspect and repair, perhaps reducing losses from a billion to a few millions.

To his credit, Minsky was acknowledging that one of the goals of AI, the use of machine intelligence to autonomously direct the behavior of robots in real time in complex situations, was still a long way in the future. We don't have the computing power, nor the theories of knowledge representation, nor the algorithms, to control autonomous robots in real time. But we do have remote manipulators that can imitate human actions in inhuman places like fires and nuclear reactors, and we do have crude means of controlling the manipulators through human gestures. Minsky saw that finding a method for closely coupling human perceptions and reactions to the actions of remote semiautonomous robots was the key to making the scheme work.

Minsky had the technical knowledge to see how remote-controlled robots had the potential to solve real problems that neither humans nor robots were able to do safely or economically alone. Although he gave a name to the experience of telepresence, Minsky didn't claim to have invented the idea, an honor that most science-fiction experts attribute to Robert Heinlein, in his 1940 novel, *Waldo,* about a rather flabby being of the future who had the power to put his hands into remote operation devices and direct the movements of powerful mechanical puppets known as "waldoes." In the notes to his telepresence proposal, Minsky acknowledged Heinlein's contribution to the conceptualization of the technology-to-be: "This proposal is dedicated to two friends. The late Ray Goertz developed in 1954 the first electric force-reflecting teleoperator. My first vision of a teleoperator-based economy came from Robert A. Heinlein's 1940 novel, *Waldo.* The first such instruments were built in 1947. Forty years after writing *Waldo,* Heinlein helped me sharpen up this paper."

It would be an intriguing challenge for one of today's science-fiction writers to imagine for us what the world might have been like if anybody in the US government had taken Minsky up on his "twenty-year, billion-dollar" project. The field of teleoperations and human interfaces for teleoperators—the external and internal aspects of telepresence—has only recently converged with the development of stereoscopic head-mounted displays. The first teleoperators were the expensive mechanical hands developed to handle radioactive materials; early researchers and nuclear power technicians put their hands into an exoskeletal controller that conveyed their hand motions to a robot arm in a shielded room. One of these devices originally developed for handling radioactive materials, the Argonne remote manipulator (ARM, of course), ended up in UNC's laboratory in Chapel Hill, North Carolina, decades later, when enough computing power came along to make direct force-feedback possible. Today, pharmaceutical chemists use the ARM to wrestle 3D molecular models into place—a combination of simulation (the molecules' predicted behavior when brought together) and telepresence (the translation of molecular forces into physical resistance through the ARM).

In 1954, Ralph Mosher at General Electric developed the "Hardiman" a heavy, complex exoskeleton that locked around the user's arm and enabled the user to control a remote robotic arm through natural movements. In 1958, the Philco Corporation built a visual telepresence system by mounting a CRT in front of the user's eyes and controlling the viewpoint of a remote camera through head movements. And then progress in the field retreated to an incremental pace. The intriguing but esoteric field of teleoperators crept along for years of dry but vital human factors research—what some cognitive scientists call "keystroke counting"—and decades of careful engineering and reengineering of the mechanical elements, while the enabling technologies of telepresence control systems matured. The field is languishing no longer if one is to judge from the prototype teleoperated firefighter I saw in England this year, the full-scale, live-ammo working model of a teleoperated gun-vehicle I saw in Hawaii a few months before that, the telerobotic setups I saw in Tsukuba and London designed to augment the abilities of the disabled, the microtelepresence research at IBM that enables people to feel the surface of molecules with their fingertips, and other telerobotics-telepresence projects now springing up around the world.

TUNING IN TO TELE-EXISTENCE

It has long been a desire of human beings to project themselves in the remote environment, i.e., to have a sensation of being present or exist in a different place other than the place that they really exist at the same time.

S. TACHI et al.,
"Development of an Anthropomorphic
Tele-existence Slave Robot," 1989

When I first began to hear about the Japanese "science cities" like the famous one in Tsukuba and the new one going up in Kansai, I projected my own mental images of what I was going to see. I must admit that I daydreamed about postmodern highrise laboratories, set (in my mind's eye) amidst Zen rock gardens, staffed by scurrying technicians in matching jumpsuits, supervised by research managers in white coats. Pulling into Tsukuba Science City via auto, after an hour and a half drive from Tokyo, I was reminded more strongly of the semidesolate industrial parks outside Phoenix, Arizona, in the 1950s. Not that the sexily designed PARC-like laboratories are entirely absent. They are there, but they are still surrounded by huge empty lots. There's something vaguely unhealthy about it, as if the place is waiting for something that isn't here yet, or haunted by something that used to be here, or, like much of Japan, hoodooed and driven by both the past and the future.

Tsukuba in 1990 looked more like a collection of enclaves that were planned to expand toward one another over a matter of decades than the tastefully designed, densely packed city of efficiency I had imagined. Blocks of apartment complexes along "Science Avenue," pastel brick and cement, five to ten stories, separated by empty lots and highway strip culture, look more or less like modern college dormitories. It was a warm morning when we arrived, and bright futons were aired out on almost every balcony in the apartment buildings.

I was driving with the *Asahi Shimbun* crew again. I remarked about the disparity between my mental images before and after seeing Tsukuba, and Hattori told me that the slight aura of creepiness I perceived clinging to the place is well known in Japan. In fact, there is something called the "Tsukuba syndrome." For reasons still unknown, Tsukuba has the highest suicide rate in Japan. At lunch, he told me that Tsukuba was originally built on the site of three old villages. The place where we were eating lunch was a slick Sheraton-type hotel for visiting scientists, overlooking lovely landscaping that gave way abruptly to scrub fields. The original site of the hotel and surrounding labs was an ancient village called "Sakuramora," which means "cherry blossom city." The only cherry tree I saw in the place, in a carefully tended garden in front of the MITI mechanical engineering lab, was blooming.

MITI is a world of its own. It is the kind of government agency that could not exist in the United States, for better or worse. The Ministry of International Trade and Industry has the power to gather Japanese industry and academia and government together in one room, decide on a grand plan to create a new technology or set of interlocking technologies for a specific national goal, then give their objective a run for its money. It doesn't work exactly like that. It is done more through a series of shared values among leaders of many interlocking hierarchies in finance, government, industry, and academia. Thinking of MITI as a "technology czar," the way an American might think of it, is not Japanese enough. MITI has the resources and respect to bring many autonomous competing forces into alignment regarding a specified goal, and that is where its greatest power lies. Even if the highly competitive info industries in the United States could be persuaded to consider the notion of cooperating with one another for the good of the national economic welfare, it would be a violation of an entire library full of antitrust laws for anybody to try to do it—especially the government.

Sometimes the chaotic market of hypercompetitive empires battling in a free-for-all—the American model—can produce astounding results. It has become clear that cooperation cannot be imposed from above. Centrally planned economies such as the Soviet Union's and those of eastern Europe failed to manage the new technology-driven industries by enforced cooperation. MITI doesn't try to force powerful actors to work together; rather, it makes it attractive, profitable, and logical for the diverse actors to line up in the same direction. MITI's strategy of aligning autonomous industrial powers toward mutually profitable goals paid off handsomely for Japan's global economic status in the fields of consumer electronics, microelectronics, and automobiles. The strategy didn't work so well in the "fifth-generation" artificial intelligence projects that were supposed to give Japan's computer industries world dominance in the speech-understanding, "knowledge-processing" workstations they set out to build in the early 1980s for the mid-1990s. The projects failed to meet their goals within that time frame because the hard scientific problems still haven't been solved, even though the enabling technologies have progressed as predicted. Indeed, many MITI projects fail. But the ones that succeed and the spinoffs that enable other technologies to accelerate their development are the ones to watch out for.

MITI sponsors facilities for basic research in robotics. Again, the image in my mind was of disciplined legions of researchers, methodically laying down the basis for Japan's future industries. Dr. Susumu Tachi, who declined to wear his white jacket, has been working on telerobotics for more than fifteen years, and although many of his colleagues certainly do intend to create the basis for future Japanese industries, Dr. Tachi has been dedicated to the use of cybernetic technology for augmenting the abilities of the disabled. In the late 1970s and early 1980s, he worked on small robotic "artificial guide dogs" for the blind. Frederick Brooks would call robot guide dogs an excellent "driving problem" for robotics and control theory. Keep in mind that

Tachi's teleoperator is not an autonomous robot but an intelligent prosthesis that is supervisorily controlled by a human. The robot doesn't have to haul around all the silicon and software necessary to perform those perceptual and cognitive functions humans accomplish so well without thinking about them consciously—like recognizing patterns, navigating obstacles, making evaluations, etc. From that perspective, it makes sense that Tachi and colleagues began working on "tele-existence," or the creation of a strong sense of remote presence as a means of operating robots under remote supervisory control.

This relationship between human and closely coupled, partially intelligent, mobile robot appears to be an example of human-computer symbiosis of the kind predicted by J.C.R. Licklider in 1960, although of a different flavor than the path pioneered by ARPA. The kind of intellectual augmentation Licklider and his ARPA researchers pursued was considered a minor technological skirmish at a time when the big money was on artificial intelligence. The AI researchers wanted to leapfrog over human capabilities and build machines that could think. Part of the effort set up to achieve AI goals quickly branched off into the field of robotics. (And one of the key elements that distinguishes robotics from pure AI is the engineering discipline of "control theory," which is concerned with the nuts and bolts of building physical systems that depend on communication of information through systems to perform actions in the world. Control theories are now crossbreeding with virtual world hypotheses in the interdisciplinary VR hothouse.)

The idea of remote presence, linking human intelligence with less intelligent but highly responsive robots, is somewhere between augmentation and AI. The possibility wasn't overlooked at the time, although it was strictly a speculation about a possible future. Twenty years after Licklider had been instrumental in developing computers as mind amplifiers, his MIT colleague over at the AI lab, Marvin Minsky, made his 1979 proposal to the US government to fund a ten-year research and development effort in telepresence technology. And ten years after Minsky's proposal was generally unheeded in the United States except for specific military uses, MITI seems to be aligning key researchers and manufacturers toward a multidisciplinary telepresence research effort.

Aligning powerful forces, funding scores of researchers, and promulgating grand visions can be a successful formula, but it does not guarantee success. It is far easier to envision how an artificial arm might move in response to tiny fingertip movements, or sketch out the possibilities of teleoperated robots in nuclear reactors, than it is to build machines you'd be willing to lock onto your head and arms. The obstacles impeding teleoperator research are gaps in our knowledge of the psychological side of telepresence, and engineering problems involved in building channels for communicating information between humans and machines. The human part of a teleoperator partnership provides the cognitive-perceptual expertise. The mechanical part of the partnership provides a sense of presence to a human operator linked with remote sensors capable of probing and manipulating environments that may be hostile to human bodies. The hard part of linking the human and mechanical systems lies in creating the right kind of communication match at the interface. From the human side, a successful interface is evaluated according to the kind of *experience* it evokes; a teleoperator interface must evoke a feeling of presence in the human that is strong enough to efficiently guide the robot in real-world tasks—without mismatching the human's other sensory input and causing simulator sickness. The sense of presence, the *sine qua non* of virtual reality, is a core issue of teleoperator research, as well.

I had met Dr. Tachi in Santa Barbara, a couple of weeks before I visited Japan. He is a quiet man, polite and somewhat formal. He and I were attending a week-long conference on "Human Interfaces for Virtual Environments and Teleoperators." When I arrived at his laboratory in Tsukuba, several weeks after our first meeting, he gave me a print of a photograph he had made in Santa Barbara, showing VR researchers from the United States, Canada, Japan, Great Britain, and Germany gathered in the parking lot outside the Santa Barbara Sheraton. Tachi had shown slides of his setup during his presentation in Santa Barbara. When I faced the same setup in person, the robot was smaller and the room was bigger than I had envisioned, judging from the slides he had shown. The room itself was built for the purpose of positioning and repositioning heavy machinery, with tracks set into the concrete floor and huge steel rings embedded in the twenty-foot-high ceilings. In one corner was a three-wheeled robot car with a set of binocular lenses in the driver's seat. It was another one of those moments when I was forced to reflect on the odd places this project had propelled me—especially when I eyeballed a metal chair that occupied a prominent place. It dawned on me that I was going to strap myself into that device and let it have its way with me before I left that laboratory.

The robot itself looked like the bottom half of a large electric light pole or a fire hydrant: its body is a red metal cylinder about nine inches in diameter. The upper torso began to get anthropomorphic, with one arm, a two-finger gripper where the hand ought to be, a neck that is cantilevered the proper distance behind and below the visual sensors and that turns the same directions a human neck turns, and twin video cameras mounted where eyes ought to be. It looks like a machine. The humanoid resemblance doesn't come from the hardware, but in the way the thing *moves* when a human is operating it. It was a strange sensation for me, because I was very aware that I was observing a piece of hardware. The weird thing about the lubricated stainless steel joints, electric motors, wire bundles was how they moved in a way that I am accustomed to seeing humans move and I am not accustomed to seeing machines move.

The human operator sits in a control station that is a little scary to observe and a little scarier to operate. I sat in something that looked uncomfortably close to a no-frills dentist's chair. The display is head-mounted, but it is a "sword of Damocles" display that uses a mechanical device, a *goniometer,* to track head movements, rather than a magnetic sensor like the Polhemus. Along with all the mechanical linkages, there is a long arm with a counterweight attached to the head-mounted display (HMD). The HMD is clunky and it clamped onto my head fairly tightly; the eyepiece was secured with a chinstrap. When I was seated and helmeted, I stuck out my right hand and the arm-hand control was locked around my wrist and forearm in a manner uncomfortably reminiscent of handcuffs. At that moment, I was blind to the world and free to move my neck and arm but nothing else. When I moved my hand, arm, or fingers, the movements were reproduced almost simultaneously by the robot across the room. I wasn't able to transmit motion in every dimension I am accustomed to use; the arm has seven degrees of freedom, the neck has three degrees of freedom. But the constraints didn't carry as much weight with my awareness as the possibilities. Just being able to pick up a wand with my robot body and thread it through a hoop twelve feet away and feel as if I did it myself was far more impressive than the fact that I couldn't dropkick the wand through the hoop.

When the apparatus was switched on, I began to look through the eyes of the robot. The world looked like the world would look if I was located twelve feet to the left of my body, where the robot was located. I reached out my arm, craned my neck, opened my gripper, and picked up a small rod. It took me about five seconds to figure out how to poke the rod through two-inch-diameter hoops that were set around me at various intervals. It took about ten seconds to be completely comfortable with my control of the robot. However, every time I heard the mechanical apparatus racheting behind me, I began to get images of flailing mechanical linkages hurling my head against the wall. When I squeezed the handlebar-brake control, the robot gripper opened and dropped the rod. In order to pick it up again and remember not to drop it, I had to remap my short-term memory to do exactly the opposite of what I've done all my life: in the teleoperator, it is necessary to squeeze once to open, squeeze twice to close. Once again, this strange protocol seemed almost natural after a couple of tries. I could observe myself adjusting to the new protocol, but I can't say how I did it. "How easy it is for a well-tuned human to adapt to a machine," I wrote in my notebook, as soon as they unstrapped me.

The strangest moment was when Dr. Tachi told me to look to my right. There was a guy in a dark blue suit and light blue painted shoes reclining in a dentist's chair. He was looking to his right, so I could see the bald spot on the back of his head. He looked like me, and abstractedly I understood that he *was* me, but I know who me is, and me is *here*. He, on the other hand, was *there*. It doesn't take a high degree of sensory verisimilitude to create a sense of remote presence. The fact that the goniometer and the control computer made for very close coupling between my movements and the robot's movements was more important than high-resolution video or 3D audio. It was an out-of-the-body experience, no doubt about it.

TELE-EXISTENCE AT TSUKUBA

There are over 67,500 quadriplegics in the United States today, with an estimated 2,400 to 4,000 new injuries occurring each year. . . . Due to advances in medical treatment (antibiotics and skin care), these individuals now have a relatively normal life expectancy. They live, virtually, through the actions of others. Perhaps, life would be more fulfilling if they were also offered the opportunity to directly control their environment through telerobotic tools for independent living. Virtual telerobotic environments for the rest of us could bring reality to these individuals.

LARRY LEIFER et al.,
"Telerobotics in Rehabilitation," 1990

The final version of the tele-existence system will consist of intelligent mobile robots, their supervisory subsystem, a remote-presence subsystem and a sensory augmentation subsystem, which allows an operator to use robots' ultrasonic, infrared and other, otherwise invisible, sensory information with the computer-graphics-generated pseudorealistic sensation of presence. In the remote-presence subsystem, realistic visual, auditory, tactile, kinesthetic and vibratory displays must be realized.

SUSUMU TACHI et al.,
"Tele-existence," 1984

From remotely operated machine guns to systems for liberating quadriplegics from the prisons of their bodies, teleoperator technology exemplifies the full range of potential applications for VR interfaces, from the darker to the higher sides of human capabilities. Dr. Tachi's work is another part of the many convergences of needs and technologies that seem to be occurring in cyberspace these days. In the 1970s, Tachi initiated another MITI project, the Guide Dog Robot, which, as the name implies, was a concerted effort to create the basic technologies for augmenting the mobility of the visually impaired. As in many robotics projects, the Guide Dog Robot, known as MELDOG (Mechanical Engineering Laboratory Dog), made definite progress toward the goal, and at the same time it revealed with greater clarity the serious problems that remain to be solved before the laboratory prototypes can offer real assistance to blind people in the untidy, chaotic, complex world outside laboratories. Two of these problems—sensory interfaces that can transfer a sense of presence from machine-sensed data to a human, and methods of enabling the human to maintain supervisory control over remote robots—led Tachi into the field of *tele-existence*, a term he believes encompasses both interface and control aspects of teleoperation.

We left the robotics laboratory and walked through the mint-green corridors of MITI's Mechanical Engineering Laboratory. Large windows overlooked most sections of the laboratory, and graphic displays in almost every window described (in both Japanese and English) the projects under way. One hallway led to the biorobotics division, another to the cybernetics division. Through one window, I saw a variety of prosthetic arms; through another window, I glimpsed a set of mirrors being ground. Dr. Tachi, the *Asahi Shimbun* crew, and I retired to his office. Without a word being spoken, one of the postdoctoral students working in the laboratory, the only woman, got up from her computer workstation and made us tea. First, Hattori interviewed Tachi in Japanese. Then Dr. Tachi and I conversed in English, with only a little help from Hattori. I recommend the exercise of speaking through an interpreter for anybody interested in honing their communication skills; it forces you to express yourself with information chunks of the right size. I asked him how he got from robot guide dogs to teleoperated robots. What was the path from prosthetics to virtual worlds?

"In designing truly effective mobility aids to the blind," Tachi replied, "we found that it is necessary to determine what kinds and how many pieces of information are necessary and/or sufficient to mobilize humans." In other words, in order to build a machine to help humans get around, it was necessary to learn as much as possible about the way humans navigate. This general problem requires close attention to the perceptual components of moving a human-jointed body in three-space, a trail that leads directly to virtual reality: from DataGloves to shape-acquisition cameras to responsive environments, one thrust of VR research has been to design machines capable of detecting and transmitting human movements, gestures, expressions. The first step to transmitting body language is to learn how we steer our body models in cyberspace, and thereby how to build more humanlike and responsive models. Mobilizing virtual puppets is essential to entertainment as well as augmentation, but the stakes are more serious when that mobilization is the only kind of mobility some people might ever know. In the physical world, their bodies are confined to beds or wheelchairs; in the virtual world, the same people might soar and glide and, in partnership with teleoperators, have effects on the world that would be impossible any other way. The key problems to be solved in each of these different fields, albeit in simpler form, were wrapped up in efforts to build mechanical guide dogs.

Tachi first focused his attention on three specific aspects: the actions of information streams from sensors, the computer programs that compare the input streams with internal maps of the environment, and the display systems that would enable a semiautonomous robot to communicate a sense of location to the human operator. If a robot dog is to be a guide, it must have a sense of where it is, and it must communicate that sense of location to the human operator in a way the human operator can understand in real time, while trying to navigate. The same information-processing and control-theory problems to be solved in MELDOG later turned out to be applicable to other areas of VR as well as telerobotics. That is one of the ways convergence can accelerate technological development: toolbuilders in many different fields can accelerate progress in multiple convergent disciplines, and in the case of calculus, printing presses, or personal computers, they can amplify the power of almost every other discipline. And now, with VR, everybody is amplifying everybody else at a dizzying rate.

The phase that often follows technological convergence is hybridization. After the implosion of diverse interests in a shared focus comes an explosion of interdisciplinary crossbreeds. Like stereoscopists and computer graphics programmers who found themselves in cahoots with roboticists and psychologists to build the first cyberspace systems, teleroboticists who want to use human interfaces to affect the physical world find they are challenged by the same design problems faced by virtual world builders who want their human operators to interact with a completely computer-contained cosmos. Both engineers-turned-teleoperations-specialists and computer-graphics-wizards-turned-virtual-world-builders find they are sharing problems with systems programmers who are studying about cognitive psychology in order to build the next generation of computer interfaces. The Santa Barbara conference at which I had met Dr. Tachi was a symptom of this multiple convergence, not a cause of it, although the decisions made at scientific and engineering conferences and the personal relationships forged there will certainly have their effect on the future of these convergences.

The second pathway from the robot guide to the teleoperator, according to what Dr. Tachi told me, was the need to find the best way to code and display the acquired information to the human operator. MELDOG navigated a rigidly controlled environment according to a map in its computer and an ultrasonic-mediated sense of where it is in the environment. A blind person cannot use a visual display, of course, and audio displays have their limitations, particularly in loud public environments. The method Dr. Tachi and his colleagues used to display information from MELDOG to human operators was "electrocutaneous"—a method of tactile communication delivered in pulses of low-voltage electricity applied to external electrodes on the skin of the operator. The operator feels a mild buzzing sensation on the surface of his or her arm or leg. Different voltages and frequencies create different kinds of buzz that operators learn to discriminate. The use of electrocutaneous feedback by Tachi and other prosthetic designers is part of another technical convergence that makes strange bedfellows: feedback devices for augmenting the visually impaired, robot gloves with a sense of full manual presence for fighter pilots, scientific visualization tools that enable investigators to probe molecules with their fingers, prototypes for sex-at-a-distance devices of the future, might all benefit from new knowledge and new technologies related to the tactile dimension of VR.

Tele-existence is part of a large-scale project on advanced robot technology, established in 1983, undertaken by Japan in cooperation with researchers from other nations, "under the international cooperation framework established at the Versailles and Williamsburg summits," as Dr. Tachi, a soft-spoken gentleman in his forties, reminded me. He mentioned the teleoperation work in Germany and at the US Naval Ocean Systems Laboratory (NOSC). When he explained that he had very good relations with NOSC researchers, Tachi added: "MITI research is only for peaceful purposes, not for weapons." When I put my hands on the teleoperated machine-gun at NOSC's high-security facility in Hawaii, a couple of weeks later, I remembered the look on Dr. Tachi's face when he mentioned NOSC. I know from his long history of dedicated work that Dr. Tachi was pointing out that although he respected his colleagues in military-related research projects, his own research was altruistic, not lethal.

Under the larger framework of the JUPITER project ("JUvenescent PIoneering TEchnology for Robot" is how MITI publications explain the acronym), Tachi expanded his goals to encompass not only augmentation systems for people with disabilities, but systems for commanding flocks of remote robots, for extending human reach into hostile and remote environments, and for expanding human perception into regions presently invisible. Tachi's research is an attempt to answer the question: "If a funding agency wanted to take Minsky's telepresence memo seriously, what kind of research would they support, what knowledge would their researchers set out to gather, what engineering goals should the research support?" One of the interesting aspects of Tachi's approach is the way his prototype projects have involved proof-of-concept for real tools designed to accomplish tasks in the physical world and at the same time have provided accurate psychophysiological measurements designed to help us understand the human components of tele-existence. The system I had experienced was both a prototype and a test-bed for trying out features of future prototypes; the Polhemus Navigation Systems spatial sensor operates at a maximum location-sampling speed of only 60 hertz (cycles per second), while the Tsukuba apparatus can measure position and velocity with 500–1000 hertz response range.

The development of the semiautonomous robot that triggered my out-of-body experience was the first stage of a project to actually build such control systems for special-purpose robots under development in other parts of the MITI robotics effort. As such, the first prototypes were consciously designed to be laboratory instruments. One of the reasons for building the experimental apparatus was to systematically study characteristics of the stereoscopic displays and reactions of human perceptual systems, then attempt to increase the power of the telepresence experience by calibrating the displays and other feedback devices. One of the first findings of this phase of the research was that the phenomenon of binocular convergence, which can be augmented by experimenting with optical systems, is a key component in human perception of three dimensions. Next-generation systems at Tsukuba will feature improved systems for augmenting the sensation of binocular convergence, which occurs when both eyes focus from slightly different positions on the same object.

Next, Tachi wants to make a master-slave teleoperator system that can move. "Our next goal is to make a full-size, mobile version of the anthropomorphic telerobot," Tachi said. "Our longer-term goals are to extend human abilities with microrobots, giant robots, robots and displays that can show humans infrared and microwave information." In this sense, Tachi is a colleague of psychologists such as Nathaniel

Durlach at MIT, who uses VR equipment as psychological probes for mapping the way human senses model the world. Durlach, who started out in Cambridge, Massachusetts, studying bats, finds himself using the same equipment and contemplating the same research issues as Tachi, who started out in Tsukuba building robot guides for the blind. Both, for example, are interested in the causes of simulator sickness— the feeling of nausea and vertigo that can accompany prolonged use of a simulator that is either too good or not good enough at conveying a sense of presence. Tachi has concluded, on the basis of his JUPITER and other research, that it is important to keep objects at the same size and apparent distance they would be in the physical world in order to prevent simulator sickness. "Keeping the visual parameters close to those in the physical world enables people to use the system for two to three hours without motion sickness," he claimed when we talked.

It is a natural step from augmenting the power of disabled people to move and affect the world to augmenting the capabilities of the nondisabled. The human body has its limitations: we can rarely lift our own weight, we are sensitive to heat, cold, and radiation, and we don't move very fast or see well enough to perform microsurgery without magnifiers. Tachi's system is a step toward building the kind of system that would enable a person to don a lightweight suit and steer a remote robot into a nuclear reactor, lift a two-ton component and replace it. The next stage prototype at Tsukuba will involve a testbed for the kind of controls needed to step up lifting and grasping power. The idea of using special body gear wired to powerful or impervious robots is an old dream revived by new components. Tachi also noted his intellectual debt to the attempts of General Electric to build the Hardiman exoskeleton in the late 1960s; he pointed out during our conversation that twenty years ago it was far more dangerous to wear an exoskeleton that was more likely to malfunction in physiologically nasty ways, and space inside the machine was at a premium because of the need to use computers, controllers, cables, and energy sources within the mobile structures themselves. Miniaturization enabled new breakthroughs in regard to the number of components that could be packed into a pseudohuman elbow joint. Dr. Stephen Jacobson at the University of Utah, leader of the team that developed the Utah Arm, has been leading research for many years in the engineering discipline of prosthetic design.

Before we left for Tokyo, we bid goodbye to Dr. Tachi and drove over to yet another campuslike complex of buildings that turned out to be an actual college campus, the University of Tsukuba. We dropped by at Dr. Tachi's suggestion to see a younger colleague, Dr. Hiroo Iwata, who was studying VR technologies at the university's Institute of Engineering Mechanics. The weather had been warm all week, so I brought only a thin windbreaker, but I regretted not wearing a sweater when we walked down the halls. I don't know if it is possible to be colder inside a building than it is outside the building, but the utilitarian gray halls of the university's engineering laboratories were cold enough to make a late March day feel like mid-February. Mr. Iwata was one of the people who won my heart from the beginning by asking me to autograph an exceedingly well-thumbed edition of *Tools for Thought*. His was not a high-budget operation, but he did have a small laboratory with a color graphics workstation, a Kubota mini-supercomputer, and some intriguing-looking hardware. Mr. Iwata is interested in generating a tactile sense that corresponds with visual models of virtual objects—a subject that eventually would draw me to Cambridge, Massachusetts, London, and southwest France. Unlike many of the other tactile feedback research projects I saw elsewhere, Iwata's funding appears to be limited, so one of his objectives was to build relatively inexpensive input devices.

Where Dr. Tachi was rather formal, dressed in a three-piece suit, Mr. Iwata was shirt-sleeved and I would describe him as "shy" rather than "reserved." As a graduate student in computer science, Iwata explained, he grew interested in the kind of human interface research that had been conducted at the Architecture Machine Group and Media Lab. In particular, he wanted to pursue ways of choreographing the movements of virtual objects with hand gestures and increasing the degree of kinesthetic and tactile feedback possible with VR systems.

The large color graphics workstation in Iwata's rather small laboratory was connected to a TITAN graphics supercomputer capable of performing 32 million instructions per second—enough computing power to provide colored, shaded, high-resolution graphics. Computationally, his setup was far more powerful than those I had seen at VPL, if less powerful than the parallel processing setup at UNC. Iwata and his colleagues have built a compact nine-degree-of-freedom manipulator as a tactile input device for desktop operation. It looks like a kind of erector-set object with highly articulated, lightweight, metal joints, shafts, and gears, about a foot high, taking up about as much space as a shoebox. I sat down in a chair and placed my hand in the device. My thumb, forefinger, and palm could manipulate various levers by squeezing and relaxing aspects of my grip, and my wrist could rotate and move within a sphere of about a foot diameter. From my seat in the chair, I peered via an angled mirror at the screen, which filled much of my field of view, but did not totally immerse me in the virtual world.

On the screen was a red sphere resting on a blue table. An articulated hand, the by-now-familiar persona of the operator in cyberspace, floated above the sphere. I moved my hand in the metal apparatus and felt it move with me while I watched my pseudohand approach the pseudosphere. When the virtual fingers of the hand on the screen made contact with the surface of the virtual sphere, the motors and sensors in the manipulator began to resist, just the way a real solid ball would resist if I tried to grip it. The sensation was not what I would characterize as powerful, but it was palpable—more of the Kitty Hawk effect than the full-blown conversion experience. If Iwata or others succeed in optimizing the slight feeling of solidity created by this prototype, however, we might soon see the day when CAD designers can not only stick their head into their designs, but reach their hands in and feel objects. One of the immediate applications of such a system interests Japanese manufacturers of cameras—the ability to simulate the feel of a hand-held instrument like a camera would be a tremendous boost for designers who now must work with mockups that take hours or days to create and modify. Ultimately, Iwata believes his prototype will evolve into a micromechanism-laced glove.

On the way back to Tokyo, Hattori and I wondered aloud about the benefits of freeing paralyzed people to move their perceptions through cyberspace, and the dangers of creating an interface to powerful machines. Just what we don't need right now are powerful machines doing things to the world before we have discovered just what it is we ought to be doing to the world. To the medical diagnosticians or surgeons who want to float their vision and manual dexterity and a microscalpel right into an artery or a cornea, teleoperated robots are a miraculous aid to alleviating human suffering and saving lives. To those who turn rain forests into plywood, semiautonomous megadozers are an ideal instrument for efficiently transforming biosphere into greenbacks. The families of bomb-squad experts and firefighters who risk their life and limb might see impervious telerobots as a gift of life. And to those who would rather not risk their own flesh but don't mind spewing shrapnel into the bodies of other human beings, teleoperated gunships are the way to go. The virtues or terrors of VR often seem to depend on who does what to whom with it.

CHEAP AND CHEERFUL TELEROBOTICS: VR IN GREAT BRITAIN

Telepresence is a specialised form of teleoperation whereby the operator of a remote system is presented with "workstation" facilities capable of creating the illusion that he is actually present at a remote, hazardous worksite. Historically, research in this area has been somewhat piecemeal, with only specific display and control facilities being provided for operators of telerobotic systems (for example, stereo, or head-coupled displays, or exoskeletal controllers). There have been very few programmes addressing the integration of these technologies in an attempt to produce a true telepresence test bed.

Robert Stone,
"Human Factors Research at the ARRC," 1990

I slipped on the "balloon-o-glove," and felt various vague but definite sensations in my fingertips. It felt like a glove with tiny, dynamic, moving lumps in it. The demonstration took place in an exceptionally well-equipped garage, but a household garage nevertheless, outside London. When the thing was fine-tuned, James Hennequin, the inventor, assured me, I would be able to use it to reproduce the feeling of grasping a teacup. This demonstration required credulity on an order of magnitude greater than that of a Kitty Hawk experience. This was like visiting Orville and Wilbur in their bicycle shop while they were assembling one of their earlier, doomed prototypes of a flying machine. Hennequin's companion, also an inventor, had a stereo CCD video camera that cleverly and inexpensively emulated the way human eyes tend to converge on whatever object they are focusing on. What Robert Stone and James Hennequin had to show me, half the world away from Tsukuba, several months after my out-of-the-body experiences in Science City, was not yet a prototype ready for testing. What I saw was a test-bed-under-construction. Indeed, I ended up traveling a considerable distance between components of the test-bed—a glove in a garage north of London, a teleoperator setup in Culham, a small mobile vehicle laboratory in Manchester. It's what they might add up to, when they are all plugged together—and the sheer ingenuity of the way it was put together for very little money—that makes the UK telerobotics research effort particularly worthy of attention.

Robert Stone takes justifiable pride in his ability to build "cheap and cheerful telerobots," as he sometimes puts it, on the relatively meager budgets available in Britain. One of his secrets is to find genius contractors to build the vital components of the VR test-bed Stone is putting together for the European Space Agency, the European Atomic Energy Authority, and various bomb squads and fire departments.

I found out about Robert Stone, and was snagged by the puppet connection, during a foolhardy field trip that threw us together at the beginning of a long scientific conference. On a rather more steep than expected hike into the dry country outside Santa Barbara, California, the previous March, Stone mentioned that one of his VR contractors was working for a London company known as "Spitting Image." Upon hearing that, I had the unnerving sensation—like a photograph from the future suddenly flashing onto my inner screen—of realizing that I would have to add a trip to London to my research itinerary. I remember standing there in the desert hills, huffing and puffing, sweating, and probably flushing half as red as Robert Stone in his suit and tie and his biological clock still in the Cotswolds. Although dire need of air-conditioning and severe lack of iced coffee had dulled my thinking ability, I realized that the same conversation would continue again in England, not too many months later, because the story was too good to ignore. Coincidentally, it was beastly hot in England, too, when I got there in July—so hot that Big Ben stopped working for a couple of hours not long after I left.

Like other odd strands of the VR drama that only began making sense to me when they started intertwining with other odd strands, my exposure to British telerobotic research started years before I began investigating VR—via my first encounter with automated satirical puppetry. Eight years ago, on a trip to England, I visited a pair of jolly, outrageous artists with a political bent, Roger Law and Peter Fluck. Roger Law looks like Blackbeard the pirate. He's six foot something, with some heft to go with it. He has wild black hair and an unruly beard, both now shot with gray, perpetually pink cheeks, melodramatic gaps in his grin, and eyes that twinkle halfway across the room. It's easy to see Law "hurling my share of dustbins through barroom windows," as he once described his New York days.

Peter Fluck, Law's coconspirator since their days at Cambridge College of Art, is as tall as Law, but is slim, weathered, sardonic. Roger tosses epithets. Peter arches an eyebrow. There is no room for any doubt that they use art as a weapon—a poke in the sensibilities as well as a tickle to the ribs—once you see what it is they create. When I first visited them in Cambridge in 1982, they had just finished constructing a puppet that looked like either Alfred Hitchcock or Queen Victoria, depending on how it was costumed; at the same time, they were turning out porcelain teapots in the shape of Margaret Thatcher's face, with the Prime Minister's nose as the spout. Automated simulacra of Margaret Thatcher, in fact, are what drew Peter Fluck and Robert Law into the force-field of VR hardware development.

Fluck and Law's workspace at the time I first visited them was an old stone building that had formerly housed the chapel of a Methodist temperance society—an irony that didn't fail to amuse them as they fueled their schemes of world media conquest on hours of bloodthirsty satire, endless pints of stout at the no-frills pub twelve steps down the lane, and loud rock and roll late into the night. (I have a persistent memory of hearing Law conversing at the top of his lungs in the other room while I lurched around the loo, trying to find the urinal, which turned out to be a moist wall with a gutter at floor level.) Fluck and Law sketched and sculpted, chortling at the raw viciousness of a Reagan puppet's jowels, or howling in self-induced hilarity at the pornographic motions they had devised for a Margaret Thatcher puppet. They ranted about building legions of their wicked three-dimensional political cartoons. They dreamed of unleashing their sick creations on entire nations.

"Spitting Image" was the name of the benevolent monster that grew out of Fluck's and Law's efforts. Spitting Image itself has always been a mutant cousin of virtual reality. They had pursued the honorable but not very secure art of political satire for years by sculpting and photographing caricatures of political figures for the covers of magazines from *Time* to *Der Stern*, to the *London Sunday Times*. Law had been artist-in-residence at Reed during my undergraduate career there. His greatest delight, even in 1968, has always been to create his filmed puppetry worlds of artistic revenge, where all the political figures he despised could be seen by all and sundry in the cruel light of caricature. By the early 1980s, they had concocted a plan to vault directly to international prime-time weekly television, in partnership with a respectable production company, with the financial backing of a reclusive computer magnate, and in collusion with a few kindred spirits in Britain's television industry.

Apparently their plan was more inspired than insane, for when I returned to their studios one shirt-soaking hot day in July, 1990, almost a decade after our first encounter, Spitting Image had moved from "the chapel" to large remodeled buildings in the East End of London, where several scores of artists and technicians were turning out a small army of movable, even nastier political puppets for their multinationally syndicated television series.

VR came into the Spitting Image scenario, and drew me back to England, by way of their quest for ever-more hideously lifelike caricatures. They already used three human puppeteers to maneuver each puppet for the camera, but they wanted fully mobile facial expressions and lip-synching. Only some kind of internal mechanism, something that required as much engineering as craft, could animate their creations the way they wanted. Spitting Image Engineering was born. Over the past two years, they've spent hundreds of thousands of pounds trying to apply cybernetics to satire. An inventor Peter Fluck had found and funded, a fellow by the name of Jim Hennequin, proved he could do absolutely magical things with tubes and hoses and cannisters of compressed air. With Fluck's puppetry experience and Hennequin's inventive talents, they discovered that not only could they make their puppets move, but they could distort their features grotesquely, liquidly, precisely, hyperrealistically, thanks to one of Hennequin's devices. These devices, I could see immediately, were going to be even more important to the VR world.

It took a half hour for Roger Law, as amiable and foul-mouthed as ever, to show me around their London studios. They had rebuilt the interiors of a series of light industrial buildings in the East End, down the street from a bell foundry that had remained on that site since 1570. Inside the studio, rooms full of young puppet makers, caricaturists, moldmakers, engineers, and production specialists were creating entire casts of puppet productions. The heads of state of every nation, prominent British political figures, each and every member of the Royal Family, British and American cultural icons were visible in various stages of construction—the Alfred Hitchcock/Queen Victoria prototype from Cambridge days had multiplied itself manyfold. And the Spitting Image puppets of the 90s were moving. Eyebrows were ticking, cheeks were inflating, lips were blubbering. Peter Fluck took me around the corner to the engineering building, where the pneumatic magic that made their puppets twitch and spit was developed.

The blast of sound—loud jazz, drill presses, and the "spiff, spiff, spiff" of compressed air escaping through valves—was the first thing to greet me. I realized that "around the corner" designated an entirely different culture from the one found at the original site. Unlike the Spitting Image main studios, which was a three-floor warren of offices and workshops, the engineering shop was more or less one huge space. Naked moving machines and fully garbed puppets ran through their paces on platforms all over the room. Huge puppets were moving their necks, hands, wrists, wattles, oversized ears, in grotesque gestural choreographies, over and over again, as various configurations were torture-tested. My next impression was the sight of Jim Hennequin—it could only be him—issuing instructions, checking gauges, a moving vortex of activity in a white lab coat, the very picture of the modern Prometheus. I remember thinking that after years of visiting research laboratories, he was the first person I could recall who actually wore a white lab coat. Later on, when Hennequin chirped away about "low-cost, clean pneumatics—none of your messy hydraulics," I wondered whether he wears the clean white coat for effect. I remembered how careful they were at the Naval Ocean Systems Center in Hawaii when they fired up their "green man" and other telerobots—if the system wasn't working perfectly, hot hydraulic fluid under pressure might squirt across the room and hit you in the face.

Hennequin delights in slipping the term "LBO" into his interviews with the press, along with other engineering acronyms. Jim Hennequin has ginger-colored hair and beard, both more or less recently trimmed. He is a devout man who delights in nonblasphemous epithets, gazes at you over metal-rimmed bifocals the way bifocal-wearers are wont to do, and spouts so many quotable sentences that I found myself filling pages of my notebook with his epigrams. When I got home I found that several of the phrases he passionately declaimed and which I dutifully jotted down also turned up in the clippings from various British publications he gave me before I left. That's fair, I guess. He has a few points to make, so he makes sure he makes them quotable and repeats them whenever the press comes by.

Low Bullshit Option is that LBO represents, and it speaks for Hennequin's attitudes toward academics and most institutions. He's a man who makes things work. Along those lines, one of his favorite sayings is "If you want to make something, you go and bloody make it. Get a piece of metal, stick it in the vise, cut it off. That's my way to do it. Most places, they spend their time having meetings." I get the feeling he loves to say those words over and over. He's the type of fellow who lives to accomplish feats that people in other fields believe impossible. If Charles Babbage had hired an engineer with Hennequin's attitude, the British might have invented the computer in the nineteenth century. Like many engineers, however, he is so highly focused on his task that the big picture isn't what he's looking for when he zeroes in on making something work.

"Inflate a flat hose and it goes round," Hennequin explained with a grin, pointing at a huge spool of the Flexator's key component—flat fire hose. That sentence explained the basic physical principle underlying the apparatus that surrounded us. He showed me one of the internal motion mechanisms visible through the head of a half-constructed Ronald Reagan puppet. A piece of flat fire hose, cut and crimped with steel clamps to a precise length, was twisted in an exact manner around a tube of specific shape, connected to a mechanism that looked like part of a trumpet or tuba, from which tubes of smaller diameter snaked toward a tank of compressed gas. The precise way a flat hose becomes round, and the different ways it deforms when it is twisted around a metal base of one shape or another before inflation, provide the dynamics for imitating human muscular movements. The actual movements are accomplished by air pressure: 100 pounds per square inch through hose that is tested for 800 psi. But the sequences of movements, the way they work together to animate the puppets, are electronically controlled. Considering the way electromechanical robots are built, or even hydraulic "anthropomorphic" robots, the idea of building a skeleton out of metal, cutting the right length of fire hose, and animating it by pumping it up with air has a certain elegance.

Once a set of Flexators has been created for a puppet, and the elements of facial or hand movement are put into place, then various combinations of those motions can be programmed in software. The animated Margaret Thatcher, always a favorite target of the Spitting Image gang, was the *pièce de résistance.* Her very gaze is enough to wilt a nearby bunch of flowers every time she swivels her neck and goggles her eyes at it. The lips are synched with her speech, and caricaturesque expressions of astonishment, cupidity, fear, and outrage can accompany her prerecorded statements. Hennequin waits until the proper moment of astonishment during the Thatcher demo to remark that they rent her out for £2500 (around $5000) per day, which means "she earns more than the real one!"

Proceeds from the Thatcher demos, which apparently are popular at meetings of large companies in London, go to Inventaid, a nonprofit company headed by Fluck, Hennequin, and Dr. Robin Platts, a medical specialist whose patients are helping them develop and test Flexator-driven aids for disabled people. Like Dr. Larry Leifer at Stanford VA hospital and Dr. Tachi at MITI's Mechanical Engineering Laboratory, the Inventaid crew think telepresence is a promising solution to the problems of people who have heretofore had very little to hope for. The idea of robotic wheelchairs with easily controllable arms has been around for some time, but robotics has not proven to be easy, reliable, or safe, and the cost of providing such equipment to the people who need it has been prohibitively high. Hennequin, in an interview with the *London Observer,* declared: "There's nothing on this earth more irresponsible than designing and building devices to solve the problem of a disabled person and then telling them that they cost £20,000. Our aim is to do this thing for £2000 per wheelchair, and I'm damned sure we will."

The Inventaid prototypes were the next things Hennequin showed me. Around a corner from the perpetually shrugging, bat-eared Prince Charles puppet is a wheelchair. Attached to the arm of the wheelchair is a mechanical gripper similar to the kind that are seen on expensive, powerful industrial robots. In place of the usual heavy mechanisms that give the gripper its precise degrees of freedom are Flexators. The Flexators provide the muscle. Over at the other end of the workshop, dismantled on a bench, are some of the mechanisms devised to control the arm. Inventaid's electronics expert, Matthew Steinberg, has been building a tongue-control system for people who can move only their tongue and eyeballs. By cutting the costs of the mechanical muscle and inventing better control devices, Inventaid hopes to extend the capabilities of people with perfectly functional minds trapped within paralyzed bodies.

One of the intriguing features of the pneumatic muscle system is the possibility of using it to build inexpensive motion platforms, another VR problem of such expense that most VR systems avoid it. The most expensive flight simulators are built in movable cabins so that the physical motions of the platform match the perceived motions in the flight simulation. Without a matching disturbance of the motion-detection system in the inner ear, the realistic simulations of cyberspaces have a tendency to make people nauseous—simulator sickness. However, the cost of building a motion platform that can move a chair or a desk or a room around quickly and precisely to match the motions of the world model has been extremely high. Electronics are cheap, but mechanical devices, especially large, powerful, precise ones, are expensive.

"Sit on that," Fluck commanded, after Hennequin whirled out of view, that day in London. He was pointing at an ordinary office chair that appeared to be bolted to a piece of plywood. He grinned and pushed a button as soon as I was seated. The chair immediately started lurching. If I had been looking at a simulated roller coaster through a head-mounted display, or viewed a dune-buggy ride like Morton Heilig's Sensorama demo, the chair's motions might have elicited a scream of verisimilitude, like the one Disneyland's Star Tours motion platform does. When it stopped moving, I stood up and stepped off the platform. Fluck kneeled, grabbed an edge of the plywood, and turned the platform on its side. On the bottom were several strategically placed Flexators. "It took us two hours to build and less than $100 worth of parts," he said. No wonder he was grinning. He gestured at the larger version in the hall, a platform large enough to convey several seated adults through an amusement park ride, complete with rolling waves or bucking broncos or accelerating space-flivvers.

Hennequin whirled through the workshop again, stopping long enough to agree to meet me the next day, when I planned to join Robert Stone for a whirlwind tour of the scattered components of his VR/telerobotics test-bed-under-construction. Fluck and Law and I retired to the nearest pub for the informal part of the interview.

The next day, Robert Stone and I found ourselves talking about teleoperators and VR test-beds once again (and, coincidentally, enduring another hot spell together), this time at the Didcot train depot, about forty-five minutes from London. At least we weren't on foot. His car, unfortunately, was not air-conditioned. We picked up the conversation where we had left it in Santa Barbara, four months previously, while he drove us through the tawny fields of truly rural Britain. It definitely wasn't suburban Silicon Valley, the urban heart of Akihabara, or downtown Anaheim. The hedgerows were high, the roads were narrow, the pubs had names like "George and Dragon," and the stone farmhouses had that kind of mossy multicentury patina the Japanese call "sabi." A half hour's drive from Didcot station, you could see the power lines converging, then the chainlink fences became visible just before an earth-toned building complex loomed into view. No patina here. The word "sprawling" comes to mind. We stopped at the security office and I showed my passport. Badges clipped to wilting shirts, we entered the United Kingdom Atomic Energy Authority Culham Laboratories.

The big news at Culham Lab that day was that the air-conditioning was down. Stone and I exchanged a look. Would our next rendezvous take place in Arizona? Before I took a look at the telerobotics setup, I needed some kind of liquid. We found a vending machine in the cafeteria. There was a sign on the vending machine. I remembered thinking that this was a very British kind of sign: "Please Understand This Machine Does Not Take 10 p Coins." An American translation would be: "NO 10 p COINS." The Japanese version would express sorrow at the inability to accept 10 p coins and proffer apologies to the consumer. A German machine would have been repaired by now.

Because of the geographically dispersed nature of the UK telerobotics effort, Stone spends a fair amount of time driving around the green and pleasant British countryside, which was more brown and scorched than green and pleasant the day I joined him. The European Space Agency (ESA) has been conducting studies in teleoperation and control through a British Consortium consisting of British Aerospace, the United Kingdom Atomic Energy Authority (Culham and Harwell Laboratories) and, most recently, Stone's home base up north at the UK Advanced Robotics Research Centre. By nightfall, I realized that Robert Stone's normal circuit of his VR contractors is several hundred miles long.

The Culham Lab robot room was fenced to keep humans out of the roaming area of the semiautonomous machines they were building. They had an industrial robot (a European make from Asea, I noticed) with a redesigned control system for telerobotics, complete with video cameras and lights. A wheeled platform gave it limited mobility. In the control room, the operator has a six-degree-of-freedom ball controller, direct voice input to a speech recognition system, twin joysticks, color graphics and video monitors, and within the next several months if the ESA funds them they plan to add DataGlove Model 2 units and a head-mounted display with the capability of switching between video and computer graphics or mixing them together. The day I was there, they were testing a predictive graphics system that showed the operator a computer model of the robot, with a delay between the operator's movements and the robot's response—simulating the kind of delays that might be expected if the operator was on the ground and the robot was in orbit. The mission scenario for the ESA was based on applications in which human operators on the ground or in orbit would assume command of telerobots in space—a similar scenario to the original NASA vision presented in Fisher et al. Indeed, Scott Fisher's demonstration of the NASA system in 1988 had been the "conversion experience" that had turned Stone into Britain's foremost VR research organizer.

After Culham we motored down to see the more modest garage workshop where Hennequin had concocted the technology I had seen in the lavish London Spitting Image Engineering workshop. Another one of Stone's contractors, David Watkins, had promised to bring along the automatic convergence stereo video camera he was building. Between the various UK agencies and private enterprises who funded his work, Stone had £100,000 annually to get his telerobotics test-bed in operation. That isn't a lot of money, especially considering the capabilities of the system he had specified. Hennequin and Watkins were Stone's secret weapons in his battle of the budget. I started to perk up when I noticed the road signs to Bletchley. It turns out that Cranfield, the town where Hennequin's house and workshop are located, is a few kilometers away from Bletchley Park, the very place computer pioneer Alan Turing had cracked the German code during World War II.

It was an ordinary suburban home. The Swedish submersible vehicle in the driveway was one clue that something other than the average suburbanite lived there. We found Hennequin bustling about the small, well-equipped workshop that shared the garage with a lovingly restored antique Riley automobile. David Watkins was there too, a somewhat quieter fellow who was clearly from the same "go and bloody make it" school of innovation. A representative of Johnson and Johnson showed up as well; he was interested in Flexator applications in rubber glove manufacturing. The lack of friction-prone sliding valves was one of the aspects that made it an interesting prospect for industrial applications. "Great enemy, friction," is the kind of thing Hennequin is likely to say in the same tone you would imagine him saying "jolly good pudding." I tried on the tactile glove. The "balloon-o-glove" he called it. A little old man in Devon made the gloves out of Lycra to Hennequin's specs. It seemed like the start of something.

"Soft robotics," Hennequin intoned, when he saw me eyeing various inflatable sacs that were clearly too big for gloves. He was also working on the idea of inflatable robots that could be fit through small holes in nuclear pressure vessels and commanded via telepresence to check welds and perform maintenance chores.

The stereo camera looked like a stereo camera ought to look—two lenses, mounted on a multiple-degree-of-freedom aluminum "neck," with miniature color video cameras. "Fourteen to one zoom ratio, very fast response speed," Watkins explained when he saw my attention attracted to his apparatus. Most ingenious was the automatic convergence apparatus. As Dr. Tachi had noted, convergence is a key element in building telepresence systems that don't make people sick. When we look at an object, our eye muscles turn each of our eyeballs inward so that our lines of gaze converge on that part of the environment where we center our visual focus. Stereo cameras, unless they are designed to do so, don't converge. Watkins had built a "cheap and cheerful" but highly accurate version of what Tachi had built in his laboratory. By using the autofocus component from an ordinary camera, Watkins was able to activate two high-speed motors to converge the stereo lenses wherever they were focused.

We left Cranfield and headed north for the Cotswolds. Eager to give me a properly British experience, Robert Stone had secured dinner reservations at Thornbury Castle, and had asked Iann Barron, the founder of Britain's foremost microprocessor company, to join us. The castle had been partially completed half a thousand years ago by an ambitious young noble who was executed by the king for not getting the proper permits to build crenellations and other military touches to the architecture. So it had been half a castle for several centuries until an American bought it and turned it into the kind of restaurant one would expect in such a setting. Hot butterscotch pudding. Calvados in the drawing room. Quite a different world from hot sake and live fish or fastburgers and Jolt Cola in other corners of the VR universe.

Barron, a quiet, gentle, perpetually smiling, thoroughly computer-hackerlike fellow, had a lifelong interest in computation; one of his instructors at Cambridge had been Charles Babbage's grandson. Barron's company, INMOS, had spun off an interesting startup company called "Division." Their product was a kind of massively parallel computer architecture known as "transputer architecture" which happens to be uniquely suited to virtual reality systems. And Division is located within ten miles of Robert Stone's house. Stone had found out about it when visiting VPL. Noticing a familiar accent in Silicon Valley, Stone had struck up a conversation with the wife of a VPL employee. She had come from Manchester, where Stone's main laboratory is located, and asked Stone if he knew that there was a VR-related startup company in Chipping Sodbury, just down the lane from his own home. When he returned from the United States, Stone found another nest of geniuses, extraordinarily young ones at that, right under his nose.

I spent the night as a guest of Stone and his wife Helen, a perception researcher for British Aerospace who is working with simulator-related issues, at their 400-year-old remodeled farmhouse. After dinner, we spent part of the evening looking through reprints from ATR and research reports from NASA, the way other people might look through travel brochures or family albums. The next morning, Robert and I started out by visiting Division. It was another stunningly hot day in the British countryside. I was expecting a cinderblock industrial park and was pleased to see an old stone building next to a creek, a small two-story structure with all the doors and windows open. There seemed to be about a dozen people, all told, none of them over the apparent age of twenty-five. I was briefed by an articulate young man by the name of Charles Grimsdale, who resembles Bill Gates, the young billionaire who founded Microsoft, the personal computer software company, in his teens.

Like Gates, Grimsdale has the habit of punctuating his presentation by pushing his glasses up onto the bridge of his nose. Wielding a whiteboard and ink marker like an old pro, he proceeded to explain clearly and simply what a transputer is and why a transputer company is interested in VR. It made sense by the time he was done. He started out by drawing a square, with a short line extending out perpendicular to each of its sides. The square is the microprocessor, where one entire functioning processing unit is squeezed onto a chip. The four lines are where information can come in or out. Then he drew more boxes at the ends of the four lines, and more lines. A transputer. A network of computers, each one on its own chip, intercommunicating and weaving the results into world models and high-resolution graphics. Massive parallelism is a fashionable, apparently powerful, but by no means proven architectural paradigm that supercharges the right kind of computation onto a whole new level of power.

Like Pixel-planes, the massively parallel reality rendering engine Henry Fuchs and John Poulton are developing at the University of North Carolina, transputer architectures are a way of building computing devices that use many microcomputers working in parallel, rather than one big powerful computer that processes information one step at a time. Iann Barron had a vision of massive parallelism which he encouraged at INMOS. Ten years ago, they made the fastest microprocessors in the world. Although INMOS doesn't sell as many computing chips on the world market as Fujitsu or Intel, the ones they sell are very cleverly made. The way computer elements are laid out on a chip, and the methods by which it accomplishes computing tasks in circuitry, constitute the computer's "architecture," and INMOS architectures are its strength. Transputing is, above all, an architecture. And dramatic advances in the architecture of reality engines is bound to drive dramatic advances in the state-of-the-art VR systems.

Transputers are to massive parallel architectures what microprocessors are to computers—a transputer is a chip that has both computing and communication capabilities (the squares and lines Charles had drawn on the whiteboard). Because a system built into each chip manages the switching of information in and out of the central processor, it becomes possible to plug these modules together like building blocks to build a community of closely linked computers that can divide labor and exchange information—a transputer. If you have a task that can be divided among semiautonomous modules—like painting a complex graphical shape on stereo CRTs or updating a million relevant sensory details of a world model—you can gain a lot of computing power by hooking up a large number of relatively cheap, relatively puny computing chips. In Pixel-planes, the output of each one of the parallel processors is one picture element; an entire computer is devoted to keeping track of whether that pixel is on or off, how bright it is, what color it is. The young transputer architects at Division see VR as a field that is hungry for what they have to sell—a system for handling massive amounts of data very quickly.

The idea, as Charles explained it, is to think of every different part of a VR system as an intelligent object that needs to exchange certain kinds of information with certain other parts of the system, but doesn't necessarily have to keep track of the big picture. Another part of the system is responsible for keeping track of which information needs to go from which module to which other module. Every module does its part, and the big picture simply emerges seamlessly. The streams of data from the glove must somehow end up translated into a series of accurate positions in a computer model that will be displayed on a CRT for the users to evaluate. And the streams of data from the 3D audio system must be integrated with the changes in the user's head position. The eye-tracker, if there is one, contributes more information going into the computer. The visual and acoustic and tactile displays coming out of the computer add to the computational load.

If you think of the accumulated processing and communicating requirements of each part of a VR system as a huge mass of computation that has to happen very quickly, you come up against reality barriers. It takes as many millions of operations per second as computer science knows how to throw at a problem to simulate the reality model our biosensors generate every second. But if you think of a VR system as a kind of community of systems that have to do smaller, specialized tasks, in perfect synchrony, very quickly, the problem can be made more tractable.

Abandoning the whiteboard for the computer room, where the heat had just conked out their DataGlove, I put on a set of EyePhones— not a highly recommended option for a clammy day—and grabbed a Polhemus electromagnetic sensor in lieu of the glove. The virtual world was a simple one. A drumset floated in space. When I moved my hand, a virtual drumstick moved in synchrony, and a sound came back to me when the drumstick intersected with the drum. With a tactile glove with some force-feedback capabilities, I suppose I would have felt the wooden instrument vibrate and bounce off the taut drumhead. A binaural earphone played back the sound of a virtual snare drum two feet in front of me, a virtual brass cymbal a foot to the right of that at shoulder level, a tom-tom to my left. The object of the young men who put together this system was not so much to create whizzier VR systems, but to build a monstrous reality engine that will blow all the other computers off the track when it really gets cranking. The VR systems are just ways of testing their product's capacity. The product itself is housed in a matte-black cube, with the word "Vision" emblazoned on the top.

Everything I knew about computer architectures I just explained in the paragraphs above. I'm far from a qualified expert. But if you hear people talk about computer architectures you will quickly notice that everybody has an opinion, often fervently held. And I think this transputer idea looks and feels like a very promising way of thinking about the kind of information-weave needed to achieve high-level VR. If one makes the large assumption that the scientific and technological barriers to building total-verisimilitude VR input-output devices can be overcome, and the 3D is perfectly realistic and so are the sound, the motion, the feel of the virtual world, then the final barrier to making the illusion work will require a computer architecture capable of coping with the information-processing explosion you create when you plug all those devices together and let a human nervous system loose with it.

Yet another enabling technology is converging on reality systems that have not yet been assembled. Ten years from now, Division might be a giant player in a new industry, or it might suffer the fate of the majority of high-tech startups and fade from view as some already formidable player or a different startup group gets some breaks, wins some battles, and occupies the computer architecture niche in the emerging VR technecology.

Now that I had seen the pieces of the puzzle he is assembling—the gloves, the cameras, the telerobots, the transputer engines—Stone and I motored up to Manchester to see one of the places where all the pieces will come together. Stone's laboratory was small, well equipped. A brick building about a fifth of a block square, two stories, housed a number of robotics, computation, and human interface laboratories. From the look of the place, remotely piloted vehicles seemed to be their area of specialization.

"In contrast to the remote manipulation bias of other international establishments," Stone pointed out, "our research is concentrating on the field of remote driving." By early 1991, ARRC hopes to combine in its Virtual Human-System Interface Demonstrator a remote navigation vehicle, a head-mounted display, speech recognition and synthesis command system, a DataGlove Model 2, a stereo camera and microphone pair mounted on a high-speed pan-and-tilt head mechanism, incorporating a closed-loop laser ranging sensor module for controlling camera convergence. Their aim is to combine stereo video pictures of actual landscape with graphics windows and overlays. The system, when assembled, will constitute a test-bed for performing precise human factors research into issues of transfer of operator control between remote driving and remote manipulation, integration of dextrous end-effectors with hand gesture input, force and tactile feedback, sensor-driven world modeling. All this on a total budget that is an order of magnitude smaller than the price of one Connection Machine massively parallel computer at ATR. With a small budget, ambitious but sharply focused goals, and a pool of independent geniuses who concentrate on making things work rather than attending meetings, Stone may have more ingredients for success than he realizes when he calls the UK effort "the poor cousin" of the VR world.

Fire rescue machines, roboguns and microsurgery are sober subjects, but virtual reality applications are not limited to matters of life and death. Entertainment and education, two areas of human activity that seem to be related, are both areas of intense study by influential companies in Japan. Nintendo is creating systems for Mattel's PowerGlove. Fujitsu, one of those only-in-Japan vertical giants, which makes everything from microchips to mainframes, AI to home entertainment, appears to have targeted the entertainment and educational potential of cyberspace. The day after my visit to Tsukuba, I paid an eye-opening visit to Fujitsu's huge research and development laboratory in Kawasaki, a Tokyo suburb. Some of the visions of the future of entertainment and education I saw at Fujitsu sent me back to what may be the oldest historical roots of cyberspace, as well. This new technology, it appears, might be the latest and most potent, but hardly the first, manifestation of a very old, little known, never more pivotally important cultural tradition.

Chapter Thirteen

THE ORIGINS OF DRAMA AND THE FUTURE OF FUN

Whereas film is used to show a reality to an audience, cyberspace is used to give a virtual body, and a role, to everyone in the audience. Print and radio tell; stage and film show; cyberspace embodies. . . . Whereas the playwright and the filmmaker both try to communicate the idea of an experience, the spacemaker tries to communicate the experience itself. A spacemaker sets up a world for an audience to act directly within, and not just so the audience can imagine they are experiencing an interesting reality, but so they can experience it directly. . . . Thus the spacemaker can never hope to communicate a particular reality, but only to set up opportunities for certain kinds of realities to emerge. The filmmaker says, "Look, I'll show you." The spacemaker says, "Here, I'll help you discover."

RANDALL WALSER,
"Elements of a Cyberspace
Playhouse," 1990

Computers are theater. Interactive technology, like drama, provides a platform for representing coherent realities in which agents perform actions with cognitive, emotional, and productive qualities. . . . Two thousand years of dramatic theory and practice have been devoted to an end which is remarkably similar to that of the fledgling discipline of human-computer interaction design; namely, creating artificial realities in which the potential for action is cognitively, emotionally, and aesthetically enhanced.

BRENDA LAUREL,
Computers as Theatre, 1991

A young man in EyePhones walks across his living room. His hands are thrust out in front of him, as if he were holding something. In VR land, revealed as a thought-balloon above the young man's head, it is clear that he wields a magical sword. On the horizon, volcanoes fume. In the virtual middle distance, Godzilla looms. It was all carefully hand-illustrated on a large piece of high-quality cardboard that was resting on an easel. The scene drawn on the display, it was explained to me, depicted Fujitsu's vision for the first step into the future of entertainment. I was sitting in a conference room on the top floor of Fujitsu's huge Kawasaki R & D laboratory, a guest of the chief engineer (and beneficiary of happenstance). They didn't roll out any hardware to show me, but they didn't lack for flip charts, overhead projections, videotapes, and inspirational oral overviews by project directors. The presentations left no doubt that Fujitsu is serious about getting into the VR business. This wasn't a short-staffed skunkworks or coalition of enthusiasts within a vast firm. These were the top research managers for the only company to beat IBM in the mainframe business—not an assignment for a middleweight.

The ability to assemble just about anything from in-house technology is one advantage of being the kind of conglomerate Fujitsu is: various parts of various laboratories were fabricating experimental HMDs and graphics chips and 3D modeling software, according to an overall plan for a VR laboratory. While they were awaiting the completion of their VR test-bed, the director of Fujitsu's VR effort, together with his colleagues, articulated a vision of where their company hoped to create a niche in tomorrow's cyberspace marketplace. The idea of the chief engineer and top executive of a company promulgating a scenario of a cyberspace initiative is reminiscent of John Walker's vision, as he articulated it in 1988; one key difference is that Autodesk does roughly $100 million worth of business annually, while Fujitsu grosses tens of billions of dollars. Although the depiction of the home reality play-house I saw in Kawasaki was just an artist's rendering of what a team of researchers wanted to accomplish, these researchers are in a position to make visions like this a global reality.

The artist's rendering of cyberspace as home entertainment was part of a presentation that the Fujitsu staff made to me in the course of an afternoon. Fujitsu's top research managers saw the future of VR in a very specific manner. At ATR, they had a vision of televirtual communications. NTT researchers wanted to build a "visual, intelligent, and personal" human-computer interface. Tsukuba, Manchester, and the US Naval Ocean Systems Center on Oahu were targeting teleoperator applications. Fujitsu aims to use VR-based media to position themselves in the fun market. VR as theater, not just as microscope, communication medium, robotic controller. CAD, entertainment, and education were the three markets Fujitsu planned to address.

If all the world's a stage, as Shakespeare declared, and if all the world is in the process of being wired into a global electronic media space, as McLuhan prophesied, and if mass media reality networks of the future might enable anybody anywhere to experience directly any role in any cyberspace they can create or afford, as some VR researchers believe, then what will people do for fun? And how might the global fun business influence the shape of a future VR society? Fun is a serious subject when you consider that the two largest blocs of users of new information and communication technologies in the 1990s will be the global finance and global entertainment industries— money and fun are spinning the wheels of the world economy. The electronic machinations of instantaneous global markets created by webs of electronic transactions, along with the sitcoms, movies, videos, double-platinum CDs, talking toys, network news, video games, and theme parks that constitute the experience industry today, are the largest driving forces behind tomorrow's information technologies.

If the idea of an "NTT RealityPhone" sounds strange, how would you feel about "Fujitsu Fantasy Worlds"? Or "Fujitsu-Disney Cyberpark"? Would it surprise you that the education industry of tomorrow might be revolutionized by MIT . . . and Nintendo? Although the three hypothetical product titles listed above don't exist yet outside my imagination, and are not based in any existing alliances I know about between these particular companies, the trend toward global partnerships among media conglomerates and research institutions brings generally similar scenarios into the realm of real possibility. The unlikely but actual Nintendo-Media Lab partnership will pop up again in the realm of VR "edutainment," a critically important cousin of cyberspace amusements.

You can't get more Japanese than Fujitsu, nor more American than Disney. Whether or not they are aware of it, these apparently different businesses are moving in on chunks of the new territories opened up by the possibility of cyberspace entertainment products. Fujitsu and Disney appear to be converging on the VR entertainment business of the twenty-first century from two different directions—home entertainment and theme parks. If you forget about modern technology for a moment and think back to bread and circuses, Punch and Judy, carny barkers and Coney Island, the mass-marketing of manufactured experiences in virtual spaces is an old tradition in the preelectronic experience industry. The present multibillion-dollar worldwide experience business is jumping into new electronic media to such a degree that its low-tech origins are obscured, but it originated and evolved from circuses, carnivals, roller coasters, and funhouses into miniature fantasy kingdoms set inside vast parking lots. Fantasy-world building today is a business for supermanagers such as Disney's Michael Eisner, no longer a game for P. T. Barnum or Samuel Goldwyn. Or so I began to think when I first noticed the Matterhorn peeking out between a couple of buildings in Anaheim.

After walking out of Autodesk's first presentation on "Virtual Reality Day," June 7, 1989, at the Anaheim Convention Center—the one where Timothy Leary in the flesh and in the virtuality proclaimed the opening of a whole new psychoanarchic territory to a rather somber convention of bewildered but enthusiastic computer-aided design specialists—I trekked across a Sahara-sized parking lot to Disneyland with another cybernaut, in order to pay homage to the original shrine of virtual entertainment world building.

Brenda Laurel, who was also following developments in VR technology, had been on the same plane from San Francisco to Anaheim. Like Scott Fisher, Jaron Lanier, Michael Naimark, and other infonauts who have emerged recently as influential figures in the VR world, Brenda Laurel came to my attention through the Atari Research Laboratory during its brief heyday in the early 1980s. We worked on other projects together, became friends. Now she is deeply entangled herself in the course of the technology, and I still try to look where she points, because she seems to be seeing where the electronic experience industry is headed, a direction she succinctly described as "first-person experiences." After the Autodesk cyberspace presentation, Dr. Laurel and I decided to ditch the conference for a couple of hours and dive into some world-class first-person VR.

The seamless surreality of the scene was irresistible—leaving head-mounted prototypes of virtual realities behind to venture through Anaheim, itself a completely synthetic zone, on the way to the grandfather of all large-scale artificial environments, seemed to be an appropriate ritual for the occasion. As it turned out, Anaheim was the first small road trip of my cyberspace odyssey. Outside the Atomic Age Motel, gardeners with chain saws were shaping pyracanthas into topiary Mickeys and Minnies. The sun was Venusian in intensity at high noon in June. Anaheim in the summer was not created for mobile creatures who do not come equipped with air-conditioning.

The "Star Tours" ride at Disneyland is a two-dimensional simulation of a wild ride through outer space and through various Lucasian mythical future architectures at eye-opening virtual speeds. The long line for Star Tours ends up at something that looks like a very modern subway car. People exit through the opposite side, so the sense of anticipation is magnified by the uneasy perception of watching all these other people disappear into the initiation chamber, scream in unison precisely two minutes later, then disappear. The interior features a wall-size, high-resolution screen, with a few dozen comfortable, heavily seat-belted chairs located in a Disneyesque futurethentic simulation of a Star Wars space flivver. The sound system is stupendous, loud, and surrounds the audience. The most powerful VR effect is invisible and inaudible—a motion platform that tilts the whole room a few feet in one direction or another and is so efficient at producing the sensation of dropping into the interstellar space projected on the screen that it is possible to stand outside the ride and time the synchronized involuntary screams of the passengers to the second.

You definitely don't need stereoscopic three-dimensional graphics to convince people that a simulation is frighteningly real. High-resolution graphics on a large screen in a small room, synchronized with total surround sound and a motion platform, is quite sufficient to induce an altered state of consciousness similar to the adrenaline rush one can find in surfing, whitewater rafting, automobile racing, or skiing. Star Tours is a roller coaster with no actual fear of death—the unlikely but ever-present possibility that a mechanical malfunction will send your car flying into physical space. Even though Brenda and I knew the physical room itself was only pitching and yawing a matter of a few inches, the utter synchrony of the motion with the convincing illusion that we were looking out the windshield of a space flivver induced involuntary screams and equally involuntary adrenaline reactions. It was a totally noninteractive experience, however; we were just along for the ride. I don't know if my poor heart could have handled the strain if I had actually tried to steer the ship through the same wild ride the preprogrammed adventure blasted at us. It occurred to me later that the experience was a direct test of one of Brenda's pet hypotheses about the architecture of cyberspaces: *mimemis*, Aristotle's term for the psychological resonance that enables dramas to move people emotionally, is the reason our stomachs dropped when the flivver "fell" out the port door of the mother ship. Where Jaron Lanier would point out that our cognitive systems try to fill in the gaps between the simulation and the reality, Brenda Laurel would point out that our hearts, our hormones, our most visceral systems can be enlisted in the aid of the right kind of drama.

Dazed from that experience, we had to navigate by our own power only a few hundred surreal feet through the Magical Kingdom to the "Captain Eo" line. Captain Eo takes place in a theater that seats several hundred at a time. Disneyland in Anaheim and Tokyo, and Disney World in Orlando, are the only places in the world where it is possible to see this big-name, high-budget fifteen-minute film. Lucasfilm, Francis Coppola, Disney animation "imagineers," and Michael Jackson teamed up to create a state-of-the-art three-dimensional video that makes you think you can reach out and grab the cute little weird creatures boogeying fourteen inches in front of your face. Disposable cardboard viewing glasses—like the red and green glasses handed out at the original 3D movies—were distributed at the door.

A few minutes before we walked into the theater, we had been upstairs at Autodesk's semisecret "demo room," where selected cognoscenti were granted ten-minute flights apiece through cyberspace. For hours, we had been watching people troop into the demo room, don the DataGlove and the "face-sucker," fly through a virtual floor plan by pointing their fingers and craning their necks, and emerge with that now-familiar ecstatic, far-focused gaze. Timothy Leary was in one corner, raising his eyebrows and grinning every time an engineer or architect or business manager took off the HMD, and the first thing the fresh cybernaut refocused on in the "real world" was— Timothy Leary.

Fresh from observing the essential VR conversion experience, Laurel and I were ripe for the kind of thinking that a direct encounter with Disney cyberspace was bound to trigger. The idea of combining Star Tours and Captain Eo into a 3D, motion-platform, cyberspace amusement leads directly into fantasies of "Disney Realityland." There are problems with anything but the lightest, most durable, least expensive head-mounted displays, however, in a place where the financial bottom line depends on extruding people at an optimum rate through a series of simulated experiences. It seems that the kind of sensations that only a very high-budget, highly physically architectured cyberspace can provide—huge, high-resolution screens, massive sound systems, and motion platforms—are on a different track from the notion of evolving today's television sets, video games, and telephone networks into home reality networks. Public reality spaces and personal reality spaces, as Myron Krueger, Randal Walser, and others have suggested, might be one boundary that eventually will distinguish two distinctly different approaches to VR technology.

FANTASY BY FUJITSU

Fujitsu Laboratories has two centers near Tokyo. At Kawasaki, telecommunications, space, information processing and personal systems are studied. At Atsugi, the focus is on electron devices, electronic systems and materials. Both laboratories lead Fujitsu's R & D and are working on advancements in fields ranging from systems to nanometer technology.

Fujitsu Laboratories develops technology for new products by working closely with various departments in Fujitsu, and independently researches future technology by designing the next generation of Fujitsu products. This work is the key to Fujitsu's future and fulfills Fujitsu's slogan: "What Mankind Can Dream, Technology Can Achieve."

FUJITSU LABORATORIES,
Research and Development, 1989

The personal electronic entertainment market has been dominated by Japanese companies, from video games to VCRs, Walkman and Watchman personal media to camcorders. When I heard that Fujitsu was sponsoring cyberspace research, I called Dr. Yasunori Kanda, the chief engineer of research and development for Fujitsu, who had been a fellow speaker at the Hypernetwork conference I had addressed a few days previously, and arranged for a visit to their facilities about an hour's drive from Tokyo.

I was beginning to get used to the "honored guest" treatment, but the red carpet at Fujitsu was rolled out with an elaborate ceremony that was mildly disconcerting at first. Katsura Hattori was on another assignment, elsewhere in Tokyo, that morning. My guide for the day was Yo Miyawaki, a reporter in his early twenties who works for *Asahi Pasocon*, the largest general purpose personal computer magazine in Japan. He is a fairly jaded journalist, considering his youth. He was accustomed to the limousine that picked us up. But when they sent four managers to greet us at the door of Fujitsu Research Laboratories in Kawasaki, escorted us to the walnut-paneled meeting room on the top floor, and distributed laser-printed copies of the agenda for my visit, I could see that even my hardboiled young friend was taken aback. So was I. Authors, artists, and teachers, I discovered, have higher social status in Japan than in the United States.

Fujitsu is a knowledge-age vertical company that covers everything from microcircuits to communications satellites. They create products and services in the information-processing, telecommunication, and electronic device markets. In 1988, they had net sales of $15.2 billion. Big boys. Fujitsu R & D's chief engineer told me early in our conversation that Japan is the only country where IBM does not have the top market share. There are two Fujitsus. The facility I visited, one of two large research and development laboratories, was in Kawasaki, a suburb of Tokyo and Yokohama. Unlike ATR or NTT, this future-factory is not located in rolling hills or a new science city, but in a mixed-use light-industry and residential neighborhood. There's a Denny's Restaurant two blocks away.

Fujitsu microcircuits using Fujitsu miniaturization techniques are part of Fujitsu supercomputers that communicate through Fujitsu telecommunication networks to Fujitsu home multimedia machines that add video and sound capabilities to Fujitsu personal computers. If you combined the American companies of Intel (microchips) and DEC (computers) and one of the Bell companies (telecommunications), you would still have to add the equivalent of Hughes Aerospace to match the technoindustrial breadth of Fujitsu.

The Fujitsu lab in Kawasaki looked to me to be about one and one half PARCs in size: six stories of stark white modernist building, of the same recent vintage as ATR. The R & D laboratory also overlooks a connected fabrication facility about half its size, devoted to producing experimental components for the different R & D divisions. The product fabrication plants are elsewhere. The chief engineer of Fujitsu Research and Development, Dr. Kanda, is revered in Japan because he cracked the main entry barrier to the Japanese in the digital age: he invented a Japanese word processor that creates kanji files—the Japanese pictographic writing adapted from Chinese characters—without a 4000-key keyboard. The national significance of finding a way to use the old language with the new keyboards should not be underestimated; this software was an important bridge between Japanese communication capabilities and the kinds of interfaces that now exist for computers.

As Dr. Kanda explained, kanji writing, based on Chinese pictograms, has a pictorial meaning along with the layers of abstractions that have been added to it. A character might represent a certain phoneme that could be part of a word for computer or canteloupe, but the visual representation of that character still might have a lingering visual reference to a rock or a person. The English alphabet is meaningless on the iconic level. That means that Japanese and English readers use different parts of their brains when they read a language. Reading Japanese characters requires more visual thinking than reading English alphabetic characters, which have no other layers of meaning beyond their strictly symbolic encoding. It made sense to me that this is one of the reasons visual communication is an important goal at ATR and NTT.

I ate lunch in the executive dining room with Dr. Kanda, Mr. Miyawaki, and Mr. Koichi Murakami, the manager of Fujitsu's "artificial reality" projects. It was one of those fancy silverware lunches with more waiters and wine stewards than diners; I realized that it was the first time I had used anything but chopsticks in two weeks. I remember thinking that the American equivalents of these people—the research managers of ATT or IBM or Apple, Disney or Media Lab—would never think of inviting a writer of popular books about the future of technology to a scene like this. The future is not as interesting to most American managers, and academics or pundits aren't always welcomed in the paneled chambers of decision-makers in American businesses.

The reason for their elaborate greeting ceremony became clear when all the Fujitsu people brought their well-thumbed copies of my book to be signed. I was soon to discover that Fujitsu research directors not only had read *Tools for Thought*, they had used chapters as part of their blueprint for future research. The book had sold an extremely modest number of copies in North America. Books about technology, I discovered, have an active audience in Japan, especially to technology managers. I had served as an unwitting bibliographer for Fujitsu R & D when they started thinking about formulating a vision for future computer technologies; it wasn't so much that my own ideas directly shaped their thinking, but the largely unnoticed American computer pioneers I introduced in the book gave Japanese technologists a guide to their customary first step into any new territory—finding out what the best thinkers in the rest of the world have done. Since they regarded me as an authority simply because I had pointed out the significance of Licklider, Engelbart, ARC, PARC, and Atari Research, I delivered a streamlined, lunch-table version of the speech I had been evolving as I traveled from one research laboratory to another—about virtual reality being too big for companies and too big for countries. Then we went back to the meeting room. They didn't have any directly VR-related technology to show me yet, but they did have some detailed plans, and one intriguing enabling technology—a neural-network-based "world model" and "user model" that "learn" how to present information to individual users.

In comparison with the prototypes I had tested elsewhere in Japan, and the laboratories under construction I had visited in the United States, the most interesting thing about Fujitsu's VR research as of March, 1990, was not what they had to show at that time, or the equipment they had assembled, but what they planned to do and the way they planned to do it. To my knowledge, the VR-related projects at ATT and IBM and DEC are still small-scale efforts, championed by a few true believers off in various corners, although the DEC project seems to be growing at the moment. Wonderful work is conducted at UNC and NASA, but the US government is hardly aware of these efforts. At ATR, NTT, MITI, and Fujitsu, however, VR research is integrated into the high-level plans Japanese industries have for future computer interfaces, communication media, and entertainment products. The key distinction between VR in Japan and the United States, as I see it, is that VR is integrated into Japanese industrial policy, and the United States does not have an industrial policy.

FROM NEURO-DRUMMERS TO FIRST-PERSON FANTASIES

Music has highly complex theories of harmony and chord progression. . . . Many researchers have tried to process musical information in terms of rule-based systems using these theories. Music, however, has a close bearing on human sentience.

This factor leads us to a seeming dead end. Tastes in music vary from person to person and there is no theory that explains such variation. With neural networks, attempts have been made to simulate intuition and sentience. We attempted to use a neural network for the sentient processing of music . . . and focused on drum playing. We think that rhythm is closely related to human sentience, and that it is easier to process rhythm than harmony. In our work, we planned to have a computer hold a musical dialog with a person using drums. The drummer plays the drums, and the computer replies to that rhythm. The computer's reply must not be random, but must be a rhythm pattern satisfactory as a reply. Judgement of whether the reply is satisfactory varies from person to person. The computer is thus taught the taste of a particular person so that it can reply with rhythm patterns which reflect that taste.

MASAKO NISHIJIMA AND YUJI KIJIMA,
"The Neuro-Drummer," 1989

Ms. Masako Nishijima made the first presentation about something that seemed at first to have only peripheral relevance to VR—Fujitsu's "neuro-drummer" product, which used neural net technology to create an artificial "partner" for a human drummer. When later presentations showed how neural nets were used as part of the "world models" and "user models" for planned Fujitsu VR entertainment systems, I began to see the direct relevance of neuro-drummer to cyberspace fantasy systems. A neural network is a kind of computer program that gains expertise at a task by testing its own performance against human-specified criteria over a number of trials, and thus, in a limited but powerful sense, the program "learns" how to perform that task by trial and error.

Neural network software is yet another technology that was originally a spinoff from artificial intelligence research but now appears to have useful applications in building personal VR systems. These programs do not operate like "rule-based" AI systems in which rulelike chunks of knowledge ("rooks move in straight lines," "two plus two equals four") simulate human reasoning processes. Rather, these programs are set up as self-modifying networks of connections, similar to the way some "neural nets" in the human brain process information. Neural nets in brains and computers do not come equipped with sets of predetermined rules, but with a network of weighted relationships between input and output variables; the weights of the relationships can be adjusted after trying a set of inputs, and thus the system can begin to "learn" how to emulate a human voice or solve mathematical problems. Neural networks are "trained" in specific ways to accomplish computational tasks that are often difficult if not intractable by normal computer programming methods—particularly pattern recognition and other attempts to emulate elements of human perception and cognition.

There was a reason Fujitsu was showing me a neural network in response to my interest in VR. The "responsive environments" pioneered by Myron Krueger twenty years ago, the full-sense fantasies prophesied by Aldous Huxley and patented by Morton Heilig, the "knowledge agents" that Negroponte's colleagues have pursued for the past decade at MIT's Media Lab, and the "Visual, Intelligent, & Personal" interfaces under construction at NTT today, all point toward future personal information systems that are very closely tuned to the perceptual, cognitive, and physical characteristics of each individual user. Cyberspace might not be just an empty space. It might have the potential to learn what you need from it, and how you prefer to communicate. In order to give computers power to communicate with us, some user-interface designers are giving computer systems little chunks of "personality" that learn about our quirks and preferences in limited but strategic domains. This is another critical inflection point: the artificial personalities that interface builders are planning as our "knowledge agents" and the artificial personalities that entertainment vendors are planning as cyberfantasies are converging on the general question of responsive environments.

Instead of trying to create a user interface that is universally usable—perhaps an impossible task—why not explore the idea of building human interfaces that learn how to fit in with each individual's intellectual and attentional style? Enabling technologies seem to be making the other elements of early visions possible now, so it isn't surprising that old dreams of dream-makers have revived. The "first-person experiences" that Brenda Laurel and her infonautic friends at Atari sought to build in the early 1980s were originally conceived as 3D graphics worlds with artificial characters that respond to each individual human user like an actor in an improvisation exercise, supported by rule-based systems that included dramatic expertise based on Aristotle's theories of drama. They were taking a stab at foreseeing the kind of computer-based knowledge representation schemes first-person fantasies might involve. Neural networks weren't important then, but now that they are, it becomes clear that deeply personal systems presuppose a system capable of learning intuitively, by trial-and-adjustment, the way neural nets are now known to do quite well.

Music is a hybrid of mathematical relationships and aesthetics, prediction and intuition, and thus offers a highly tunable instrument for exploring human-computer duets. While sound can be synthesized easily, and patterns of sound can be recorded with great accuracy, the truly difficult tasks for computers are related to the fuzzy evaluations involved with harmony and rhythm. Fujitsu researchers are interested in whether computers can learn to be aesthetic partners—can a human musician play a pleasing, improvisational duet with a computer? Based on well-known ideas, with next-generation technologies, the Fujitsu effort is aiming to make the visions of yesterday's cyberprophets a reality. Myron Krueger's work is fundamental, but his experiments were based on much earlier generations of computer technology. There's some ongoing research into musical partnership at Media Lab and elsewhere.

Fujitsu's neuro-drummer is a synthetic drumming partner that takes its instructions from a human partner via neural network software. The input is a rhythm pattern played by a human drummer. The output is a rhythm pattern played by a neuro-drummer. The human drummer can adjust the weights of the network by selecting among the outputs offered by the computer. The drummer riffs. The computer riffs back. The drummer replays the computer's riff the way he or she personally would have done it. Then the human drummer plays a different riff and the neuro-drummer ventures a reply, based on all it has "learned" to that point.

Listening to the neuro-drummer play after one, five, twenty, and forty learning trials was like listening to a baby learn to talk in ten minutes; neuro-drummer is a quick study. The specific rhythmic tastes of an individual human constitute the language this silicon novice was learning to speak. The machine and the human conversed, and from that conversation, the machine began to mold itself to the shape of the human's mind. If a different human drummer wanted to create a drumming partner, it would be necessary to train a new network.

Neuro-drummer is incorporated in a consumer product recently introduced by Fujitsu, the subject of much discussion in Japanese and American multimedia circles: the "fmTOWNS" machine offers a personal computer with the latest, most powerful digital processors, special sound and video chips, and a built-in CD drive that made it possible to pop in encyclopedias and video libraries. Apparently, fmTOWNS is the elementary module of the personal interactive fantasy systems Fujitsu intends to build—the "front end" to a more elaborate, powerful system that is under construction.

My mental gears shifted when yet another technocrat in a three-piece suit walked up to the overhead projector and put on a transparency with two quotations, both familiar, but surprising in their juxtaposition, in standard drop-shadowed Macintosh presentation software style:

Illusion—Seeing Unseen Worlds

The screen is a window through which one sees a virtual world. The challenge is to make that world look real, act real, sound real, feel real. [Sutherland, '65]

See me, feel me, touch me, heal me. [The Who, '69]

That first screen woke me up. Mr. Murakami seemed to be thinking the same thoughts I had been thinking after talking to VR enthusiasts. But I'm a writer and Mr. Murakami is a senior researcher for Fujitsu. Mr. Murakami declared that it was now possible to begin creating the kind of magical windows described by Ivan Sutherland, and to create the kind of new mass-entertainment medium envisioned by The Who. There was no mention of Aldous Huxley, but it was clear that Murakami was saying that the era of the "feelies" was on the horizon. In terms of enabling technologies, I agreed that it is time to start the R & D that might lead to true full sensory virtual experiences twenty years from now. The difference between Mr. Murakami and the earlier effort at Atari Research Laboratories was that Atari Corporation itself was an out-of-nowhere instant company that the founder-entrepreneur had sold to Warners, another global entertainment giant, whose managers were impressed by the sudden cashflow from video games (more money in the year of Pac-Man than Hollywood and Las Vegas combined) but unschooled in the process of technological research and development. Fujitsu, however, has been around for a long time, and when they decide to back a line of research, it tends to be supported for the five, ten, twenty years it takes. Ask IBM about Fujitsu and the supercomputer business.

The next screens Mr. Murakami displayed, outlining Fujitsu's VR strategy, were cause for further wonderment. It is one thing to understand the way Japanese researchers seize upon American innovations that are often overlooked by American industries; it is another to find out that one is part of the medium by which the process occurs. *Tools for Thought* had a chapter, "Brenda and the Future Squad," which referred to the elements of the future interactive entertainment media Brenda Laurel and her colleagues at Atari Research had planned. In the seven years since then, Laurel has expanded the research that began with her college thesis into a book, *Computers as Theatre*. The main direction of her research into what it takes to create an artificial world that "pushes back at you" is described by the title of her book: she sees the ancient arts and philosophies of drama as the source for future inspirations about human-computer interactions and computer-mediated human interactions. Interacting with a truly intelligent interface is a dramatic act, not strictly a perceptual or logical one, she believes. Laurel's Ph.D. thesis was a hybrid drama–computer science project that demonstrated how interactive media such as video games and computer adventure games are evolving toward the kind of dramatic interaction described by Aristotle in his *Poetics*.

Laurel was the one who first pointed me toward the past, to the origins of drama, as a place to find the framework for making sense of the future. The machinations of imagineers in Anaheim and Kawasaki are only the latest, most technology-intensive, perhaps least self-aware manifestation of what may be a very old cultural tradition indeed. Fun might be more serious than previously thought, and the connection between ecstasy and cyberspace might be older than anyone suspected. The future of cyberspace might be decodable only by those who understand its past.

DIONYSIAN CYBERSPACES AND THE BIRTH OF THEATER

There was a metal bowl associated with the initiation which has been mathematically reconstructed. The concavity of the bowl was such that a young man looking in expecting to see his own face, would see instead the face of an old man, or the mask of an old man held up by the candidate. The shock of realization, the death and old age within youth, represents an opening of the mind to a logic dimension of his own existence. Not becoming fixed to this particular moment of life, the initiate is awakened to the whole course of life. Out of that there will be associated restructurings as to the sense of it all. This kind of shock would not be experienced if the young man had been told by a friend who had gone through the mystery. That is why it was regarded as criminal to betray anything of the mysteries. Now, if things like this also were associated with a slight hallucinogenic situation in the mind, you can imagine what kind of illumination would come through.

JOSEPH CAMPBELL,
"Day of the Dead," 1988

The Festival of Dionysius was an annual event that celebrated the symbolic death and rebirth of the god, and hence nature. Several plays were commissioned for performance at each festival as contestants for a prize for the best drama. The theatrical people who were involved in the production of the plays (including actors, musicians, and costumers) maintained a strong connection to the Dionysian religion, eventually forming a guild whose head was usually a Dionysian priest.

Early Greek drama sprang from the intersection of philosophy, religion, and art. The occasion was ostensibly religious, and there is reason to believe that at least some of the actors felt themselves to be "in possession of the god" as they performed in the festival that honored him. The subjects chosen by the great tragic playwrights for theatrical representation at the festival were matters of serious import, depicting the evolution of Greek philosophy through their dramatic treatment of known myths and stories such as the tragedies of Agamemnon, Orestes, and Oedipus. They communicated philosophical and religious ideas as well as providing the occasion for the collective experience of emotion. Quite simply, Greek drama was the way that Greek culture publicly thought and felt about the most important issues of humanity, including ethics, morality, government, and religion. To call drama merely "entertainment" in this context is to miss most of the picture. The Greeks employed drama and theatre as tools for thought, in much the same ways that we envision employing them in the not-too-distant future.

BRENDA LAUREL,
Computers as Theatre, 1991

Young men, secretly chosen, were given the initiatory potion just before dawn. When the active ingredients started to kick in, their sponsors took them by the arms. Then they descended, literally, into the mysteries. Their guides led the novices down the ancient steps, to the esoteric dioramas arranged for them in dark chambers. In the Eastern Mediterranean, an underground *son et lumiere* ritual took place as part of the initiation of the mystery societies. These mysteries were associated with the cult of Dionysius, an old god of wilderness, wildness, dance, song, and intoxication, whose reign of influence shrank with the shrinking of the primordial forests. The French scholar Alain Danielou, in his book *Shiva and Dionysius,* presents evidence that the Mediterranean cults of Dionysius and the Asian devotees of Shiva were both part of the same phenomenon. It was part of an older culture of small, isolated farming villages that existed in the Mediterranean and in north India until the sword-wielding, horse-riding invaders swept out of the Caucasus to impose the Hellenic and Vedic ways of life on their conquered subjects. The central ritual of both Dionysian and Shivaite cults was the same: men and women would steal away from their homes in the dead of night and gather in clearings in the wilderness, there to partake of wine or cannabis, then dance, sing, and ceremonialize. The results were often orgiastic. Ritual cannabalism was sometimes involved. In the hills of Greece, these gatherings often took place in natural amphitheaters.

The Greek theater, now synonymous with the foundations of Western civilization, explicitly evolved from Dionysian revels. The annual festival of Dionysius, celebrating and reenacting the symbolic death and rebirth of the god, the earth, the soul, and the cosmos, became the tragedies of Aeschylus and comedies of Aristophanes. The Greek theaters in which the great dramas of Euripides were performed were shrines to Dionysius, and the original purpose of the theatrical productions that took place there was the same as that of the Dionysian and Shivaite revels—simulating the release of spiritual energies and summoning mystical illumination. The wilder audience-participation aspects died down when the tribal, animistic hill-villages were succeeded by the urbane, pagan city-states, but the core of the ritual was preserved, symbolically. Ultimately, theater evolved into the art of tragedy.

Aristotle proposed that the function of tragedy was to induce the emotional and spiritual state of *catharsis,* a release of deep feeling that originally had a strong connotation of purification of the senses and the soul. According to Aristotelian theory, the emotions of pity and terror are stimulated by the art of tragedy, and those fundamental emotional components of spiritual experience trigger the desired state. The method by which these emotions are evoked, *mimesis,* a combination of vicarious participation and suspension of disbelief, will turn up again later, when the mimetic component of future virtual realities enters the story. Mimesis, like literacy, is an abstraction, a method of packing dense experience into coded information. By imitating other things, ritual and drama serve to invoke mental and emotional states through the highly developed human power of association.

At some point, several thousand years ago, the older physiological intoxicants of plant infusions, song, dance, and sexual frenzy were replaced by new abstractions—dialogue and plot, props, masks and choruses—that simulated the original Dionysian ecstasies by arousing similar feelings in the audience. Indeed, the famous masks of Greek drama are now known to have been audio devices designed to amplify the sounds of the actors' voices in the open-air amphitheaters, which were also constructed for their acoustic properties; and devices such as cranes to suspend flying figures of the gods were used to create special three-dimensional effects.

The subthemes of the older ecstasies and newer Greek dramaturgy were the same questions, the kind of questions that could be addressed only in nonordinary states of awareness. What kind of creatures are we, who are descended from beasts but dream of gods? What is the core of human nature and how is it changing? In his *Poetics,* Aristotle stated that drama gives pleasure because it is an imitation of deep needs, feelings, and ideas and helps people understand the world; entertainment and knowledge come in the same package in Aristotelian drama. But the open rituals of the theater analyzed by Aristotle were a late development, an Apollonian-influenced strain of ecstasy—the tamed public version. Certain ceremonies remained secret, and were conducted underground, for smaller groups. Until recently, little was known about the actual contents of the mysteries celebrated for thousands of years at such sites as Eleusis. One thing that was known was that initiates were forbidden on pain of death from revealing what took place in the initiation rituals. A few artifacts survived, however, and the encoded symbology eventually was deciphered and linked to the initiation rituals.

In 1988, I attended a lecture by the mythologist Joseph Campbell. During this lecture, Campbell approached the mystery initiations by discussing the fundamental complementary elements of many cultural experiences: the ecstatic and the rational. The following quote is from a transcript of Campbell's lecture that was later published as an essay:

Nietzsche's *The Birth of Tragedy* points out that the two principles in art are, respectively, the Dionysian and the Apollonian. Dionysius represents dynamism which brings things into form and then shatters them. Apollo represents the illuminated mind giving form to this process.

The Apollonian principle causes you to become attached in your thinking and in your sentiments to the form, and delight in the unique qualities of a special moment, person, or thing. On the other hand, there are moments of Dionysian ecstasy when the delight is to see and feel and hear the form as it shatters and smashes. The sublime expression of power and force that shatters all things and brings forth all things is the two points of view. One ascends the dynamism and the other ascends the formal principle. But for the work of art you must have both. When one becomes too concentrated on the form and the thinking aspect of the living experience, it dies. The other aspect has to be there. You can't have life unless something is cracking up. Living your dying is a sense of that process, and it requires a wonderful balance. Nietzsche says that the balance between the Dionysian and Apollonian principles was what was achieved in the great tragedies that gives them their special quality. Together, the two give us the experience of our own living and life.

In contrast to the Apollonian principle, the Dionysian ecstasy has a quality of drunkenness. Nature brings out this throwing away of the binding and constraining principles and brings forth the joy and rapture of being in being. Dionysius is the lord of the grape and wine. He represents youth in its wonderful, energetic moment of high beauty, which was celebrated by the poets who sang the praises of the young athletes.

Campbell explained a series of close-up photographs of a hammered metal platter that depicted a sequence of enigmatic scenes. By cross-comparing the symbols depicted on the metal platter with other, more well-known artifacts, scholars were able to reconstruct the scenario for the Eleusinian mystery initiations. The purpose of the ritual, according to Campbell, was to awaken in the initiate an awareness of the balance between life and death, Dionysian and Apollonian forces, that was presented in milder form in the public tragedies. At the beginning of the initiation sequence, the initiate passes by a priestess holding a grail of blood, a reminder of the Dionysian allegory of sacrifice and trans-formation, the death of the self that must precede the rebirth of the spirit. Accompanying the priestess is the figure of Orpheus. At the entrance to a temple is a figure holding a basket, from which the initiate takes a pine cone, the seeds of which are reminders that the temporary form of the body is not as important as the life for which the body acts as a mortal vehicle. The initiate, holding the pine cone, continues with a raven, symbol of death, riding on his shoulder. Further symbolic sequences introduce the twin goddesses Demeter and Persephone, rulers of rebirth and the descent into the underworld. It is the descent into the underworld, the realm of Pluto, lord of death and transfor-mation, that takes the initiate into a subterranean chamber in which a virtual reality technology awaits.

The metal bowl described by Campbell in the quote at the begin-ning of this section was a tool for creating a lifelike visual illusion. By seeing the face of an old man in place of his own, the initiate is shocked into an awakening of his true nature as the mortal vehicle of an immortal spirit. A virtual reality is used to attract the attention of the initiate to a deeper reality underlying the appearances of the mundane world—a reality that differs not in the form or matter or energy of its manifestation, but in the way the initiate is conscious of it. As William James wrote in *The Varieties of Religious Experience*, "Our normal waking consciousness is but one special type of con-sciousness whilst all about parted from it by the filmiest of screens there lie potential forms of consciousness entirely different. We may go through life without suspecting their existence, but apply the requisite stimulus and they are there in all their completeness." Other scholars have presented evidence that this shock, the "requisite stimulus" of the Paleolithic shamans and Eleusinian mystics, was am-plified by the use of hallucinogenic substances, possibly lysergamides derived from ergot, the source material for the modern halluc-inogen LSD.

It is possible that Plato was alluding to these initiations in his Allegory of the Cave from *The Republic*, in which he used the idea of a kind of cave-theater as a way of discussing his theories regarding the nature of reality:

"Imagine mankind as dwelling in an underground cave . . . imagine a low wall has been built, as puppet showmen have screens in front of their people over which they work their puppets. . . . See, then, bearers carrying along this wall all sorts of articles which they hold projecting above the wall, statues of men and other living things, made of stone or wood and all kinds of stuff." The objects and beings we see in this world, Plato asserted, are like shadows projected by the "ideal" objects and beings of which they are merely temporary instances. There is a real world in the Platonic cosmos, but we never see it directly. We see an illusion based on a reality, a virtual world.

One can only speculate about the content of the knowledge imparted in the Paleolithic virtual reality chambers (and the notion of where it all came from will become increasingly important in the final chapter, when the story gets around to looking at where it is all going in the future). More is known about the knowledge at the heart of the mystery religions—knowledge related to the birth of the new posttribal, equally Apollonian and Dionysian kind of person who would eventually become the "citizen" of Western civilizations. The "posttribal encyclopedia" transmitted in the mystery ceremonies can be seen as a kind of map of where humans originated and where we are going, together with a strong vision of what human beings really are: spiritual beings who have evolved from and are still embodied in animal forms. The rituals, in all their different cultural forms, share the core message that we are creatures in the midst of a process of change, and that this capability for change, learning, and growth is itself an important part of human nature.

Theater is a psychological process that uses language, rhythm, voice, myth, and perception-altering technologies to achieve a specific state of mind in the audience. Aristotle said *catharsis* was a healthy and necessary way for people to deal with the great themes of life and death. The means by which catharsis is induced in an audience, mimesis, is the emotional simulation capability that enable humans to empathize with the actors onstage and internalize the dramas presented before their eyes and ears after the curtain goes up. Understanding the human components of mimesis might become one of the foundations of the emerging psychological investigation of virtual experience. A challenge as formidable as discovering the nature of the human sense of mimesis is the thorny question of how to get the mimesis into a machine. Thorny questions tend to be very valuable on the frontiers of computer science when they furnish the "driving problems" that Frederick Brooks believes to be a powerful engine of scientific progress. The challenge of building autonomous characters in cyberspace might be the driving problem that supercharges the application of VR technology to entertainment products. It is not unlikely that future scientific and technological breakthroughs could be stimulated by spinoffs from the entertainment industry's R & D. The development of artificial theories of drama that can be encoded in algorithms, for example, might be driven by the desire to create "artificial characters" for future forms of entertainment.

In the early 1980s, one of the most exciting AI technologies was the "expert system" or "rule-based system" that captured a portion of human expertise in a set of rules that was "extracted" from human experts through a question-and-answer process. By using such systems to augment one's limited knowledge of a field such as medical diagnosis or geological prospecting or laying out the architecture of microchips or configuring telecommunications networks (all successful application areas for expert systems), many people have been able to amplify their ability to make judgments about specific problems. One of the most successful early expert systems, MYCIN, contained a data base of information about a certain class of diseases and a set of rules that was contributed through interviews with diagnosticians who discussed the criteria they used for making a diagnosis on the basis of many different sets of symptoms. Another diagnostician who is not expert in MYCIN's class of diseases could query the system regarding the symptoms of a particular case and use the system's "knowledge base" to help make a diagnosis.

Brenda Laurel and the other Atari-ites wanted to build a world you could walk into, with rocks that are hard to move, and heroes and villains and all the other characters that make fantasy fascinating. Techniques for controlling the appearance of computer-generated graphics require expertise and many hours of hard work, but at least we have an idea of how to do it. David Zeltzer and his students at Media Lab's Snake Pit are pursuing the difficult but not intractable problem of building autonomous 3D graphics characters, focusing on how they appear and how they move, but not paying attention to what they might say or how they might react to human communication. Building a sense of drama into a world, creating artificial characters that don't respond from a script, but improvise according to the rules of dramatic interaction, however, would require some kind of artificial intelligence beyond the capabilities of present technology. How, then, might such a future technology develop from what we know today? Laurel proposed in her thesis that an expert system could be applied to this problem. The expertise would consist of the rules laid out by Aristotle in *Poetics*, and the knowledge base would contain information about the artificial world and about the user-player's reactions to it.

But rule-based systems have turned out to be only part of the answer. Several different scientific breakthroughs in the past few years have shown the possibilities of fitting other programming approaches into a system that might overcome the deficiencies of rule-based systems. Chaos theory and the study of complex systems have revealed that the right set of very simple rules, applied repeatedly to the same information, can generate highly complex behavior. And neural networks have shown that it is possible to build computers that know very little but are capable of learning a lot. Perhaps the answer lies in a hybrid of rules and networks. Or perhaps the pursuit of such a hybrid will uncover another obstacle that must be overcome by yet another computer tool. Brenda Laurel, in a 1986 paper, "Interface as Mimesis," put forward a strong reason why people are bound to pursue fantasy amplifiers as soon as technology makes them possible:

> It is not enough to imitate life. Drama presents a methodology for designing worlds that are predisposed to enable significant and arresting kinds of actions—where characters make choices with clear causal connections to outcomes, where larger forces like ethics, fate, or serendipity form constellations of meaning that are only rarely afforded by the real world. Dramatically constructed worlds are controlled experiments, where the irrelevant is pruned away and the bare bones of human choice and situation are revealed through significant action. The predispositions of such worlds are embodied in the traits of their characters and the array of situations and forces embedded in their contexts. If we can make such worlds interactive, where a user's choices and actions can flow through the dramatic lens, then we will enable an exercise of the imagination, intellect, and spirit that is of an entirely new order.

The reason I found myself thinking about Brenda Laurel's ideas while I was sitting in a meeting room at Fujitsu was that Mr. Murakami invoked her name early in his presentation, when he started talking about the "highest level" of the "artificial reality" (AR) Fujitsu planned to create. Myron Krueger was mentioned, but Mr. Murakami was more interested in personal systems than responsive environments, at least for the immediate future. The three levels of AR, as Fujitsu sees it, consist of the *device* level, the *system* level, and the *art and application* level. The device level is where the DataGlove, EyePhone, DataSuit, and other input hardware reside. Future eye-tracking and gloveless gesture-sensing technologies will fit into this layer when their design is optimized. The hardware systems that link the human and the computer are controlled and mediated by a software system that includes, according to the Fujitsu model, computer graphics, speech and gesture recognition, world modeling, and user modeling capabilities. The highest level consists of the applications of such a system to operation of remote tools, to communication systems, and as a means of training people in various skills. The other half of the highest level is the "artistic" level, and artificial, first-person dramas—soap operas and action movies the user can enter and participate in—appear to be the most important component of Fujitsu's research in this area.

Fujitsu sees the art of VR as something worth paying attention to. The three application areas they are aiming for are entertainment (their viewgraph subtitled this "computer-based tools for a lot of funs [sic]"), CAD, and education. Each of these involves a key human element that Fujitsu is concentrating on now in their own laboratories and in research they support at Carnegie-Mellon University in Pittsburgh and elsewhere. The cyberspace system is designed as a growable collection of interlocking components that will come from convergent R & D efforts. The EyePhone, DataGlove, and fmTOWNS machine constitute the part of the present development system that the user sees, the "front end." The "back end," the powerful computers and software that work behind the scenes to maintain the reality level of the virtual world, includes a Pixel Machine, a graphics supercomputer optimized for fast 3D models, a SUN workstation, a neural-network-based back end to the DataGlove that can be taught gesture recognition, a real-time graphic rendering system, and a world model system. The user puts on the glove and repeats key gestures until the neural network interface binds the gestures to specific commands. If you feel like pinching your nose to fly, instead of pointing your finger, the interface can be trained to do that.

The fmTOWNS machine seems to be designed to be the personal portal to a much larger virtual world of software and services that Fujitsu plans to develop in the future. The back end in particular involves several key assaults on driving problems. The Pixel Machine will take care of real-time rendering, but the details of the virtual world's appearance and behavior will be mediated by a world model system that is also linked to the user by a "neurosimulator." The world model of the future will include today's standard physical model for the virtual environment and nonliving objects in it. For artificial characters in the virtual world, a model will consist of rule-based and neural network models for behavior. Something they call "scenario control" will also be involved in the back end. In other words, they want a world that you can walk around in, that will react to you appropriately, and that presents a narrative structure for you to experience.

One of Mr. Murakami's depictions of what Fujitsu planned to market to the global teenagers of the next century was the cartoonlike sketch of a young male described at the beginning of this chapter. This hypothetical VR entertainment consumer of the future appeared to be approximately twelve years old, with a futuristic HMD and a pair of futuristic gloves linked to a familiar-looking television set with two black boxes under the monitor. A thought-balloon emanating from the young man's head showed the same young man in a fierce primeval landscape. In front of the armor-clad hero, Godzilla approached, making menacing sounds and gestures. In the hands of the user-player-actor-hero of the drama was a glowing sword. Presumably, the Godzilla module of the world model knew to bleed and die when it was stuck in the right place, and to breathe fire when the actor-model let his guard down. And presumably the scenario module would be capable of learning that this user wants more Godzillas to fight than maidens to rescue, or prefers exploring outer space to engaging in combat with mythical video creatures. Perhaps, with an additional Nintendo edu-module, the same world will become a geometric puzzle or historical simulation with educational content to be mined through the user's explorations.

The attempt to create the "personalities" of artificial characters in cyberspace is taking place at Carnegie-Mellon University (CMU), one of the long-time bastions of AI and robotics research, by computer scientist Joseph Bates, based in large part on Brenda Laurel's ideas and Dr. Bates's previous work in AI. Dr. Laurel is a collaborator on the project. Bates knows how to build knowledge representation structures in LISP, Laurel knows the rules of dramatic interaction. "Project OZ" is the name for the CMU project, and the principal funding comes from Fujitsu Research and Development. I exchanged e-mail with Dr. Bates, which was appropriate for the present level of his project, which is demonstrable only as text on a screen. While others such as Zeltzer at MIT disregard questions of content in order to concentrate on building the form of a virtual character, Bates is using a text-only, content-oriented prototype, with the understanding that the "intelligence" of the system he and Dr. Laurel and their CMU colleagues are building eventually will be linked with animated characters in three-space. How these characters might act and respond is easily modeled and tested in text; and as anyone who has ever played a text-only adventure game can attest, ample mimesis can be triggered by interaction with a well-constructed narrative even in the form of words on a screen.

Indeed, a kind of text-based cyberspace phenomenon seems to have broken out all over the Internet. Because Internet nodes are connected with very high-speed data communication channels, thousands of computers all over the world are available to those who have the right passwords at their command. If I have the permission to use a computer in Oslo to create a little world model, I can log onto an Internet node in California, type onto my keyboard the proper incantation, and have the eerie knowledge that a light is blinking on a computer in Sweden, where my command has triggered a disk access operation. Because certain areas of the computer's memory can be set aside as a kind of three-dimensional puzzle, full of rooms and treasures and gremlins and swords, computer users have always built adventure games inside them. In the Internet era, however, the latest fad adventure cyberspaces are known as "MUDs." MUD stands for "multi-user dimension." A player logs onto a computer via a packet-switching network, and from that computer uses the right passwords and communication services to reach a computer that contains a MUD.

The first thing to do when you reach a MUD is to incarnate, to create a character. Usually, a welcome message instructs you on how to do that. There are guest characters for those who are just roaming through the network, checking out MUDs here and there. There are public MUDs, open to anybody with a modem and password access, and there are very private MUDs where the wizard has to decide who can enter or not. Long documents explaining the mores and rules, powers and economies, topographies and mythologies of various MUDs are available. Many MUDs are self-constructed, where newcomers gradually learn how to gain the power to actually build new rooms into the MUD. "Wizards," sometimes called "tinkers," are the system administrators who actually run the computer programs that maintain the MUDs. "Robots," often called "bots," are programs that run around a MUD and perform services. MUDs are the subjects of newsgroups, mailgroups, zines, and conventions. And when goggles and gloves and protocols for transmission of presence make it possible to jump right into a graphic MUD, there will be a population of thousands of sophisticated architects/players. While volunteers on the Internet with computer time on their hands build cyberspace from the ground up, and researchers at CMU and Media Lab search for ways to interact with artificial characters, Fujitsu seems to be positioning itself to take advantage of the research results when they come in.

Fujitsu owns Habitat, a kind of MUD where participants command cartoon representations of themselves and interact with one another via text messages in real time. It seems clear that they are planning to converge their telecommunication, immersion, and game technologies sometime in the future. Right now, the hardware and software prototypes for a fantasy-oriented cyberspace are being assembled in Kawasaki as a development test-bed for the different parts of the artificial reality system, with the assumption that when the hardware and software systems are working, they will be ready to integrate the findings of the work Bates, Laurel, and others have undertaken into computational theories of drama. Those theories of drama should be testable by trying to build the dramatic infrastructure of a do-it-yourself computer-based fantasy world. Simple systems are already running at CMU, and increasingly intelligent generations are in the development process. It will take months to get working prototypes, years to discover how to make them work, and more years to integrate with the input and output technologies. Considering how long it is taking to develop effective 3D cinema, the idea of 3D movies that play along with your actions in real time is not in the category of something we can expect to happen this year or next. If we all meet in the land of cyber-OZ, seven or seventeen or twenty-seven years from now, we'll know whether Fujitsu or Disney or somebody entirely new succeeded in building interactive fantasies in cyberspace. And then it won't matter, because we'll all be elsewhere, swallowed by a combination of mimesis and mesmerism.

Maybe we ought to think about where this is all leading, before we find that we've been led somewhere by the technology's economic imperative, the way we were swept away by commerical network television.

The last thing I heard about VR research in Japan was more than six months after my visits to Japan's R & D laboratories. Izumi Aizu sent me the following message via electronic mail:

> In the September 10th issue of the *Japan Economic Journal,* an article appeared on a virtual reality study commission currently being organized by Japan's Ministry of International Trade and Industry (MITI). Dr. Tachi, a MITI section director in charge of biorobotics and an assistant professor at Tokyo University, is to be the chairman of the newly formed virtual reality commission. The study commission's steering committee will be composed of two personnel from an annex of MITI, one from each of the Advanced Telecommunication Research Institute (ATR) and NTT's Human Interface Laboratory, and professors from several of Japan's leading universities. One of MITI's section chiefs and Professor Takemochi Ishi of Tokyo University will form the advisory board of the fledgling commission.
>
> The study group will meet monthly to share information gathered on international virtual reality projects and maintain open lines of communication with overseas researchers. For an annual fee of 300,000 yen, anyone may become a member of the group. Although there are no plans to design and develop virtual reality equipment

and software at this stage, the formation of the MITI-affiliated study commission is an indication of the growing interest in virtual reality in Japan.

The first study group meeting will be held on October 19, 1990. The session will open with a keynote speech entitled "The Social Impact of Virtual Reality" by Professor Ishi, followed by a presentation on the present and future of virtual reality. Mr. Tachi and various members of the steering committee will then lead a panel discussion. The meeting will conclude with a party. Free for members, nonmembers will be charged 50,000 yen to attend the half-day event.

From telemicrorobots to Aristotle-on-a-chip, the future of the VR effort in Japan might have profound effects on the rate and scale of growth of the VR industry worldwide. If VR devices and VR networks could be installed on a large scale, cyberspace might become more than just a new toy for those who can afford access to the latest high technology. But the rest of the world would do well to ask the Japanese consumer reality vendors the same question I posed: Is reality going to be what the communications industry calls an open system? Will citizens and competing companies be able to trade realities over the network, to exchange and barter and buy objects and worlds and points of view? In other words, will there be a reality marketplace? Or will there be a de facto monopoly built into the constraints on the kinds of realities accessible through the mass market gadgets, in which everybody will be forced to buy either Sony or Fujitsu brand reality?

The marvelous illusions I saw and heard in Mountain View and Tsukuba and Chapel Hill are not all there is to virtual reality, I discovered. Seeing is believing, and hearing something that syncs with what you are seeing is better yet, but feeling something with your hands entails a deep, visceral kind of *knowing* that is normally associated with only the most "rock-solid" realities. I got a taste of force-reflection feedback at UNC and in Dr. Iwata's laboratory in Tsukuba. In Cambridge, Massachusetts, however, I was gripped by the powerful illusions of texture. In Cranfield, England, I put my hands into the first step toward a device capable of transmitting the way something feels in your grasp, as well as the way it looks from your point of view. And in Grenoble, France, I played a virtual violin with a virtual bow. The feel of things to come.

Chapter Fourteen

THE FEEL OF THINGS

Touch is the oldest sense, and the most urgent. If a saber-toothed tiger is touching a paw to your shoulder, you need to know right away. Any first-time touch, or change in touch (from gentle to stinging, say), sends the brain into a flurry of activity. . . . When we touch something on purpose—our lover, the fender of a new car, the tongue of a penguin—we set in motion our complex web of touch receptors, making them fire by exposing them to a sensation, changing it, exposing them to another. The brain reads the firings and stop-firings like Morse code and registers smooth, raspy, cold.

. . . Research suggests that, though there are four main types of receptors, there are many others along a wide spectrum of response. After all, our palette of feelings through touch is more elaborate than just hot, cold, pain, and pressure. Many touch receptors combine to produce what we call a twinge. Consider all the varieties of pain, irritation, abrasion; all the textures of lick, pat, wipe, fondle, knead; all the prickling, bruising, tingling, brushing, scratching, banging, fumbling, kissing, nudging. Chalking your hands before you climb onto uneven parallel bars. A plunge into an icy farm pond on a summer day when the air temperature and body temperature are the same. The feel of a sweat bee delicately licking moist beads from your ankle. Reaching blindfolded into a bowl of Jell-O as part of a club initiation. Pulling a foot out of the mud. The squish of wet sand between the toes. Pressing on an angel food cake. The near-orgasmic caravan of pleasure, shiver, pain, and relief that we call a back scratch. . . .

DIANE ACKERMAN,
A Natural History of the Senses, 1990

"What does it feel like?" Margaret Minsky was sitting in front of her workstation in the Snake Pit. I was sitting beside her, my hand on a small joystick that looked like the control device for a video game. Think of a metal chopstick with a Ping-Pong ball on top, like an automobile's gearshift. The bottom of the chopstick protrudes from the top of a metal shoebox, and the shoebox is plugged into the computer. Like the ARM, this gadget pushes back. I stirred the joystick around. I could move it forward, back, to the side, "shift" it, "stir" it. The rod slid easily in some directions, resisted strongly in others, and strong resistance often collapsed suddenly, as if an obstacle had moved out of the way.

"Slippery, hard chunks of something fairly light, packed loosely," I reported.

"That's virtual ice in a virtual bucket." She adjusted a slider control on the screen: "And this is virtual ice and molasses." I felt the molasses, instantly, in my hand. I'll always remember that as a particularly weird moment in my personal history of reality.

I gooped the joystick around in the virtual substance that suddenly had gained viscosity, in addition to the feeling of jostling a multitude of smooth, irregular, rigid objects packed into a container. It was hard to think of the invisible substance at the other end of the joystick as anything but ice in molasses. *Where* the end of the joystick was, precisely, was a question that kept hitting my awareness like a koan knocking on the door, asking for admittance. If I put my attention in the shoebox part of the apparatus, it did indeed feel as if there was a bucket of ice cubes directly under the shaft of the joystick. When I tried to move the shaft, the joystick resisted and slipped with just the right degree of force, in just the right relationship to my own actions, to convey the impression that I was moving the physical object through a physical substance.

"This is sandpaper," Minsky announced, summoning a patch of black and white texture to her display screen that, yes, looked like a television close-up of a piece of sandpaper. I moved my chair closer to her monitor; she moved the joystick in front of me. When I pushed the joystick, a cursor moved across the patch of texture on the screen. At the same time, a pattern of slightly vibratory microforces, felt by my fingertips and wristbones through the joystick, conveyed to me the strong impression that the joystick was really a pencil and the end of the pencil was scratching a piece of sandpaper. I could feel that sandpaper out at the end of the joystick the way you can feel real sandpaper at the other end of a physical pencil or a screw at the end of a screwdriver.

"Where is a virtual object located in physical space" is a quasi-philosophical question that gets mind-itchingly concrete when you literally grasp something in your hand. The question of the location of the virtual object became more acute. I could look at the screen and watch the cursor move across the 2D black-and-white simulation of sandpaper. Watching the cursor caused the apparent location of the virtual sandpaper to switch to the surface of the screen. If I diverted my visual attention back to the shoebox, I could reconvince myself that the end of the joystick was there in the box, not at the cursor on the screen. But if I scratched my virtual pencil over a patch of pattern on the screen again, the apparent origin of the stimulus switched back to the screen. Locations are more fickle in cyberspace. In a way, haptic illusions like the ones Margaret Minsky demonstrated shook my reality sense far more than the cartoonlike visual worlds I had explored. That sensation of contacting something solid, with a recognizable texture, combined with something as visually crude as a black-and-white, flat-screen 2D display, hit that part of me that assumed, until that moment, that when I feel something with my fingertips, I'd better believe it.

Minsky showed me how to alter different parameters in the computer's texture simulation models and thus change the way the test patch of texture appeared on the screen, simultaneously altering the way the joystick reacted when I tried to move the cursor over it. There were several grades of sandpaper. Several black and white, smoothly shaded depictions of what looked like arrays of metal cylinders were arranged in a neat array on the screen, as well. It felt like running a pencil over an array of polished metal tubes packed close together, like organ pipes. There was even a fractal option: I could tune the parameters and create an irregular pattern on the screen that looked like a close-up of a piece of granite. The mathematical model of natural irregularity felt so lifelike it bordered on spooky—I could have sworn I was trying to write my name on a rock. Minsky smiled. She knew that the demo was the place to start. After one stirs virtual ice cubes and molasses with one's hand, and writes on a simulated rock, the theoretical part of the presentation seems less academic and more dramatic.

The Snake Pit is the informal name for the computer graphics laboratory in the basement of MIT's Media Lab in Cambridge, Massachusetts, the place that made "demo or die" a computer science cliché. Ten-foot-long green cloth snakes, the laboratory's media-age totem, were woven into the bundles of communication network cable strung uncosmetically across the ceiling. I kept one hand on the joystick while we talked, and with the other hand instructed the computer to change my sensations from sandpaper to molasses and back again. Tactile doodling. It was almost hypnotic, the way her ideas about simulated textures made deep-down sense while I was trying to sign my name on a piece of virtual glass.

I had missed her by a few weeks when I visited Chapel Hill. Minsky had returned to Cambridge in order to finish writing her doctoral thesis about the research she had conducted at UNC. Jaron Lanier told me years ago that she had been one of the early backers of his effort to use a glove to create a visual programming language. And I had seen her presentation at SIGGRAPH 1989, including that video of an extremely realistic architectural walkthrough of Sitterson Hall I experienced directly at UNC. So I sent Minsky e-mail before I went on tour and she told me to drop by to try "virtual sandpaper." I had no doubt, after the brief demonstration, that she was onto something. The problem of how one adapts the joystick's technology to the task of feeling virtual worlds with your hands in useful ways is an engineering problem, and a hard one. But engineering is based on solid science, and sometimes scientists cause engineers to solve hard problems in order to build instruments for extending scientific knowledge. Minsky was using a cleverly engineered device to probe the core scientific problems involved with simulating the feel of reality.

Margaret Minsky came into the VR act, and into the realm of haptic feedback, from several interesting directions at once. She has grown up among computer terminals and computer programmers in the exact center of the computer science world because her father is Marvin Minsky, one of the founders of the branch of computer science known as "artificial intelligence" (AI). Her Ph.D. advisers are Nicholas Negroponte and Frederick Brooks, two of the deans of contemporary computer science. Margaret is a computer scientist, but she is from the Atari generation rather than the mainframe generation, and games, not military or business concerns, were her early driving problems. Indeed, her first collaborators in the joystick realm were young inventors and programmers from Atari's Cambridge laboratory in the early 1980s. And the texture-sensing joystick I held in my hand like a ballpoint pen, unlike UNC's equipment, traced its lineage not to remote manipulators for handling radioactive materials, but to the force-reflecting steering wheel of "Hard Drivin,' " Atari's commercial driving-simulation arcade game.

"You want to play Hard Drivin'?" Minsky asked, after she showed me her basic demos. "That game and my research have a few things in common," she added, as we left the Media Lab with a few Snake Pit folks. Atari had donated a game to the Boston Computer Museum, and after a telephone call from Minsky, they agreed to open up the part of the museum that was closed that day to let us do some hands-on research. We talked about the genealogy of joysticks on the way across the Charles River. Fall was beginning to happen; the air was still hazy and warm, but the trees were turning red and gold. The Boston Computer Museum is located very close to the spot where the Tea Party happened.

Minsky had chosen the sensation of texture as her first driving problem in pursuit of virtual tactility. "I didn't think that shape was a good place to start," she explained. "It's a compound of bulk and surface properties that aren't easy to split apart—properties like texture, elasticity, and viscosity."

Minsky's first exposure to joysticks that push back was in the early 1980s. At the same time that other infonauts were dreaming about the distant 1990s in Atari's California laboratory, Atari had been funding another small research laboratory in Cambridge, Massachusetts. Max Behensky, who Minsky knew from MIT, was working with a consultant by the name of Doug Milliken to build a two-degree-of-freedom joystick with force-reflecting motors connected to the joystick via gears. Minsky was around when Behensky and Milliken hooked up their first working model to a powerful computer graphics workstation. "One of the lessons of that first model was that it was exciting to feel continuous forces like inertia and vicosity." That feeling of excitement was Behensky and Milliken's goal. The Atari lab was trying to develop an arcade game that would extract change from people's pockets by providing a viscerally exciting automobile simulator with a steering wheel that responds the way a real steering wheel does. It took years for them to come up with a manufacturable product, but when they did, "Hard Drivin' " turned out to be a hit for Atari.

Minsky and I wandered around the Boston Computer Museum, conversing among the displays of such cybernetic antiques as Altairs and Altos, after taking our turns at the wheel. I sat in a driver's seat, with a standard steering wheel, gearshift, and foot pedals. The wide-field-of-view screen and high-fidelity sound create the standard video-game low-level illusion of looking through the windshield of a vehicle moving through toon-land. It wasn't as 3D and immersive as Sensorama, but the steering wheel was the convincer. The way I could almost feel the asphalt in the road when I maneuvered the steering wheel is the reason for the game's popularity. The way a physical automobile responds to different kinds of motions at different speeds on different kinds of roads is modeled and transmitted with striking verismilitude through the force-reflecting steering wheel.

When she started out as a grad student at Media Lab, Margaret Minsky was working with others on Alan Kay's "Vivarium," a long-term research project that involved building artificial creatures that children could experiment with. Minsky wanted to build a joystick that would help children experience something of the artificial creatures' point of view. So she asked her old friends from Atari to build a device with the same basic design as their first joystick. "A graduate student named Megan Smith designed a version that used motors and brakes together," Minsky recalled. Minsky and Smith were joined by another mechanical engineering graduate student, Massimo Russo, and together they built a three-degree-of-freedom prototype. By that time it was the late 1980s, and word of Ming Ouh-Young's work at UNC was getting around. "Frederick Brooks was always a leader in haptic research," said Minsky. "Batter and Brooks used their two-degree-of-freedom joystick in 1972 to see if physics students could use them to sense simple fields—truly landmark research." And then there was the ARM. So Minsky moved to Chapel Hill for a while and teamed up with Oliver Steele, a UNC undergrad, to design texture-simulating software for the 3-DOF joystick.

How do you fake solidity or smoothness accurately enough to feel with your fingers? The key to 3D graphics lies in certain tricks that have been discovered over years of research into visual display apparatus—the use of shading to simulate light sources, stereographic imaging to simulate binocular parallax, head-tracking to simulate motion parallax—but Minsky and her colleagues didn't have decades of research into haptic illusions to guide them, the way visual illusion-builders do. For that reason, one of her goals for her Ph.D. project was not only to demonstrate a method for simulating texture, but to create a personal computer-based workbench for perceptual psychologists to study the basics of haptic perception.

The way in which the UNC team simulated texture takes advantage of a haptic illusion that you can simulate as a thought experiment. You'll need a pencil and a ridged surface, like a piece of corrugated paper or the edge of a deck of cards. Run the pencil over the ridges and concentrate on the feeling in your fingertips. Now imagine a highly magnified schematic cross section of the ridged surface, depicted as a series of "bumps"—"hills" and "valleys." Now imagine the way the pencil seems to resist your efforts to push it up the leading edge of a ridge—the part where the pencil point is climbing up the hill in your mental schematic diagram. When the pencil gets to the top of one ridge/hill, there is a moment of equilibrium, and then the resistance seems to fall away quickly as the pencil point slides into the valley before landing between this ridge and the next one.

One approach to simulating texture would be to simulate the way the point of a probe changes the way it resists finger movements as it moves over a surface that is composed . . . for the sake of simplicity— of bumps and valleys of various sizes and slopes. The bumps and valleys can be depicted graphically. Think of a regular sine wave, again for the sake of simplicity, wiggling from the left to the right side of your mental blackboard. Now imagine a straight line drawn horizontally above the wavy line depicting the bumps. If you were to draw straight vertical lines down from different points on the horizontal line, you could imagine that each of these vertical lines represented the amount of force exerted by a virtual spring at that position, either resisting strongly as the pencil tip climbs a bump or resisting very little as it slides down into a valley. If the motion of the pencil tip can be modeled in the computer, and the cross section of the texture can be modeled, then the output of that model could determine the amount of power to give to tiny motors that push back against the joystick.

The interesting thing about using motors to produce forces that simulate the actions of springs is that the reactions of springs can be described mathematically. In fact, the properties of any physical system in motion can be simulated, in principle, from the proper combination of three different kinds of idealized objects: springs, dampers, and masses. Therefore, the local position information generated by the joystick, combined with the information about that position in the world model, can be plugged into an equation that will create the sense of virtual springs.

When we got back to the lab from the Computer Museum, Minsky enabled me to feel what she had been talking about in regard to springs and dampers and masses. I used the joystick again, and this time the simulation gave me the feeling that I was moving a lever that had a weight on the end of a spring attached to it. I had a strong sense of a virtual weight flying around in the air. Then I found I could change the mass of the flying weight or even change the air it was flying through to water or something thicker. By adjusting on-screen "slider" controls, I was able to test the springs, dampers, and masses together, and get a sense for myself how the feel of textures can be closely mimicked by creating just the right mixture of these idealized forces. I could fiddle with the damping forces and feel the tip of the probe moving through an increasingly viscous fluid. The difference between molasses and ice, bricks on a smooth slope and bricks on a rough slope, a lasso and a spring that is trying to move your hand around is simply a matter of the "recipe" you choose.

The simulation of virtual springs, dampers, and masses—an approach that was developed in parallel by a group in Grenoble, France—might end up as a false lead on the quest for lifelike haptic simulation, or it might be the Rosetta Stone of haptic illusions. Minsky and her colleagues noticed something else, besides the relationship between physical ideal forces and the elements of haptic textural illusions: As researchers into tactile perception had suggested, people recognize objects through their fingertips by making a series of stereotyped actions. Just as we unconsciously gauge the distance and depth of a visual stimulus by moving our head and eyes around to see it from different angles, we determine the feel of the surface of an object by rubbing our hands back and forth across it—or move a transducer like a pencil or a joystick back and forth across the virtual surface. Haptic perception, like vision, is not a purely passive process, but involves active exploration.

The tactile component of haptic perception—the relatively low-energy surface-level perceptions concentrated on the fingertips—is the subject of another convergence of interests. Just as there are researchers in parallel computing architectures looking for the right kind of processing problem, and 3D graphicists looking for position-sensing technology, I felt certain that there were roboticists who were thinking about tactile perception from the machine side. I met one, a few

months after Margaret Minsky started me thinking about the future
of fingertip virtuality. M. A. Srinivasan and I spent five days together
at a scientific gathering in Santa Barbara, California, together with
several dozen other unlikely colleagues, who were discovering that
virtual reality was the place where optics converges with robotics, cognitive science gets down and dirty with biomechanical engineering,
computer graphics programmers confer with control theorists, and
Bell Laboratories meets Disney and Lucasfilms. Srinivasan, who was
based at MIT's Neuman Laboratory for Biomechanics and Human
Rehabilitation, briefed the rest of us on the fundamentals of tactile
perception. I brought some of his reprints home with me and discovered that his written report, "Tactile Sensing in Humans and Robots,"
offers a compact, only slightly technical, almost poetic description of
the informaton flow involved with surface touch:

> Human tactile perception is the culmination of a series of events.
> When the compliant skin comes in contact with an object, its surface
> conforms to the surface of the object within the regions of contact.
> The associated distortions inside the skin and its substrates cause
> mechanosensitive nerve terminals embedded within to respond with
> electrical impulses. While each impulse is almost identical to another
> (50 to 100mV magnitude, about 1 ms duration), the frequency of
> impulses (up to a maximum of about 500/s) emitted by each me-
> chanoreceptor depends mainly on the intensity of the particular com-
> bination of the stresses and strains in the local neighborhood of the
> receptor to which it is responsive. Since these stress and strain fields
> within the skin are directly dependent on the mechanical stimulus at
> the skin surface, the response of the population of receptors repre-
> sents a spatio-temporal code for the applied stimulus. This code is
> conveyed through peripheral nerve fibers to the network of neurons
> in the central nervous system, where appropriate processing enables
> us to infer the surface features of the objects in the contact areas and
> the type of contact by touch alone.

In other words, the pressure sensors embedded in your skin play
only one note, but each sensor can play that same note over and over
again, very rapidly under strong contact or less frequently under more
gentle pressure. (Habituation can occur, in which the sensor will stop
responding after a stimulus has stayed steady long enough—the reason
you don't feel your shoes unless they are too tight or you think about
them.) Higher-level processors look at the differences in stimulation
at adjacent areas of the skin and gain motion information about stimuli.
One technical fact in Srinivasan's statement above contains an important clue for haptic research. One of the people I know and respect

from the robotics research community noted, when I mentioned tactile VR, that any kind of tactile display system is going to have to achieve a frequency of about 500 cycles per second in order to match human tactile acuity. Roboticists would love to have mechanical hands capable of feeling surface features of materials. Even if you are able to create thousands of tiny vibrators per square inch of skin, each one vibrating at speeds up to 500 cycles per second, all those cycles and all those sensors sending data back to the reality engine add up to a monstrous computational problem. When I talked about the upper ranges of fingertip tactile sensitivity—the normal human ability to discriminate between tissue and bond by holding a piece of paper between thumb and forefinger, for example—he mentioned that surprisingly small features on smooth surfaces can be detected by humans. What we call "smoothness" is a kind of biological signal filter applied to a network of exquisitely fine-tuned receptors.

We learn about the world and create an internal data base that serves as a matching template for the patterns of impulses received from our peripheral sensors. The on-off coding of mechanoreceptor signals that are triggered by feeling a soapy windowpane or a piece of bark is matched by knowledge in our experiential data base. As Srinivasan explained it to me, the way our mechanoreceptors and nervous systems work together to detect such qualities as "slip" on a surface offers clues to the way "microfeatures" of surfaces are associated with stored neural codes. Once again, thinking about ways to make VR work forces one to think about the way humans work: A sophisticated matching of world knowledge and real-time reality sampling enables me to close my eyes, reach out my hand, and determine whether the fender of an automobile is polished or dusty. Part of that perception is here and now, reflecting an accurate analysis of what the fingertip sensors are doing at the moment. Part of it is abstract, as the stored memory of previous encounters with the textures of the world.

Haptic perception—the feel of reality—is another one of those labyrinths of research and philosophy that can swallow you if you step off the main path of VR. Clearly, neuroscientists as well as roboticists and VR researchers must be thinking about issues related to the way humans transduce and process tactile information. Before Minsky et al.'s devices can evolve to credible haptic illusion machines, or alternatively an entirely different approach to haptic simulation emerges from a different group, a whole body of "human factors" knowledge must be gathered and validated by painstaking experiments. Minsky directed me to K. O. Johnson of the department of neuroscience's biomedical engineering division of Johns Hopkins University. When I talked to Johnson, he reminded me that there is a difference between having a detailed understanding of human sensing mechanisms and knowing enough to fool those mechanisms. Just as "pictures" are ways of visually encoding informaton to create a kind of illusion of depth and lighting, devices such as force-reflecting joysticks help create what Johnson calls "tactile pictures." Tactile illusions come from a basic knowledge of how to trick mechanoreceptors.

"To recreate high-fidelity tactile virtual reality," he told me during a telephone conversation, "you don't need so much knowledge of how the brain works as you need knowledge of how to make a believable representation. If technology were no problem, and there was a way to vibrate and indent the skin in precise patterns, tactile representation could be of a high quality." As my roboticist friend had tipped me, it's the specifications that make the engineering task more crucial than the scientific theory. Fingertip receptors, Johnson reminded me, can detect bumps as small as a millimeter, vibrations up to 500 cycles per second. The moving component of skin indentation—the way objects feel when we rub our fingers across their surface—gives us information about the surface details of the object. "The way a screwdriver gives you a picture of what's on the other end," is the way Johnson put it.

"The problem is that no output device today comes close to those specifications," he explained. Input devices are less of a problem— even today's crude gloves give relatively accurate information about the positions and motions of fingers. But creating a glove or grip or joystick that returns dynamic haptic information to the operator is, Johnson suggested, a task equivalent to the invention of the telephone. Even as Alexander Graham Bell and others needed to understand just enough about human hearing to create a device that transduced the spatiotemporal encoding of sound waves, future engineers will have to build a perceptual encoder-decoder for tactile senses. It might be embedded in a body glove. Jets of air or tiny balloons might be involved, or arrays of flexible ultraminiature piezoelectric cells. I met a man in California who is building arrays of fingertip buzzers the size of ballpoint-pen tips out of special alloys that change their shape when electrically stimulated. One of these devices might work, or it might take a material or technology that doesn't exist yet.

The potential of haptic perception is real, but it is too early to tell whether we will soon see a technology for exploiting it fully. We might be limited to feeling things at a distance, through joysticks or other hand-held devices. Or tactile feedback might be extendible to the entire body. One of the more provocative implications of full-body tactile feedback is explored in a later chapter, in the section on "teledildonics." However, it is clear that the addition of real-time haptic feedback, even limited, through a device that is gripped or felt with the fingertips or palms, to a 3D graphic display serves to augment the feeling of reality, particularly when it is combined also with audio representations. Seeing those molecules fit together visually, feeling their atomic forces in my fingertips, and hearing the bump forces when they collided worked together to create the strong conviction that I was dealing with something real. I had no doubts about the something. It was the "real" part of my lexicon that was beginning to change.

PUPPETRY, PROSTHESES, AND BEYOND

Indeed, successions of jabs or other mechanical gradients on the skin surface can come to mean anything you want them to mean; ingenuity in coding provides the only limit. A practically infinite variety of cutaneous patterns, representing particular collocations of spatial, intensitive, and temporal variants, offers a vast multitude of sensory symbols to be coded. The Copper Eskimos welcome strangers with a blow of the fist on the shoulder. Polynesian men rub each other's backs in greeting, while Andaman Islanders of the Gulf of Bengal blow on each other's hands by way of saying that "parting is such sweet sorrow." Cutaneous discriminations of size, shape, locus, intensity, or duration can be made to say these things, or report weather information, or detail daily trading on the stock exchange, or provide the data essential in gun-laying or uranium prospecting.

FRANK GELDARD,
"Adventures in Tactile Literacy," 1957

A year ago, I was halfway convinced that cyberspaces where you can experience the sensation of hefting a brick or squeezing a lemon probably won't be feasible for another twenty or thirty years. A month ago, I saw and felt something that shook my certainty. When I tried the first prototype of a pneumatic tactile glove in inventor Jim Hennequin's garage in Cranfield, an hour's drive southwest of London, I began to suspect that high-resolution tactile feedback might not be so far in the future. The age of the Feelies, as Aldous Huxley predicted, might be upon us before we know what hit us.

After surviving the exoskeletal contraptions that I locked around my hand, somewhat uneasily, in Tsukuba, the chopstick-in-a-bucket tactile joystick I wielded at the Media Lab, the fingertip buzzers I tested in TiNi's Emeryville, California, workshop (described in a later chapter), and that force-feedback arm with its ominous deadman switch I grasped in Chapel Hill, the elegance of Hennequin's concept was immediately evident. I put my hand into a glove. Sandwiched between layers of the glove were strategically placed air bladders that were

rapidly inflated and deflated. The inventor claims that the size, location, and pressure of the tiny air bladders are the critical variables
needed to create tactile illusions. By inflating the right patterns of
bladders in the glove to the right amount of pressure, Hennequin
hopes to be able to replicate the sensation of grasping an object and
recognizing its shape.

Keep in mind that this isn't the same kind of computerized glove
that has dominated the VR story thus far, although the ultimate goal
is to combine a tactile glove with a gesture-sensing glove. The
DataGlove and PowerGlove are *input* devices to the computer that tell
the computer what the human is doing. The kind of glove that Hennequin and a few others are trying to build is an *output* device that
can transmit information from the computer to the human via the
sense of touch.

"I want to store in the computer the feeling of holding a teacup,
and I want you to recognize it as a cup when you put on the glove
and I replay it," Hennequin explained. I tried the glove the second
time we met, the very afternoon all the systems were hooked together
for the first time. I presume he had finally got it working in the wee
hours, and mine was actually the first demonstration after the inventor
made sure he wasn't going to make a fool of himself. It was as close
to the source as I was likely to get—raw, uncut technology. Hennequin
inflated and deflated parts of the glove, and I could feel various kinds
of pressure that could conceivably, in the right combination, begin to
feel like edges or surfaces. Maybe the first computer graphics looked
this crude but this promising. Some stimuli felt like something soft
was pushing against my fingertip. Other stimuli felt like something
hard was pushing against the palm of my hand. Like the TiNi memory
alloy stimulator, some of the stimuli felt distinctly pleasant, like a good
massage or scratching an itch.

Nobody knows whether the initial promise of this approach will live
up to the reality test. Time will tell whether Hennequin and his crew
will succeed in using his "Air Muscle" successfully to duplicate the feel
as well as the look of physical reality. Just as "photorealism" is the goal
of one branch of computer graphics, "tactile realism" has the potential
to grow into a general goal for cybernetic clothing designers. Hennequin and his crew had just completed a "proof-of-concept" prototype when I met him, in the summer of 1990, and faced months of
tinkering and tuning ahead. But from the look of other, more fully
developed applications that were going through their paces in his
workshop, Hennequin has a fighting chance to achieve his goal.

Until I showed him the research reprints from laboratories in California and Japan, Hennequin wasn't more than mildly aware that his

invention, if it works as well as he hopes, could be the solution to technical problems VR researchers around the world have been trying to solve for years—and the beginning of a whole new set of social problems. The same technology that could liberate the trapped minds of quadriplegics and tetraplegics around the world could also be applied decadently to teledildonics or malevolently to brainwashing. As one of William Gibson's characters said in *Neuromancer*, "The street finds its own uses for technology." Simulated tactile sensation could have profound impacts on our body sense, our morals, our culture. Important scientific questions must be answered through careful, time-consuming research before we can know exactly how tactile illusions could work; it isn't easy to predict how fast the long-dormant study of tactile perception can gear up to provide data for realiticians. And building tactile perception capabilities in a lightweight apparatus is a significant engineering problem.

Hennequin might be creating an important piece of the puzzle. His pneumatic devices might provide or point to the key breakthrough. Or maybe TiNi's David Johnson, in his Emeryville workshop, is on the right track with shape-memory alloy. Or somebody else none of us ever heard about will reapply an obscure old theory or invent a new technology. Especially in the realm of peripheral devices, VR technology is allowing the long-endangered breed of garage inventors to become real contenders again. Hennequin's very attempt at making the tactile perception glove—and earlier versions by VPL and the US Air Force contractors—in the absence of a highly evolved scientific theory of tactile perception raises the possibility that someone might hack together a tactile feedback device before neurophysiologists or human factors experts understand how such an apparatus might work!

By the time I packed for London, I had long since acquired the habit of carrying a folder full of reprints along with me on my travels. I began to feel like an informational honeybee, bearing select pollen-grains of knowledge in my briefcase during my pilgrimages from one laboratory to another in search of VR experiences. The Japanese had been interested in what I knew about American research, and so were the British. The Dutch researchers wanted to know more about Japanese research, and part of the British effort was converging on the same general problem that the people I met in Grenoble, France, had been approaching for twelve years. None of this was secret information. These people simply did not yet know about one another. I carried a few reprints away from each laboratory I visited, and let the people at the next lab make copies of the reprints I brought along with me. And I told them how to gain electronic access to Sci.virtual-worlds. In a historically minor way, this has been one of the strangest relationships between a nonexpert author and the field of expertise he is covering.

Just as the control theorists, the computer graphicists, and the stereographic cineastes took decades to realize they were working on different approaches to the same goals, many other VR theorists worldwide in 1990 were relatively unaware of each other's existence. They had started out years ago to build puppets, prosthetic devices, or musical instruments, to design robots or make art, to advance the state of telecommunication or understand the nature of human perception, and it was only beginning to dawn on them that somebody else, halfway around the world, in a previously unrelated discipline, might have just the tool or fact they needed in their own quest. In Hennequin's case, he had invented the technology for other reasons before he became aware of the possibilities in VR and teleoperator technology.

Whenever I think about how difficult tactile feedback technology will be to create, I wonder how many Jim Hennequins are loose in the world, creating pieces of tomorrow's VR systems.

RÉALITÉ ARTIFICIELLE AND THE ART OF THE VIRTUAL INSTRUMENT

The means available to man to make his presence felt to his outside environment are vocal and gestual. Verbal utterance is essentially used among his fellows while gesture is exclusively employed on objects, in space, with or without tools. We can therefore say that in natural communication situations gesture is multiform. It is vigorous and motor, silent yet visible, minute but audible, and in all cases omnipresent.

Since the computer is playing an increasingly meaningful role in our environment, the obvious question that arises concerns the communication *conditions between the former and the latter. . . . Among the others, the question of "gesture and computer" thus exists in its own right. . . .*

Today's media are essentially "audio-visual." In other less "mediatic" disciplines techniques relative to gesture processing have been developed for a long time now. Take, for example, telemanipulation. *Thanks to appropriate organs, the gestual phenomenon in its turn can be captured, tele-transmitted, restituted, processed, and memorized.*

It is quite clear that this is exactly what occurs when a gesture stroke is captured by an alphanumerical keyboard. In fact, this happened in a very reduced form with the Morse key manipulator, but what precisely had not then been understood (and which still remains to be seen even today) is that gesture is a very subtle and rich phenomenon. . . .

The instrumental gesture can become an object of research just like sound and image if we have suitable organs that are not only appropriate but also efficient enough.

Claude Cadoz and Christophe Ramstein,
"Capture, Representation and Composition
of the Instrumental Gesture," 1990

The lights were dimmed in the *salle de manip*. The keyboard glistened. I put my fingers in the loops as directed, moved them tentatively, and almost jumped—even though I was expecting it—when the sound of a violin string, inexpertly bowed, issued from an impressive array of speakers. More startling than the sound was the feeling in my hands, my fingertips, the bones in my arm. I had traveled for days just to get my hands on this device, the Modular Feedback Keyboard, the technology that made the virtual violin and other virtual instruments real enough to feel. When I first heard about the device and wondered whether getting my fingers on it justified an extra fifteen hours of train travel, there seemed to me to be little connection between gloves-and-goggles VR research and this violin bow that looks like a keyboard. But I trusted my source, Margaret Minsky, and made the trip. The moment I felt that unexpected violin bow vibrating in my fingers, I knew these folks had been stalking the feel of things from an entirely different direction. The ACROE researchers looked to the arts for their driving problems, embedded in a pragmatic engineering approach, within a hard-core AI research institution.

It was deliciously cool and dark in the *salle de manip*, and ecstatically chilly next door, in the *salle de machine*. When you get enough computer power assembled in one place, the electrons tend to generate a bit of heat in their microcircuit picosecond marathons, which means that if you want to find the one room in a building that is guaranteed to have some kind of air-cooling appliance, look for the place they keep the expensive computers. At LIFIA, the *Laboratoire d'Informatique Fondamentale et d'Intelligence Artificielle* (Laboratory of Fundamental Informatics and Artificial Intelligence), they call that room the *salle de machine*, the "machine room." I found the presence of cool air to be an enormous personal relief, after my time in England. The heat first hit when I was in London and didn't relent for a moment during the two weeks I traveled the European VR circuit. I started to wilt in Amsterdam, abandoned most of my wardrobe at a friend's house in Paris, and by the time I arrived in Grenoble, home of France's civilian

VR research, I was down to a pair of knee-length paisley-print shorts, matching tee-shirt, and the black baseball cap I had hand-painted with mystical designs in metallic copper ink the night before I left California. I had been saving this particular combo for my day off in Paris.

I washed the cotton shorts and tee-shirt each night, while soaking my feet, and hung the wet clothing on my hotel room balcony, where they would catch the morning sun; the garments were dry by the time my room started to heat up in the morning. Heat is hardest on one's feet, especially when traveling, so I figured that if I was going to be informal I might as well be utterly informal, and therefore ·decided to complete my ensemble with the DayGlo green plastic thonged sandals I carry for hotels and long train rides. I remember giving a two-hour seminar to a dozen researchers from Institut IMAG, the *Institut Informatique et de Mathématiques Appliquées de Grenoble* ("Institute of Informatics and Applied Mathematics of Grenoble") on my second day there, still clad in that costume (sans chapeau). The research staff must retain some strange mental images of the writer who came so far, bearing news of VR research, showing slides of marvelous instruments in Chapel Hill and Tsukuba: the American with the paisley shorts, Johnny Realityseed. Deciding on the right garb for each of my appearances, which increasingly require me to give presentations as well as stick my head into demos and conduct interviews, had emerged as a minor subtheme of my VR quest; I learned that people were universally interested in the information I had for them, so I began to feel that there was no need to be terribly sartorially orthodox. I had the advantage in most cases of knowing that Jaron Lanier had preceded me, so minor idiosyncrasies of dress like mine don't come as a complete shock in VR circles. The folks at NTT probably still remember the author in the gray suit, conservative tie, and van Gogh shoes. Thomas Furness asked where he could get a tie-dyed shirt, after I wore one while lecturing to his virtual worlds class in Seattle.

But the heat had taken priority over propriety by the time I got off the train in Grenoble, and the French VR team of Claude Cadoz, Annie Luciani, and Jean-Loup Florens seemed to be interested enough in the scientific reprints I brought, as well as the videos, the slides, word of mouth about what James Hennequin and Margaret Minsky were doing, the latest news about the Super Cockpit and VPL, to ignore the informality of my garb. Speakers and color CRT monitors lined the walls of the *salle de manip,* the "manipulation room." The place where demos happen. Foam baffles in odd corners indicated that someone wanted to be certain the acoustics in the room were right. An Evans and Sutherland graphics workstation with a huge color monitor took up most of one table. But the visual focus of the room was the "keyboard," a beautiful electromechanical clockwork of brass and brushed aluminum, its innards visible through a clear lucite case, the result of more than twelve years of research into "gestural control."

Except for the plastic case, the Modular Feedback Keyboard looked like the kind of instrument that aristocratic scientists like Newton and Lavoisier commissioned their craftsmen to create from hardwood and alloy in centuries past. The keys themselves were the same size as piano keys, and later, when they took off the peripheral device and entered the appropriate incantation via the computer keyboard, the keys indeed felt and sounded like piano keys. A world-class pianist was part of their team of artistic collaborators. At the time I first saw it, an additional device with stainless steel finger loops was attached to two of the metal keys with small clamps; it looked like something a surgeon might use. I put my thumb and forefinger into the device and moved them. The loops moved, in tandem. A rather unskillfully bowed violin note blatted out of the speakers and my hand reported to me that I had just moved a violin bow across a tuned string. It twanged something inside me that the most vivid visual illusions alone had failed to reach.

It was the same kind of visceral response that had surprised me before, in Chapel Hill and Cambridge, wrestling molecules in cyberspace or stirring virtual molasses. I could see that I was holding metal, I knew that the sound was synthesized in real time by electrons racing around a chip somewhere in the *salle de machine*, but it *felt* as if I had been bowing a violin. I could control the pressure of the bow and hear the violin vary its pitch. I could bounce it or draw it in a long slow motion, the often unharmonious feedback from the virtual strings revealing me as an utter ignoramus in matters of bows and violins. It was the most proprioceptively alarming illusion since I stirred that bucket of ice cubes in molasses, a year previously, in Margaret Minsky's laboratory. Minsky, in fact, was one of the people who suggested that I pay a visit to Grenoble. I understood why. There's something fundamentally important about the relationship between a musician and a violin, for example, that is more or less missing from the human-computer interface, despite all the progress that has been made in audiovisual representation.

There would be more demonstrations later. As soon as they had allowed me to experience that simple simulation of a bowed instrument, we were all eager to exchange information for a while before plunging back into the hands-on if not head-in world of "gestural transducers" and "instrumental gestures." I had felt and heard enough to know that they were onto something. HMDs are one thing. Understanding and transmitting tactile and proprioceptive perceptions are another matter. These three researchers clearly believed they were onto something. So we regretfully left the coolness of the *salle de manip* and retired to the offices of ACROE, where a whiteboard and colored pens—instruments of choice for scientists worldwide—could be found. Support staff scurried around, as they had in Tokyo and London, bearing papers to the copier, videos to the dubbing deck, and slides to the nearest photo lab. I added several reprints and slides from ACROE to my portable VR library. There were three principal investigators, several graduate students, some support staff, and a respectable accumulation of computing power devoted to ACROE, which was one cell of a hive of interconnected acronyms housed in a modern building in an old city on the Rhone. ACROE is the *Association pour la Création et la Recherche sur les Outils d'Expression* ("Association for the Creation and Research into Artistic Tools"). The three core members of the French VR team have been supported by the French Ministry of Culture, but ACROE is geographically and bureaucratically nested within Institut IMAG, as part of LIFIA.

When I first heard about them from Margaret Minsky, I reached out through the Worldnet to see what was happening. First, I faxed them a note about what I was doing and included my electronic mail address. When I logged onto the WELL the next day, there was a message from O. Roualt, a postgraduate student working with the ACROE principals. He mentioned some kind of piano that could also be a violin. He said that if I was writing about the state of VR worldwide, they had something to show me in the southeast corner of France, three hours by the fastest train from Paris, that would be worth the journey. I told him I'd show up on a certain date. The next day, I got more e-mail. O. Roualt asked if I would stay an extra day and give a seminar about what I had seen at other laboratories, which is how I ended up lecturing an assortment of French information technology researchers on the state of the world of VR research. By this time, it was getting easier to make my presentatons, but it took longer with each telling: I just told the French researchers what I had said about US research in Japan, along with reports of what I had seen of Japanese, British, and Dutch research, and heard of German research.

LIFIA reminded me of ATR, albeit an order of magnitude smaller, in that it involves an integration of several overlapping research fields: AI, parallelism and neural networks, mathematics and software theory, new programming languages, computer modeling of physical systems, computer vision, robotics, and a variety of VR that sought to synthesize not just sounds, but *instruments*. There's a hotbed of VR research nestled in this pleasant old town, with the French Alps looming to the east and an increasingly cosmopolitan, unfortunately increasingly smog-bound city filling the valley. The French VR research program is small, but it is closely linked to many other research programs that just happen to be furnishing crucial information and technologies to VR research. The program at ACROE is also highly focused on a specialty. Just as UNC specializes in visualization tools, the Advanced Robotics Research Centre concentrates on teleoperated vehicles, and ATR focuses on VR as a communication medium, the French effort seems to specialize in the *transmission of gesture,* especially as a means for linking human eye-hand skills and imagination with computerized rendering tools. Virtual orchestras and teleoperated microsurgical robots are part of the same cluster of research questions.

Like ATR, LIFIA researchers are after several different kinds of fundamental knowledge about the perceptual and information-processing aspects of visual images, as well as the applied engineering disciplines, such as automatic scene recognition ("gloveless VR"), that are built from that knowledge. About seventy persons are directly involved in research at LIFIA, with ten people contributing to administrative and technical functions. The collection of state-of-the-art computers, the long-term funding, the think-tank atmosphere, all tucked away in a new building behind an old stone gate down a cobblestone street toward the center of old Grenoble, indicated a fairly high level of financial support from somewhere. Somewhere turned out to be Europe.

Some of LIFIA's funding comes from various French ministries, but about 70 percent of the research funding comes from research contracts, about half of which are for the major European cooperation programs. The continental scope and disciplinary breadth of the LIFIA research program reminded me of what Jens Blauert had said in Santa Barbara, when he talked about the nascent German *telepresänz* consortium of thirty-seven German scientists of different disciplines "who devote a major part of their research activities to the foundations of telepresence technology." Blauert, who is based in Bochum, Germany, noted that it was easier to get European funding than national funding for cross-disciplinary efforts like this. As of the summer of 1990, the German consortium Blauert had mentioned was still in the planning stage—the researchers were pursuing their separate activities, but the *telepresänz* test-bed had not yet progressed to the point of having a building and fully equipped laboratory devoted to VR. That situation has probably changed since 1990.

The VR effort represented by ACROE was small, but it had been evolving since the late 1970s, which puts them back there with Krueger, Brooks, Negroponte, and Furness as old-timers in the VR game. Just as there seems to be a heavy bias in Japan toward the use of VR as a human interface to computers and a communication medium, I sensed in the French research program a distinct appreciation of the importance of linking gesture with the visual and acoustic arts. Like Krueger, they seemed to grasp the idea that interaction is the medium of expression uniquely suited to computers, and that artists could create works of art that were also experiments in the nature of human-computer interaction. Krueger concentrated on the gross behavioral aspects of that interaction—position and gesture and movement of the arms, hands, and legs. In Grenoble, they were focused in on what it is that a pianist or violist or painter does with his or her hands and eye and brain to create something new, both as a way of extending the scope of human artists and as an appropriate framework for a comprehensive human-computer research effort.

This is what LIFIA's brochure says about the role of artists and artificial reality in their overall research program:

Physical models, gestual [*sic*] control and real-time interaction are the three keys which characterize the research at LIFIA on the most advanced confrontation between computing technology and artistic, musical and visual, creation. In this domain, there is a demand for fundamental research. First, a research on physical models, since, in sound and image, this is the only way to accept the challenge of naturalness and expressiveness, while everything relies upon the computer, a totally artificial device. But models of physical objects, which are deformable, mobile, vibrating, generate expensive simulations, and research on the optimisation of simulation algorithms is crucial. Then, a research on gestual control and on force-feedback gestual transducers, which establish the physical relation between the gestures of the operator and the reactions of simulated objects, an essential relation in the act of creation, which is expressed through the gestual channel. This leads to the implementation of devices for sensorimotor gestures, which open a new field to the dialogue between man and machines. These devices indeed make it possible to capture a gesture, to build a representation of it, and to produce, by extraction of significant forms, intelligent interpretations of effective gestures. Finally, research on real-time interaction is necessary, since gesture is only meaningful when simultaneous with the sound it produces, the image of the object being deformed, the mechanical reaction, captured by the hand of the virtual object to which it applies. This original research of LIFIA requires real-time synthesis machines for sound, image and mechanical reaction and leads to the implementation of a complex software and hardware system, at the forefront of current technology, where these machines cooperate to make present an artificial reality.

When we got to the whiteboard, however, I got the intensive version of their presentation, which Annie Luciani and Claude Cadoz conducted in tag-team manner, standing up, taking the marker from each other, erasing whatever the previous partner had written, even as I was trying to get it down in my notebook. Occasionally, Jean-Loup Florens, who was seated at a table with me while the other two performed their duet, responded to one of their invitations to explain more fully. Luciani is redheaded, animated, friendly, and brooks no nonsense. Cadoz was dressed in black, even on a sweltering July afternoon. He also has a neatly trimmed goatee, high forehead and widow's peak that give him a slightly saturnine look. Luciani and Cadoz were quick, and their English was far better than my French, although I found I could help get my questions across better if I used whatever French popped into my head. Florens's English was only moderately better than my French, so it took him a bit longer to respond. My "briefing" ended up taking more than two hours. They had been thinking about their particular peculiar avenue into VR for a long time, and it isn't easy to explain a couple decades of intellectual work in a few minutes.

ACROE started building force-feedback devices for sound synthesis in 1978, starting from a principle very similar to one Myron Krueger was espousing at about the same time: Computers and electronic circuitry had been successfully applied to the synthesis of sounds, reproducing the acoustic characteristics of known instruments as well as creating sounds that have not been heard before, and had progressed in a similar manner in the visual dimension, toward increasingly lifelike synthesized visual objects, but few people and only a handful of computer scientists had applied themselves to synthesizing instruments—reproducing the way it feels to play a violin or piano. Media Lab is working with the interactive element in virtual instruments, but ACROE seems to have the longest focused research program.

Gesture, as Krueger also noted, plays an important role in human communication and could be a key means of human-computer communication. Aware that gestural communication could be a central issue in human-computer interface design in general, the ACROE research from the beginning was based on their belief that the creation of new artistic instruments is an important and useful "driving problem" for studying the elements of gestural (or as ACROE calls it, "gestual") communication. "We were not simply aiming at improved ergonomy of gestual control in sound synthesis, but rather at a fundamentally new insight into musical synthesis itself. We were led to propose not only a synthesis of the sound but also of the *instrument*," Cadoz explained.

For Brooks and his crew in Chapel Hill, one goal was to make a molecule you could feel with your hand; for Furness and the USAF, the supreme goal was to ensure the survival of pilot and aircraft under confusing conditions; Margaret Minsky wanted to create virtual sandpaper and study tactile perception. ACROE has been aiming to build virtual violins for more than decade. In 1978, Cadoz and Florens created a one-dimensional device, in which motors exerted a resisting force to a mass that the operator could slide along a bar and feel the resistance as a virtual object with varying degrees of elasticity. By studying the way damping forces, masses, and springs operate in one dimension, they began to specify what they would need for a device that could act as a gestural transducer (a means of turning human gestures into computer data) and tactile feedback device (a means of creating tactile renderings of virtual objects). They were aiming at a device for simulating what they called the *instrumental gesture*. They knew that they would have to design special extremely compact motors with ultrarapid and accurate response and sizable power; their patented "sliced" motors, one to each key, one key for each degree of freedom, were the result of that branch of their research. In the early 1980s, they built a second-generation gestural transducer. It looked to me like a cross between a sextant and a compass, with gears and counterweights and tiny motors.

"We needed to build very high-performance transducers like these," Luciani interjected, taking the blue marker from Cadoz and stepping up to the whiteboard, "but we also have been creating languages for connecting gestures to sound and image." They had a clear focus on the instrumental gesture, the intelligent movements of the hands, arms, and fingers that guide human craft. But gesture leads everywhere. CORDIS was their ten-year effort to create a software architecture built around a very high performance force-sensing, force-reflecting input device, in ways that connect gestural expression with the synthesis of sounds and images. We ducked back into the *salle de manip* for some cool air and another hot demo. They showed me a bit of ANIMA, the graphics manipulation language they had been developing for almost as long. They were fascinated, of course, at tales of VPL's glove. Their interests in the finest fingertip skills, naturally required input devices of a much higher degree of performance.

Back in the cool dark, I settled into the hotseat, the comfortable office chair where I was able to see, hear, and feel the best quality of gestural reproduction. I used the same surgical peripheral that had been the violin bow to manipulate a complex soup of kaleidoscopic computer graphic forms in real time. There's a certain childlike pleasure in mastering these abstract but fascinating tasks, related perhaps to that excitement a video game fanatic feels. When you find a way to interface your finger-knowledge to a complex visual task, you feel, at a very low intensity, the way F15 pilots must feel with all those fight-or-flight hormones amping up their intelligence. These kinds of tasks—separating vital information from noise, processing perceptions intelligently, making rapid decisions based on changing, uncertain data—are among the things humans have been shaped by evolution to do very well.

Their goal of giving a virtual violin and a virtual bow to a master violinist served to inspire more than twelve years of hardware and software development, far from the technological centers of the world. Extending what they had learned from experimenting with the 1978 force-reflecting feedback device, and the single-key system that they built in 1981, the ACROE crew constructed the instrument that their decade of research had led them to build—a keyboard, in which each key could represent a degree of freedom, and through which the extremely fine perceptions of a musician's fingers might be able to exchange information between human and computer.

"The instrumental gesture must be genuine," Florens interjected, at one point. I recalled the amateurish sound of my violin bowing. Here was the proprioceptive equivalent of photorealism as a goal for computer graphics. The specially designed motors and delicate ma-chinery of the transducer, together with the sound synthesis and image synthesis software tools they had built, represented not just ACROE's realization of "virtual instruments," but an ideal test-bed for studying the nature of instrumental gestures and the high-performance di-mension of human-computer gestural interaction. Why stop with pro-ducing virtual instruments that sound and feel exactly like their physical counterparts? Why not create virtual instruments that can do what no previous instruments can do? Luciani and Cadoz took some time to point out how the generalized abstractions they had created for translating gestures into sights and sounds mapped perfectly onto the task of controlling a teleoperated robot or feeling one's way around a tactile virtual world. The Modular Feedback Keyboard was first pre-sented to the public in 1989, and research into the human factors side of the project—the sensorimotor skills that make the difference be-tween a maestro and a muddler—has barely begun.

Research and development in the late twentieth century is a hybrid of capitalism, science, and technology. Discoveries and technologies lead to products; the demand for products, and the money they bring in, drive research to uncover principles that lead to new technologies and improved products. Sometimes, the critical R & D is performed in giant, well-equipped laboratories. In other instances—particularly when the science and technology are new and unproven, and the products are unprecedented—the true pioneers are small businesses. In Silicon Valley, the technopreneurs who start out as "homebrew" hobbyists have been accorded a special kind of respect, ever since the days of Hewlett and Packard, Wozniak and Jobs.

Chapter Fifteen

HOMEBREW VR

So on crucial billboards in the area—at PCC, at Lawrence Hall, at a few schools and high-tech corporations—Fred Moore tacked up a sign that read:

"AMATEUR COMPUTER USERS GROUP—HOMEBREW COMPUTER CLUB . . . you name it. Are you building your own computer? Terminal? TV Typewriter? I/O device? Or some other digital black magic box? Or are you buying time on a time-sharing service? If so, you might like to come to a gathering of people with likeminded interests. Exchange information, swap ideas, help work on a project, whatever. . . ."

The little club formed by Fred Moore and Gordon French had grown to something neither could have imagined. It was the vanguard of a breed of hardware hackers who were "bootstrapping" themselves into a new industry— which, they were sure, would be different from any previous industry. . . .

It was the fertile atmosphere of Homebrew that guided Steve Wozniak through the incubation of the Apple II. The exchange of information, the access to esoteric technical hints, the swirling creative energy, and the chance to blow everybody's mind with a well-hacked design or program . . . these were the incentives which only increased the intense desire Steve Wozniak already had: to build the kind of computer he wanted to play with.

STEVEN LEVY,
Hackers, 1984

The largest wheels take the longest to turn. Although VPL, Autodesk, and the commercial side of HITL seem like the most important events in the emergence of a VR industry, there are many more smaller enterprises popping up, one or a few of which might evolve out of the crowd, the way DEC and Apple and Autodesk did. VR research also marks the return of the homebrew inventor and technopreneur.

For a long time, it looked as if the garage science and industry mythos was dead, unfortunately. The personal computer revolution was made possible by the small elite of computer scientists at ARPA and ARC and PARC, but it was carried to the masses by the Homebrew Computer Club, who started out as hobbyists and ended up as an industry. During the time I've worked on this book, I've watched a couple of living room inventors strike out on their own, create their first prototypes, find their first clients, and move into their first offices. I've also tracked a half dozen other companies who hope to sell a product or service in the as-yet almost entirely hypothetical VR industry. It might not be an Edison or an Alexander Graham Bell who kicks VR up a level from fringe science to global industry. Perhaps the person who will take the dream of the true believers to the land of the mass market will be more like Henry Ford or Steve Jobs.

Never underestimate the power of the enthusiast, particularly the American technical enthusiast. One of the most positive personae of the composite American personality is the tinkerer, the inventor, the garage-scientist, who messes around with crystal sets or homemade computers for the fun of it. The Edisons and Teslas and Wozniaks just seem to emerge at regular intervals from the pool of technological amateurs in a warehouse in Menlo Park, New Jersey, or a garage in Menlo Park, California. One of the prerequisites for such homespun R & D is that the tools and materials for tinkering be affordable by amateurs. The general trend of high technology has been away from such availability, however. You can build your own crystal set, but you can't manufacture your own microchips. You could do fundamental physics research in the nineteenth century, and much of it was accomplished by amateurs, but you need to be a consortium or a government to explore the frontiers of high-energy physics. However, when the miniaturization revolution made it possible to put all the fundamental components of a computer on a single chip, suddenly it became possible for knowledgeable amateurs to tinker in a realm that had previously been confined to the Defense Department or IBM-size corporations—computers. When the Intel Corporation began selling microprocessor chips, in the early 1970s, and MITS in Albuquerque began selling microcomputer construction sets a few years later, the "homebrew" movement was born, and so was one of the key origin legends of Silicon Valley.

The "homebrew" enthusiasts created a second computer industry in a niche that the computer giants considered too minuscule. The hobbyists who met at the auditorium of Stanford Linear Accelerator in Palo Alto to exchange parts and information later became the founders and chief engineers of Apple Computer, Osborne, and a dozen other companies that spurred the birth of the personal computer industry. Because there was a rich infrastructure of small companies who were all probing the needs of the growing industry, creating products that would accelerate development everywhere, the homebrew tradition carried over to the personal computer revolution when it began to reach interested amateurs outside the hobbyist circles. When the first Apple computers made it possible for nonprogrammers to play a wide variety of different games on their home computer, simply by changing the software, a rich population of "third-party developers" created the software that has caused millions of people to buy Apple computers. As Apple grew and IBM jumped into the personal computer game, the shakeouts came, and many of the original hobbyist-entrepreneurs went back to being simply hobbyists. The technology of personal computers grew increasingly sophisticated, the companies grew too unwieldy to innovate, and the homebrew element retreated into the background of the computer world again.

A few years from now, you might be able to write an entire book about the homebrew VR industries that are springing up in the early 1990s. I have not been able to track all the developments in the realm of small companies, but my own antennae for the pace of change are triggered at regular enough intervals, in widely separated places, to strongly suspect the homebrew VR revolution is in a very powerful incubation phase. There is certainly a pool of knowledgeable hardware and software enthusiasts, and they have superb means of communicating technical informaton among themselves, via Usenet. But can it be done cheaply enough? The first step has already been taken: the Autodesk system demonstrated that you could build a low-end VR system for about $20,000. When that price drops sufficiently, the homebrew revolution will take off. If hardware prices drop, if clever software solutions are found to otherwise expensive problems related to processing VR data, if HITL makes good on their promise to create a public-domain virtual operating system, and if information continues to circulate about how to put together a VR system, we might see the emergence of yet another powerful driving force in the reality-industrial complex.

In my efforts to keep in touch with developments, I've visited small startup companies in Vancouver, British Columbia, the San Francisco Bay Area, and Austin, Texas. Fortunately, I've been able to track the establishment and growth of a homebrew outfit within bicycling distance of my house. I've seen it grow from a living room operation into a small and growing company with a product, clients, an office and workshop. I had met the founders of Sense8 Corporation when I first started visiting Autodesk. It's hard to miss Eric Gullichsen, with his shoulder-length hair and his shin-length pants. His partner, Patrice Gelband, cuts a less flamboyant figure, but when I started talking with her it became clear that she soaked up every detail of what was happening in the room. When I visited the demo room at Anaheim, I got into a conversation with the two of them and Timothy Leary about the infinite horizons of potential that had intoxicated everyone involved in the VR industry in 1989. From the programmer's point of view, there will be a lot of work to do in order to craft products from the demos the first wave of the VR industry has produced. Gullichsen and Gelband were hard-core programmers, although Gelband informed me that she is fundamentally a mathematician who became a programmer because she is interested in simulation as well as entrepreneurial opportunity. In grand old Silicon Valley tradition, they left Autodesk to start their own VR company.

The Autodesk system had used special graphics processors in an $8000 add-on module that boosted the power of an IBM-style personal computer as the reality modeling engine and a pair of Amiga personal computers as rendering engines. When Gullichsen and Gelband set out to create something from scratch—a legal necessity when you leave one company to found another—they thought they saw a better way to do it cheaper. The eternal search for a cheaper way to build a basic toolkit is part of the homebrew phenomenon. The homebrewers themselves drive down the price of entry into their club. When Gullichsen called me and told me I would be the first person other than the creators to test their system, I went over to their house-workshop to find a scene out of homebrew heaven: the living room was given over to a welter of computer terminals, raw motherboards with their rows of chips exposed, looking like rectangular black cockroaches, bundles of color-coded cables, mysterious metal and plastic boxes with jacks and switches. They had purchased a Polhemus position-sensing system. They built their own HMD from more or less the same off-the-shelf parts that were used at NASA. And they purchased their optics from the original VR homebrewer, Eric Howlett. Howlett, a fellow with frizzy white hair and a mad-professor presentation—I heard him speak once to a stereographic engineering convention while he wrote semilegibly in grease pencil on transparency sheets for overheads—owns a small company called PopOptix in Waltham, Massachusetts. Scott Fisher had tipped off McGreevy and Humphries, who used Howlett's optics, and VPL followed NASA's lead. By 1990, Howlett was selling his own HMDs.

Until they could take delivery of a DataGlove from VPL, Sense8 was using an input device I had seen before at UNC—a six-degree-of-freedom "orb." The orb is a small sphere, about the size of a billiard ball, set in a base, like a joystick. It can be pushed, pulled, and rotated. In some ways, the orb is a better control device than the glove. The glove is indispensable in establishing your sense of presence and giving you a literal handle for manipulating the virtual world, but the orb is far easier to use for navigation in cyberspace.

I remember dropping by Gullichsen and Gelband's place a couple of times while they were getting the demonstration ready. Gelband would be doing mathematics with pencil and yellow pad in the kitchen; Gullichsen would be hacking code in the living room. Once she understood the problems that programmers face in trying to maintain complex virtual worlds and render them in real time, Gelband was able to exploit her background in mathematics to develop more effective algorithms. Like nuclear physics, which suddenly turned up uses for previously "pure" branches of mathematics that had been invented centuries before, VR software stimulates a mathematician-programmer like Gelband to look through her conceptual toolkit when she comes up against VR-related software problems.

They finally got it working in mid-February, 1990. They were about to pack it all in a suitcase and take it on a plane, so I only had a few minutes of test-flight of their homebrew reality. I put on their HMD— certainly cruder than VPL's face-sucker, but considerably less expensive. They fired up the engine. And a very grainy world swam up around me, in colored diamond-shaped pixels. The first world they had working was expectedly crude in resolution, consisting of a green plane—thirty polygons—with three pyramids. You could use the orb or the buttons on the orb and your line of sight to fly around. I returned a few months later to test-fly their next-level demo. This time they had a glove working, and my disembodied hand—the emergent hallmark of early-generaton VR systems—was floating above a checkered tabletop. Blue objects—a pyramid, a cube, an automobile—rested on the tabletop. The thrill of reaching over and using my hands like mitts to pick up cartoon-objects was losing its potency. But I could benchmark the experience: Sense8 showed me enough to believe that the low-fidelity-reality VR systems that cost tens of thousands of dollars today might be on desktops by the mid-1990s.

Of course, VPL's multihundred-thousand-dollar version is slick, and UNC's Pixel-planes worlds are slicker yet, and both of them remain far from slick enough. But the homebrew version is certainly more than a hundred-thousandth as exciting as the high-end worlds. It is now possible for people to build systems and exchange worlds, to propagate improvements, to evolve the way personal computers did. It remains to be seen whether there will be very many cyberspace homebrewers, or whether they come up with a rich set of tools, or whether they find ways to share their efforts. But Sense8's system is an existence proof. You don't have to be NASA. You don't even have to be Autodesk. You can do it in your living room, the way Gullichsen and Gelband did.

By the fall of 1990, Sense8 had developed their product, a software system that makes real-time rendering of 3D worlds easier to do with lower-powered hardware. They call their product WorldTool and plan to sell it for between $2500 and $5000. The investment for turning your desktop machine into a reality engine is still fairly high. But it is within reach of many university computer science departments or a group of hobbyists.

Sense8 wasn't the only startup VR company I encountered in my travels. "Fake Space" and its founder, Mark Bolas, also made an impression. If Mark Bolas is a typical example of the kind of entrepreneurial talent waiting in the wings, then the larger VR companies will face some real competition. In fact, Bolas's company actually did begin literally in a garage in Menlo Park, true to Silicon Valley legend. In June, 1988, Bolas started working with Scott Fisher at NASA's VIEW laboratory on the "Molly" (named after a character in William Gibson's *Neuromancer*)—a slaved binocular remote camera platform. In addition to the contract work, Fisher granted Bolas access to the VIEW system after hours to conduct his own experiments.

"I convinced the faculty of the Stanford Design Program to allow me to create virtual worlds for my master's project work," Bolas recalled when I interviewed him. The Design Program is a joint program involving the Art and Mechanical Engineering Departments. Bolas's stated goal was to "treat the system at NASA as a medium in its own right. I intended to exploit that medium to create some cool experiences for purely aesthetic reasons. This was a new thought back then— it was very hard to convince the faculty that my project was to be a series of experiences and had little to do with the hardware itself."

Bolas's first real test of the system was a series of evenings in which he forced himself to spend an hour each night in existing NASA environments. "It was boring," he recalls: "I was worried that I had found the emperor's new clothes. So I programmed and tried out a new variation and reprogrammed and tried it out again and then I made other people try it out—until we began to be satisfied with the quality of the experiences. I ended up with three art pieces and about five cool interface ideas."

One of the three art pieces was "Mondrian," an experiment in the aesthetics of virtuality. "Mondrian" was an interpretation of a Mondrian painting in the medium of the VIEW system. The final piece consisted of a perspective explosion of the painting Composition with

Line, 1918. As Bolas described it, "Mondrian felt that canvas should never try to imitate the three-dimensional world, so I took his painting and blew it up backwards. If you look at his painting from within a virtual art gallery, it looks like Mondrian's original, but if you move out of the virtual art gallery, you realize that the 'painting' is a three-dimensional line sculpture the size of a football field, and 'lines up' to look like Mondrian's original only from within the virtual gallery."

His other project with Fisher, accomplished with several colleagues who would later join his VR company, was "The Boom." The Boom is a binocular eyepiece for viewing virtual worlds that is mechanically suspended over a desktop, the way a desklamp is mechanically suspended. When you need to look into a virtual world, instead of locking your head into goggles, you reach over and put the Boom in front of your eyes, like a pair of binoculars. Because it is mechanically linked instead of position-tracked via electromagnetic means, the Boom reduces lag time, and is less expensive to manufacture than EyePhones. But Bolas, like Fisher, has an interest in the arts as well as the sciences of virtuality.

In September, 1989, Bolas founded "Fake Space Labs" (with Ian McDowall, Russell Mead, Tim Parker, and Nanci Anderson) to make "real tools for virtual environments." Bolas adds: "At the same time that we're building hardware and software for NASA and others, we are trying to sneak a little design sense into virtual world building. I'm amazed by how little time people actually spend inside virtual environments and how much time they spend talking about polygons per second."

When I found out that the contractor who had made the vibrotactile technology involved in the Super Cockpit's "tactile gloves" was within an hour's drive of my office, I paid a visit to David Johnson's office-laboratory-factory operation in Emeryville, where his company TiNi is exploring the possibility of using "shape memory alloys" to create tactile reality devices. Johnson's operation, the TiNi Company, was in a postmodern corrugated-steel light-industrial building in Emeryville, California, a formerly decaying heavy-industrial area south of Berkeley that now seems to be reemerging as a center of late-twentieth-century microtechnologies—there are software companies and futurists, genetic engineering plants, digital mapping outfits. The cafes are flush with knowledge workers, eating their low-cholesterol goat cheese lunches, speaking of algorithms and bioforms. The TiNi plant was a split-level facility, with desks and workstations in a relatively officelike area upstairs, and some light machinery like drill presses and deposition machines for creating microchips.

TiNi uses an alloy called nitinol as the basis for a small grid of what look like ballpoint-pen tips. Nitinol assumes whatever shape it is cast in, and can be reshaped; then when it is electrically stimulated, the alloy returns to its cast shape. It can be used to perform the kind of mechanical switching that solenoids do, on a smaller scale. By entering the proper command to the computer interface, the six-by-five pin array, about three quarters of an inch square, starts moving. I touched my finger to the grid and felt something like a pencil lead underneath a piece of cloth, moving across my fingertip as the rows of pins were activated in the programmed sequence; I could feel the individual pins, but I detected the edge that their pulsed arrays created. In October, 1990, at yet another VR convention called "Cyberthon," I tested the five-finger version. There was a tactile whisper of possibility for transmitting something about the feel of the world through this technology. There are hard engineering problems to solve. In order to be effective enough to create full-blown tactile illusions, it appears that up to 500 vibrations per second are necessary from the tactile actuator if human sensitivity to tactile variations is to be fooled.

The pins on my fingertip felt good, kind of tickly and soothing. My friend Flash Gordon has a chair that does something with your vertebrae in rhythmic vertical sequences that can seem obscenely pleasant. I imagine something like that for the long-term possibilities of the technology David Johnson and his crew are pursuing. Or perhaps the fellow I met later that year, in a workshop in London, with his pneumatic gloves and dirt-cheap motion platforms, was onto something better. In either case, the ingenuity of homebrewers is now an emerging possibility, on the verge of being a small force, in the VR industry. With a marketing success here or there, they could become a driving force.

The world does not lack for forces that will tend to drive the development of a reality industry in the 1990s. Where VR is likely to drive us, and what kind of creatures we might become as a result, are a complex matter, a case for wide-ranging speculation, because the answers to these questions are likely to lead to questions about the future of our most basic human characteristics, from our sexuality to our sense of identity.

Part Four

VIRTUAL REALITY AND THE FUTURE

Chapter Sixteen

TELEDILDONICS AND BEYOND

There was a young man named Kleene,
who invented a fucking machine.
Concave or convex, it fit either sex,
and was exceedingly easy to clean.

TRADITIONAL.
This version attributed to
John von Neumann.

The first fully functional teledildonics system will be a communication device, not a sex machine. You probably will *not* use erotic telepresence technology in order to have sexual experiences with machines. Thirty years from now, when portable telediddlers become ubiquitous, most people will use them to have sexual experiences with other *people,* at a distance, in combinations and configurations undreamed of by precybernetic voluptuaries. Through a marriage of virtual reality technology and telecommunication networks, you will be able to reach out and touch someone—or an entire population—in ways humans have never before experienced. Or so the scenario goes.

The word "dildonics" was coined in 1974 by that zany computer visionary Theodor Nelson (inventor of hypertext and designer of the world's oldest unfinished software project, appropriately named "Xanadu"℠), to describe a machine (patent #3,875,932) invented by a San Francisco hardware hacker by the name of How Wachspress, a device capable of converting sound into tactile sensations. The erotogenic effect depends upon where you, the consumer, decide to interface your anatomy with the tactile stimulator. VR raises the possibility of a far more sophisticated technology.

Picture yourself a couple of decades hence, dressing for a hot night in the virtual village. Before you climb into a suitably padded chamber and put on your 3D glasses, you slip into a lightweight (eventually, one would hope, diaphanous) bodysuit, something like a body stocking, but with the kind of intimate snugness of a condom. Embedded in the inner surface of the suit, using a technology that does not yet exist, is an array of intelligent sensor-effectors—a mesh of tiny tactile detectors coupled to vibrators of varying degrees of hardness, hundreds of them per square inch, that can receive and transmit a realistic sense of tactile presence, the way the visual and audio displays transmit a realistic sense of visual and auditory presence.

You can reach out your virtual hand, pick up a virtual block, and by running your fingers over the object, feel the surfaces and edges, by means of the effectors that exert counterforces against your skin. The counterforces correspond to the kinds of forces you would encounter when handling a nonvirtual object of the specified shape, weight, and texture. You can run your cheek over (virtual) satin, and feel the difference when you encounter (virtual) flesh. Or you can gently squeeze something soft and pliable and feel it stiffen under your touch.

Now, imagine plugging your whole sound-sight-touch telepresence system into the telephone network. You see a lifelike but totally artificial visual representation of your own body and of your partner's. Depending on what numbers you dial and which passwords you know and what you are willing to pay (or trade or do), you can find one partner, a dozen, a thousand, in various cyberspaces that are no farther than a telephone number. Your partner(s) can move independently in the cyberspace, and your representations are able to touch each other, even though your physical bodies might be continents apart. You will whisper in your partner's ear, feel your partner's breath on your neck. You run your hand over your partner's clavicle, and 6000 miles away, an array of effectors are triggered, in just the right sequence, at just the right frequency, to convey the touch exactly the way you wish it to be conveyed. If you don't like the way the encounter is going, or someone requires your presence in physical reality, you can turn it all off by flicking a switch and taking off your virtual birthday suit.

Before plunging into questions about whether it is ethical to build or moral to use teledildonic technology, it pays to ask how far today's technology seems to be from achieving such capabilities, because the answer appears to be: very far. Fiberoptic networks will be required to handle the very high bandwidth that tactile telepresence requires, perhaps including the kind of hybrid circuit and packet-switched technology NTT is installing in Japan as "broadband ISDN"; fortuitously, it looks like the world is going to be webbed with fiberoptic bundles for other reasons. Carrying information back and forth across town, continent, or hemisphere in large amounts, fairly quickly, will not be a problem. Until the speed of light barrier is broken, the physical size of the planet precludes a truly instantaneous on-line shared cyberspace; the larger your cyberspace is distributed geographically, the larger your system lag time is likely to be. The computation load generated by such a system is definitely a problem, too, a show-stopper, in fact, in terms of today's computing capabilities. The most serious technical obstacles that make teledildonics an early-to-mid-twenty-first-century technology rather than next year's fad lie in the extremely powerful computers needed to perform the enormous number of added calculations required to monitor and control hundreds of thousands of sensors and effectors. Every nook and protuberance, every plane and valley and knob of your body's surface, will require its own processor.

Transducers are a real problem, as well. It will take decades to develop the mesh of tiny, high-speed, safe but powerful tactile effectors: today's vibrators are in the ENIAC era. The engineering problems in building the transducers, the parts of the system that communicate in a form that people can squeeze and scratch, stroke and probe, may be formidable, but they are already the subject of focused effort on three continents. Hennequin with his pneumatics and Johnson with his shape memory alloy are not the only ones. The researchers who were showing me their demonstrations at ATR in Japan were very interested in the transmission of touch. Researchers in Italy may have just made a big step toward the kind of intimate cybergarment described above. A very crude prototype of the lightweight sensor-effector mesh has aleady been developed, according to these passages quoted from a 1990 article by Shawna Vogel, "Smart Skin":

> One of the most sophisticated approaches to this goal is being developed at the University of Pisa by Italian engineer Danilo De Rossi, who has closely modeled an artificial skin on the inner and outer layers of human skin: the dermis and epidermis. His flexible, multilayered sheathing even has the same thickness as human skin— roughly that of a dime.
>
> De Rossi's artificial dermis is made of a water-swollen conducting gel sandwiched between two layers of electrodes that monitor the flow of electricity through the squishy middle. Like the all-natural human version, this dermis senses the overall pressure being exerted on an

object. As pressure deforms the gel, the voltage between the electrodes changes; the harder the object being pressed, the greater the information. By keeping tabs on how the voltage is changing, a skin-clad robot could thus distinguish between a rubber ball and a rock.

For resolving the finer details of surface structure, De Rossi has created an epidermal layer of sensor-studded sheets of plastic placed between thin sheets of rubber. The sensors are pinhead-size disks made of piezoelectric substances, which emit an electric charge when subjected to pressure. These disks can sense texture as fine as the bumps on a braille manuscript.

These scientific frontiers provide the jumping-off point for the VR sex fantasy: Put together a highly refined version of "smart skin" with enough computing power, cleverly designed software, some kind of effector system, and a high-speed telecommunication network, and you have a teledildonics system. The tool I am suggesting is much more than a fancy vibrator, but I suggest we keep the archaic name. A more sober formal description of the technology would be "interactive tactile telepresence."

Teledildonics seems to be a thought experiment that got out of control. The quantum physicists had used this technique of imagining a certain set of conditions as a kind of mental scenario, a *gedankenexperiment*, a "thought experiment." The idea is to induce people to put themselves into an appropriate mindset for seeing the implications of a new discovery. I made the mistake of performing my gedankenexperiment on my local node of the Worldnet.

I wrote a short riff on teledildonics, not too different in content from this chapter, and posted it on the WELL; I used my modem to send the electronic version of that essay from my home computer to the larger computer a few miles from my home that stores the electronic record of conversations and publications that constitute the WELL. It's a cheap way of getting instant feedback from a few dozen respondents out of a local readership of a couple of thousand people. Anybody whose home computer is communicating with the WELL a moment after I post it, or in the middle of the night, or six months later, can tell me what they think publicly in the public conversation, or privately via e-mail. People can also do other things, like copy documents and send them places. I thought that my piece would stimulate discussion that could help me think through the various implications of sex at a distance. The weird part came when I started receiving electronic mail from around the country within hours of posting that essay. I did not take special measures to prevent anybody from duplicating the file. Apparently, one of the several thousand people who had access to the WELL had sent my teledildonics riff elsewhere via electronic mail; it only takes a few keystrokes to send an existing file on your host computer to any other computer in the Worldnet.

The piece seems to have struck a nerve. I noticed that an editor who interviewed me in Tokyo for one of Japan's largest computer magazines had a printed copy. I got calls from London and Amsterdam. People seemed to skip over all of my verbal qualifications and descriptions of technical difficulties, and almost all the people who contacted me for information seemed to believe that such a device actually exists somewhere and that I've seen it or tested it in some fulsome way. Among the other experiences my thought experiment led me into was a dinner with a German journalist. A young reporter for *Der Spiegel,* one of Germany's two largest news magazines, was traveling in search of the next computer revolution, and that had led him to VR. After our dinner, he wrote an article; it was published several months later. My name was mentioned. There was a photo. My agent sent me a very rough translation, which apparently was too rough, or else I failed to read it thoroughly enough. A month after that, I got a call from a woman in Augsburg, Germany, who was quite insistent that I should come to address a convention for NCR, within a few days. She persuaded me to speak, as specifically or vaguely as I wanted, about "future prospects for virtual reality." When I arrived in Augsburg, I was immediately whisked to a reception given by the mayor at the splendidly gilded, restored Augsburg City Hall. Then we retired to the municipal ratskeller for beer, sausage, and an interminable bilingual skit about Augsburg's history. It was at this point that NCR's German marketing communications director told me that I was a hot commodity in Germany at that time, because of what *Der Spiegel's* story had said, or what they thought it said.

"What do you mean?" I asked.

"The part where they said you were experimenting with ways to have sex with computers," the marketing communications director replied.

No wonder the vice presidents who introduced themselves to me were smiling the way they were when they told me they were looking forward to my talk. So I opened my presentation with John von Neumann's limerick.

The teledildonics story was also published in *Mondo 2000,* an avantgarde, technology-oriented "mutazine." I got even more weird phone calls after that. I can't help believing, from the reaction I've received in response to an essay I wrote in ten minutes, for fun, that the interest in this possibility will remain high. When people seem to want a technology to develop, to literally lust for a possible new toy, that need can take on a force of its own, especially given the rates of progress in the enabling technologies and the enormous market-driven forces that will be unleashed when sex at a distance becomes possible. Yes, teledildonics is a titillating fantasy, far from the serious human realities of medical imaging or teleoperated machine guns. But once you start thinking about sex at a distance, it's amazing how many other questions about future possibilities present themselves, questions about big changes that might be in store for us. Given the rate of development of VR technologies, we don't have a great deal of time to tackle questions of morality, privacy, personal identity, and even the prospect of a fundamental change in human nature. When the VR revolution really gets rolling, we are likely to be too busy turning into whatever we are turning into to analyze or debate the consequences.

One side effect of technological power seems to be that human culture is growing more mechanized. We wake up and eat and sleep and arrange our days according to the dictates of the machines that make our lives easier—or at least difficult in different ways—than our grandparents' lives. At the same time, human desires have been progressively stimulated, confused, and ultimately numbed by the barrage of provocative images, sounds, words thrown our way via electronic media; McLuhan didn't tell us that the global village would be experienced primarily by most people as an overdose of beautifully crafted advertisements, based largely on sexual innuendo, for the products of multinational corporations. Electronic media have been used thus far by a few to manipulate the desires of many, resulting in unprecedented financial profit. It is possible that telepresence technology, if linked with an inherently distributed network system such as the telecommunications infrastructure, will give this power to many, rather than reserving it for a few. Whether that is true and whether it is a good idea are both questions that remain to be settled.

To many, the idea of literally "embracing technology" seems repugnant. Computer ethicist Joseph Weizenbaum, author of *Computing Power and Human Reason*, I am sure, would consider it antihuman. And perhaps it is. We should think about those deep moral reservations of a few less than optimistic prophets very hard and very long. But there is no doubt that people everywhere in the world are fascinated by the prospect. And why not? Contemporary philosophers have pointed to progressive mechanization of human culture and the future of sexual expression as the site of a potential cultural collision of immense dimensions.

Think about a few fundamental assumptions about the way things are that might have to change if teledildonics becomes practical. If everybody can look as beautiful, sound as sexy, and feel as nubile and virile as everybody else, then what will become the new semiotics of mating? What will have erotic meaning? In the area of sexual-cultural coding, much can be learned by the way people seem to be using other electronic communication technologies to construct artificial erotic experience. "Telephone sex," in which paying customers are metered for the number of minutes they have a conversation about their choice of sexually charged topics, with a real human of the gender of their choice, might offer clues. So says Allucquére Rosanne Stone, a scholar of such matters. I met her on the net, and we knew each other's opinions pretty well by the time we met "ftf" ("face to face"), as the computer conferencing habituees say. Stone had been spending her time interviewing VR programmers, telephone sex workers, and amputees, because they all shared the experience of disembodiment and of feeling sensations from a body that does not exist physically.

When she found out about my quest into all the odd corners of VR research, Stone made contact with me via e-mail. In an electronic mail exchange, Stone sent me some provocative observations about telephone sex, part of a work-in-progress named "Sex and Death Among the Disembodied," which seemed to have direct bearing on the idea that people might use the telecommunication infrastructure for erotic gratification:

> Phone sex is the process of constructing desire through a single mode of communication. In the process, participants draw on a repertoire of cultural codes to construct a scenario that compresses large amounts of information into a very small space. The worker verbally codes for gesture, appearance, and proclivity, and expresses these as tokens, sometimes in no more than a word. The client uncompresses the tokens and constructs a dense, complex interactional image. In these interactions desire appears as a product of the tension between embodied reality and the emptiness of the token, in the forces that maintain the preexisting codes for body in the modalities that are not expressed in the token; that is, tokens in phone sex are purely verbal, and the client uses cues in the verbal token to construct a multimodal object of desire with attributes of shape, tactility, etc. This act is thoroughly individual and interpretive; out of a highly compressed token of desire the client constitutes meaning that is dense, locally situated, and socially particular.

The secondary social effects of technosex are potentially revolutionary. If technology enables you to experience erotic frissons or deep physical, social, emotional communion with another person with no possibility of pregnancy or sexually transmitted disease, what then of conventional morality, and what of the social rituals and cultural codes that exist solely to enforce that morality? Is disembodiment the ultimate sexual revolution and/or the first step toward abandoning our bodies? Whenever I think of the vision of billions of earthlings of the future, all plugged into their home reality sets, I think of E. M. Forster's dystopia of a future in which people remain prisoners of their cubicles, entranced by their media, not even aware of the possibility of physical escape. And then I think that it is good to beware of looking at the future through the moral lens of the present: in a world of tens of billions of people, perhaps cyberspace is a better place to keep most of the population relatively happy, most of the time.

Back to thought-provocative implications of telesex. If you can map your hands to your puppet's legs, and let your fingers do the walking through cyberspace, as it is possible to do in a crude way with today's technology, there is no reason to believe you won't be able to map your genital effectors to your manual sensors and have direct genital contact by shaking hands. What will happen to social touching when nobody knows where anybody else's erogenous zones are located?

Privacy and identity and intimacy will become tightly coupled into something we don't have a name for yet. In Unix computer systems, such as those used by the host computers of Worldnet, files (documents, data bases, graphics, encoded sounds and programs) and categories of users who have access to those files can be grouped into nested hierarchies by a system of "permissions," like hiding information behind doors of encryption that can be opened only by those who know the key. People who use Unix systems today often have publicly accessible file areas, in which everyone with access to the computer system has the key (a secret combination of numbers and letters and punctuation marks) to read and copy these files, and private areas for which only a small group of associates or one trusted partner knows the key. In cyberspace, if a parallel structure emerges, your most public persona—the way you want the world to see you—will be "universally readable," in Unix terms. If you decide to join a group at a collegial or peer level, or decide to become informationally intimate with an individual or group of individuals, you will share the public keys to your identity permission access codes. It might be that the physical commingling of genital sensations will come to be regarded as a less intimate act than the sharing of the data structures of your innermost self-representations.

Potential psychosocial effects of present state-of-the-art VR technology were cannily anticipated thirty years ago by Marshall McLuhan in *Understanding Media,* which seems to make more sense in the 1990s than it did at the time it was published. But future cyberspace spinoffs are getting into territory beyond the McLuhan horizon. With all those layers of restricted access to self-representations that may differ radically from layer to layer, what happens to the self? Where does identity lie? What new meanings will "intimacy" and "morality" accrete? And with our information machines and our bodily sensations so deeply "intertwingled," as Theodore Nelson might say, will our communication devices be regarded as "it"s or will they be part of "us"?

MEDIA WARP AND MYSTERY RELIGIONS IN THE 1990s

LSD is a way of mining the invisible electronic world; it releases a person from acquired verbal and visual habits and reactions, and gives the potential of instant and total involvement, both all-at-onceness and all-at-oneness, which are the basic needs of people translated by electric extensions of their central nervous systems out of the old rational, sequential value system. The attraction to hallucinogenic drugs is a means of achieving empathy with our penetrating electric environment, an environment that in itself is a drugless inner trip.

MARSHALL MCLUHAN,
"Playboy Interview," 1969

The second thing most people and all journalists want to know about virtual reality, after their curiosity about teledildonics is satisfied, is whether the technology might conceivably become a form of "electronic LSD." If you begin to question that question, to pick at the reasons it seems to pop up so often, a thread unravels. Follow the thread and you will find yourself winding through shadowy alleys in the collective unconscious of white industrial culture. If you follow it long enough, Ariadne-like, you will end up back at Lascaux at the dawn of VR and other thought-leveraging tools.

"Is virtual reality going to be electronic LSD?" I heard the same question asked, word for word, at every VR panel discussion and public demonstration I attended in 1989 and 1990. I asked the question myself, before I even experienced cyberspace. It's a good question. But it isn't the *only* question, and far from the most important one. It happens to push a hot cultural and political button, however. I began to wonder why it is that the same interesting but minor concern keeps surfacing and gravitating directly to center stage when there are so many other, more interesting questions to ask. Sex and drugs are lurid, sure, but there's more to it than that.

First of all, it is not an unreasonable speculation; it is the disproportionate attention paid to that speculation, in the context of the technology's overall potential impact, that reveals a submerged cultural neurosis. There is no doubt that something about the experience makes people note the connection. Jerry Garcia, guitarist for The Grateful Dead, house band for the acid tests of the 1960s, took a ride in virtual reality at Autodesk in 1989. He was quoted as saying: "They made LSD illegal. I wonder what they are going to do about this stuff." Indeed, the cyberspace architects at Autodesk, an otherwise somber-appearing company that seems far more devoted to CAD than cultural revolution, chose Timothy Leary to narrate their first videotape introducing their research. I interviewed both Garcia and Leary regarding VR, and they both agreed that their first trips to cyberspace reminded them, in an abstract way, of their first psychedelic adventures, that a machine that can change your worldview is similar to something like LSD in that one dimension, but not in all significant dimensions. I found that both Garcia and Leary had much more to say about many other things; I thought it equally interesting that Garcia wanted to perform virtual concerts in cyberspace, where the audience could participate in creating the performance, and that Leary called attention to the revolution in "interpersonal computing" that gave VR important cultural significance.

The media interest in the topic drew me into the plot but didn't become part of the story I was trying to cover until January, 1990, when the *Wall Street Journal* ran a front-page article on virtual reality technology with the subhead: "Electronic LSD?"

The *Wall Street Journal* reporter played on the possibilities of future ecstatic virtual realities (while failing to mention that the technology is being used today to plan radiation therapy for cancer patients). A few days later, I received a call from the first of several journalists around the world who were tracking stories on "virtual reality and other machines that make people high." Between the people who want to plug in and the people who want to stamp it out, the interest level in VR as an ecstasy machine is almost as high as it is for teledildonics. Ironically, both sex and psychedelia are much more distant possibilities than others that are less juicy but equally profound, such as augmenting the capabilities of quadriplegics or firefighters or diagnosticians or radiation treatment planners.

There is no doubt that the potential for trance, intoxication, ecstasy, or mind control is inherent in any technology that strongly affects human perceptions. That doesn't mean it is an easy task to create an electronic Eleusis or a digital brainwasher, however. We may be closer to sex at a distance, technically speaking, than cybernetic tripping. But I have a feeling that the most lurid of the wide variety of speculations about VR will continue to draw the attention of the press with irresistible semiotic power. Timothy Leary and LSD in 1990 represented not so much a person and a mind-altering substance as value-laden signs and symbols of all that is feared in the national subconscious.

Strangely, few people voice what seems to me a much more credible fear—that VR might become as addictive, energy-sapping, and intellect-dulling as *television,* "the plug-in drug" that requires the average American abuser to consume for more than seven hours a day, to the profit of those who won the battle for control of the gateway. Journalists and commentators, including myself, just can't stay away from the phrase "electronic LSD." It has word-magic mojo power.

Sure, there are sex, drugs, and even rock and roll to be found in the future of VR technology. The same could have been said of the printing press. It is important not to ignore technology-triggered changes in our culture because they involve contemporary social taboos; it is also important to refrain from obsessing on those changes. After zooming in on one attractive but less than immediate scenario, it is necessary to widen one's focus.

The problem is ecstasy, and how to handle it.

America in the 1990s has a problem with ecstasy in its original sense of *ex-stasis,* of moving out of one's daily trance for a moment, transcending the mundane particulars of mortality, shedding one's ordinary waking consciousness, to make direct contact with the numinous. Unlike the Jivaro of Amazonia or the ancient Greeks at Eleusis, we postindustrial urbanites have no socially sanctioned method of putting aside our everyday consciousness and quenching our thirst for direct experience of the *mysterium tremendum et fascinans.* Unlike most adults of the past 100,000 years, we were never initiated in fear, trembling, and joy. We never made acquaintance with our own birth and learned about our connection with death. The illusion of the self was never demonstrated in myth, song, dance, chant, and direct confrontation. The old teaching stories weren't whispered to us in moments of awe and terror. We are swept along in a digitized hyperreality that is not of our making. And now we are suffering in ways most of us don't understand.

It isn't a new story, this war between the sky religion and the earth religion, the Apollonian and Dionysian civilizations, the warriors and blacksmiths versus the healers and herbalists. But America in the 1990s has taken it to a new limit of doublespeak, disinformation, confusion, repression, and most of all—denial. We are denying so many things so strongly that many are willing to lose our reality rather than face it.

Our cultural problem with ecstasy is compounded by our cultural denial of the problem. Like the substance abusers who don't see their own behavior and deny that which everybody else knows, our official institutions conduct a War on Drugs but never stop to ask why people want to get high. We worry about the reasons our youth are hypnotically attracted to interactive electronic games, at the same time our educational infrastructure is crumbling; and except for a few crusaders, few stop to ask whether other kinds of knowledge might be made fun to learn. The use of plant substances by indigenous people on this continent has been all but exterminated. Yet alcohol kills half a million people a year, and while attendance figures at conventional churches are steadily dropping, the populations of fundamentalists and new-agers who share nothing but a sanction for direct religious experience are undergoing unprecedented growth.

The VR experience breaks the frame of everyday reality, although it does not, as yet, catapult the user into the kind of profoundly different experience that can be catalyzed by psychedelic chemicals. But that simple frame-breaking, and the vast potential for symbolic and what Jaron Lanier calls postsymbolic experience represented by today's crude VR systems, represents the possibility that someday, in some way, people will use cyberspace to get out of their minds as well as out of their bodies.

Is that such a bad idea? Maybe the problem with intoxicants in Western industrial society is not the urge to indulge, but the intoxicants themselves and the context in which they are consumed. Looking at the electronic ecstasy question from the other side is another instructive thought experiment: virtual reality, if inspired and talented people are seized by the vision and the desire to make it so, *might* become the first wholesome, integrating, nonpathological form of ecstasy capable of liberating safely the long-repressed Dionysian energies of our heavily Apollonian civilization. One answer to the "electronic LSD" question is, therefore: "Yes, VR might become a key to open the doors of perception, but only if someone has the grace and good sense to design it properly."

What will happen to the economics of both society's sanctioned and proscribed ecstasy-substitutes when a reasonable electronic facsimile comes over a fiberoptic pipeline with your nightly video selection? That causes problems in worlds where supply, demand, and profit are important. It is helpful to use sex and ecstasy, two highly charged subjects, as a way of seeing the possible effects of VR on the way people live and think. But those are not the most important implications of VR technology to worry about. Not with military development of remotely piloted lethal weapons proceeding at full steam. The application of telerobotic research to weaponry has been a matter of ongoing interest by several military services. To make matters more confused, the people who are trying to use telepresence to save lives are benefitting directly by some of the knowledge garnered by military teleoperation researchers.

VIRTUAL WARFARE, TELENANOROBOTICS, AND TELEOPERATOR RESEARCH

Anthropomorphic teleoperator systems have special advantages and are particularly well-suited for some kinds of work, such as underwater repair jobs in which the precise nature of the task cannot be anticipated. The sense of remote presence one experiences in the green man's operator station is quite strong, and there is virtually no need for the operator to think about how to translate his or her motions into motions at the remote work location.

WILLIAM R. UTTAL,
"Teleoperators," 1989

I decided to stop in Honolulu on my way back from Japan, to see if I could catch a glimpse of the green man. It is said to reside on the windward side of Oahu, a grenade lob away from a lagoon where dolphins practice underwater demolition missions. The US Navy's anthropomorphic telerobot—called "the green man," although it is the color of brushed aluminum and stainless steel—wasn't in residence, unfortunately; it was attending a conference in Utah when I was in Hawaii, but I did get a close-up look at a jeep that had binocular television cameras and robot arms in place of a human driver. Behind the robo-driver of the Teleoperated Land Vehicle, and connected to it, was a huge spool of fiberoptic cable that was designed to pay out over the landscape for as much as 300 km at a time. At the other end of the cable, in a control van, is the human driver, who sees a stereoscopic version of everything the robot sees, who moves the robot's cameras by swiveling his or her own neck, and who, presumably, controls the trigger of the 50-mm machine gun mounted forward of the optic cable spool.

I was aware of the Air Force role in developing head-mounted displays for jet pilots, and I knew that the Navy's interest in remotely operated vehicles dated back to the time they were asked to retrieve thermonuclear warheads that had accidentally fallen out of a B-52. In Seattle, I had seen one of the nodes of SIMNET, the global war game simulator. But there's nothing like a 50-caliber machine gun and a robot driver to bring home the implications of military teleoperations.

Like many other aspects of computer technology, military needs and

research sponsored by the Department of Defense were the most powerful driving forces in the development of teleoperator technology.
Durlach at MIT is interested in testing the limits of human nature.
Tachi in Tsukuba, Leifer at Stanford, Hennequin in London, are
trying to design teleoperator systems for the disabled. Robert Stone
in Manchester wants to build a teleoperated firefighter. But the first
serious research money began to spur the development of teleoperators when a couple of thermonuclear-armed missiles landed in a deep-
sea trench. Researchers in the US Navy began immediate work on a
well-funded, well-staffed crash program to develop a means of steering
a robot submarine down to hook a cable onto those missiles. If you
are trying to use a teleoperated robot arm to grip a wrench and dis-
mantle the detonator of a thermonuclear weapon in deep ocean, it
helps to have a feel for your tools. Hence, the international reputation
of the Naval Ocean Systems Center (NOSC) in Hawaii as a hotbed of
advanced teleoperator research.

At the Santa Barbara conference on human interfaces for teleoperators, Dr. Walter Aviles had exhibited an amusing videotape of
NOSC's roving vehicle. There was no machine gun, but there was the
robot with the binocular cameras in the driver's seat. Apparently, the
operator, who was using a high-performance HMD back at the NOSC
laboratory, had steered the vehicle to the edge of the NOSC property.
A woman who was jogging along a road adjacent to the robot vehicles
test area, and who clearly was not aware of this top-secret research,
did a double-take when she saw the robot driver. The operator, speaking into a microphone miles away, asked the woman if she would come
a little closer; she stopped jogging and came closer, cautiously. The
operator explained that he was a half mile away, inside a building.
The anonymous woman looked at the robot and shook her head skep-
tically and jogged away.

When I was in Japan, Susumu Tachi had spoken highly of NOSC,
so I changed my itinerary to give me twenty-four hours in Oahu on
my return flight. An exchange of faxes arranged a date for me to be
escorted through the facility by Dr. Hugh Spain, who advised me that
this was a "secured military research facility." I arrived at night. The
next day, I drove my rented car up over the mountains that look
exactly like Hawaii is supposed to look—verdant and primeval. The
lagoon next to the NOSC buildings would be a snorkeler's paradise if
it wasn't top secret. This close to Pearl Harbor, they are pretty tight
when it comes to military security in general. The military applications
of NOSC's research make it even more security-conscious. Because
everybody has to park and to verify credentials at the guardhouse
before entering the base, there's a bit of a wait. After I showed my

passport and they saw me on the list and called Dr. Spain, I was given directions. I remember the route very clearly, because in order to get to the area adjacent to the lagoon where NOSC is located I had to wait for a traffic signal to change at the edge of an airstrip. The sound of two F-15s taking off at the same time, a hundred feet in front of you and ten feet over your head, is a serious acoustic experience. A couple of minutes after they took off, I put my teeth back in my head and drove across the airstrip.

Unfortunately, the entire "green man" was in Salt Lake City at the time, where Dr. Aviles was conferring with Dr. Jacobson's team at the University of Utah—the people who had created "the Utah Arm," the most advanced artificial arm designed as a prosthetic. Again, there is a fateful connection between prosthetics and weaponry: you can make weak people strong or make strong people deadly by using telepresence to put human perceptual and cognitive capabilities at the helm of a robot. Whether the robot wields a knife or a scalpel is a matter of human intention, not mechanical capability. The green man is as close to anthropomorphic in its movements, if not its appearance, as current engineering can achieve. Films of it in operation are eerie. The operator puts on a helmet HMD and carries a harness that locks the arms and hands and fingers into a lightweight exoskeleton. Like a cobra dancing for a snake charmer, the green man duplicates with uncanny precision, very close to instantaneously, every movement of the operator. You know it's a machine, but it moves the way only humans are supposed to move. I saw the laboratory, I tried on the HMD, but the teleoperator wasn't there.

"Telepresence performance assessment" is what Spain told me, when I asked about his own specialty. "We want to lay the foundations for the future design of teleoperation systems."

"What kind of tasks can your systems perform with present capabilities?" I asked.

"They can thread a number-ten nut onto a stud, screw in a lightbulb, take cheese off a mousetrap, hammer a nail, bounce a tennis ball on a racquet," he replied. Later, he showed me a video demo of teleoperators performing all those tasks, which are easy for humans, not easy for machines, and somewhere in between for teleoperators.

A pleasant and hospitable fellow in standard Hawaiian informal civilian garb, Spain made little reference to military applications, although they were obvious. We went into another building where I put on an HMD and operated a robot arm in a standard pegboard task. Pegboards are like test patterns for teleoperators; I saw an almost identical arrangement at the British Atomic Energy Commission facility in Didcot. On the way to test teleoperator rigs, Spain and I passed through a quonset hut in which I saw the teleoperated vehicle. The idea of VR warfare takes on a different kind of significance when you look down the bore of the machine gun in the passenger seat. Spain saw me looking at it.

"Semiautonomous weapons are a politically sensitive area," he said. Even with the "semi" in front of it, the idea of "autonomous weapons"—"Robowarriors"—might indeed become a political hot potato if enough people knew about it to make it an issue. I found Dr. Spain to be as sincerely devoted to furthering the technology of teleoperators as Tachi and others, and his publications are solid contributions to the scientific literature. He seems like a thoroughly nice fellow. And I've heard arguments from others who develop high-tech weapons systems that it was just such technological superiority that convinced the Soviet Union to make significant arms reductions. The teleoperated gun vehicle gave me the willies, though.

Gazing at the teleoperator-rigged version of the High-Mobility Multipurpose Wheeled Vehicle, I remembered something I had seen months previously: SIMNET. While I was in Seattle, visiting the HIT Lab, one of the people who called Furness's laboratory-under-construction to inquire about employment was a graphics programmer from the Seattle branch of Bolt, Beranek and Newman (BB&N). He was working on something called "SIMNET" for the military. BB&N was involved with the invention of the first computer networks, and they are still one of DARPA's most advanced contractors; I decided to pay them a visit and peek at the state of the art in military simulation. What I saw was impressive at the time, sobering in its implications, but it began to make a whole new kind of sense when I saw that robot guncar at NOSC and when I found out about "Ender's Game" and the video game warfare scenario.

SIMNET is a project funded by the Defense Advanced Research Projects Agency that includes over two hundred tank simulators, located in Germany, Washington, D.C., Fort Knox, Kentucky, and a few other places. Although they are geographically dispersed around the planet, these telecommunication-linked simulators interact with the same virtual battlefield in real time. Using the highest-speed communication lines of MILNET, the contemporary military branch of the original ARPAnet, these four-person simulators make it possible to fight an entire war game in cyberspace. The tank simulators look and feel like the interiors of M-1 tanks, and through the viewports is visible what the tank crew would see of a battlefield, except the battlefield is a very high resolution computer simulation. The terrain data base of the simulation I saw represented the country around Fort Knox where mock tank-battles are sometimes conducted. They have probably uploaded, at other times, the data base for Middle Eastern battles and other possible sites of armored warfare. The other tanks that are visible during a SIMNET session are controlled by other tank crews that can be twelve feet or half a world away, in real time. The view of the battlefield changes as the tank moves, in real time.

If an enemy crew aims accurately and fires at the right time and if the tank crew on the receiving end of the missile or shell does not succeed in evading, the target tank is *hors de combat*. The other participants in such combat, primarily tank-killing helicopters and high-speed fighter-bombers, can also participate, from different simulators. Considering the high cost of training soldiers in physical war games, and the declining cost of computing power, SIMNET is an economical option. And like Furness's test cyberpilots, the tank crew members went for this training experiment with unbridled enthusiasm. Besides playing against hundreds of other crews in real time, it is possible for a single crew to play against "Semiautomated Opposing Forces." Video games for real warriors.

War games in cyberspace certainly use less petroleum and create fewer conditions for fatal accidents. It might not be an altogether bad idea. The idea gets a bit creepy when you hear about "Ender's Game." I first heard about it from a fellow I met on the net, who told me about a science-fiction story that had such serious implications that NATO had already conducted a workshop on how to prevent the scenario from happening. The Institute for Simulation and Training (IST) at the University of Central Florida is the only academic SIM-NET site. Eventually I met J. Michael Moshell from IST face to face, at the First Conference on Cyberspace in Austin. When I mentioned to Moshell that I had peeked at SIMNET, he told me the plot of Orson Scott Card's science-fiction novel: Suppose you train a crack crew in a SIMNET-like setup to use state-of-the-art virtual weaponry to destroy an entire civilization. And suppose that during one of their "training exercises," the simulated weapons at the other end of the video war wizards' reflexes are secretly switched to real weapons.

Weapons of mass destruction are, unfortunately, often technologically indistinguishable from tools to save lives: A remotely piloted vehicle that can enter a building and save its inhabitants from fire can just as easily kill them. The question is not whether such research can be prevented, but whether there will be controls on the military development of telepresence, the way there is at least some attempt at controlling the proliferation of nuclear weapons, research and deployment of chemical and biological weapons, and other technologies that should not be turned against human beings. Certainly, the degree to which warfare is automated is itself a public policy minefield. It would be ironic if public apprehension over the possibility of teledil-donics or electronic LSD were to dominate the debate over VR tech-nology, while the development of much more dangerous and more probable semiautonomous weapon systems goes undiscussed.

The increasing use of robotics in manufacturing has the potential to drive teleoperations R & D in the commercial sector, building upon and extending the military research to civilian applications. Scientific research certainly is one of the most active extensions of teleoperator technology, when the teleoperator is linked to sensors in environments that are hostile or inaccessible to humans. The "fantastic voyage" brand of medical scenarios, such as the one Thomas Furness quotes, is one possible direction in which teleoperated vehicles might be turned to human benefit. Although I did not meet them, I became aware of two groups who were actually working on the technologies needed to build teleoperated microrobots capable of swimming through the human bloodstream and performing internal surgery. Yotaro Hatamura and Hiroshi Miroshita, in their report, "Direct Coupling System Between Nanometer World and Human World," tell of using one of the most powerful devices for probing the world of the very tiny, the scanning electron microscope, as the source for stereo images of a submicro-scopic world that is also probed by highly sensitive force-sensors. This system, if it can be developed a great deal further than today's pro-totypes, someday might be capable of building microscopic robots ca-pable of navigating through blood vessels.

Telenanorobots ("nano" is the prefix used for scales at the molecular level, where "micro" is at the larger magnification of the cellular level) can be used as probes for pure scientists. A team at IBM's Yorktown Heights has succeeded in connecting a scanning tunneling microscope with a teleoperator and a force-feedback puck that enables the op-erator to feel the surface of atoms with his or her fingertips. The IBM team published a report in 1990 with the title "Toward a Tele-Nanorobotic Manipulation System with Atomic Scale Force Feedback and Motion Resolution." The possible revolutions that could stem from new technologies at the micro and nano level are beyond the scope of this book; however, if and when microrobotics and nanorobotics blos-som, teleoperation will play an important part. It's a convergence in progress.

INDUSTRIAL STRENGTH VR: TELEPRESENCE IN FACTORIES

"Charley, I think the mixture in vat 6 is cooling too fast."
"Then stick your hand in it and find out how it feels."

TELEPRESENCE overseers at a future
chemical plant.

Computers can be wonderful amplifiers of human capability, and computerization can be a soul-deadening juggernaut of alienation. Is virtual reality the ultimate form of alienation, or can it be used as a tool to diminish some of the alienation that already has occurred as the result of computerization? Consider those cases where the *nature of the computer interface,* not the use of the computer itself, has removed people from the true nature of their work. There is no stopping global computerization. But the way in which we deal with the computational membrane that is surrounding us is not entirely out of human control. The economic advantages of computerization will ensure the continued introduction of computer-mediated or computer-controlled systems in the workplace. The alienating effects of the interface—the place where people make contact with their tasks—is one area of both human and economic significance where cyberspace might, paradoxically, return people to more direct contact with their work.

Industry—the coordination of matter, energy, and human effort in the production of the ten thousand objects and energies that seem to be vital to modern life—is an area where the right kind of cyberspace application could produce high leverage. And tools tend to evolve very quickly when they are found to produce leverage in an expensive and competitive enterprise. Design—which one might regard as the "front end" of the industrial system—is a discipline in which 3D visual immersion, telepresence, and the ability to manipulate actual design data bases by reaching out and touching virtual objects could produce enough competitive advantage to drive virtual reality as a CAD interface. There are areas in the "back end" of industrial production, the molten-metal and caustic-chemical part of the system, where similar points of leverage can be found.

The control of industrial processes, for example, is a little-seen but economically vital area in which the introduction of computers has had strange effects, and in which a virtual reality interface has high payoff potential.

Process control is somewhere between a science and an art, involving a delicate, vital, and often dangerous minuet of humans and machines.

Although those who aren't involved with it don't pay much attention to the way process control is managed, the results are essential to many parts of daily life, from the pulp mills that produce the newsprint in our morning papers to the petroleum refineries that fuel our automobiles and the smelters that produce the steel in the freeways we travel, the buildings we enter, the bridges we cross. Process control used to be performed by experts who felt the heat of the smelter on their face, looked at the color of the liquid in the vats, reached right in and grabbed a handful of wood pulp and squeezed it between their fingers. Over the past twenty years, however, every major industry in which process control is a major concern has switched to automated, computer-controlled systems. And that has removed the humans from direct sensory contact, stripped them of their ability to directly interact with the process.

In her book, *In the Age of the Smart Machine: The Future of Work and Power*, Harvard-based social scientist Shoshana Zuboff investigated the way the introduction of computers changed the way people work in factories and offices. In the case of process control, Zuboff uncovered an unsettling kind of alienation. Skilled workers who had been fine-tuning industrial processes for decades found themselves in a situation where their old skills were useless, and the nature of their interface to the process forced them to learn unfamiliar skills. In some ways, it looks like human senses and intuition are better for monitoring a complex process than a highly abstracted computer control that removes humans from the factory floor and puts them in control booths, looking at screens. Zuboff interviewed skilled workers at pulp mills that were in the process of changing to a computerized system.

Zuboff saw a larger pattern, in those who worked with information and services as well as those who worked with industrial processes, involving the emergence of an unplanned set of new skills required by workers in what the author called an "informated environment." She wrote this about the pulp mills she studied:

A fundamental quality of this technological transformation, as it is experienced by workers and observed by their managers, involves a reorientation of the means by which one can have a palpable effect upon the world. Immediate physical responses must be replaced by an abstract thought process in which options are considered, and choices are made and then translated into the terms of the information system. For many, physical action is restricted to the play of fingers on the terminal keyboard. As one operator put it, "Your past physical mobility must be translated into a mental thought process." A Cedar Bluff manager with prior experience in pulping contemplates the distinct capacities that had become necessary in a highly computerized environment:

"In 1953 we put operation and control as close together as possible. We did a lot of localizing so that when you made a change you could watch the change, actually see the motor start up. With the evolution of computer technology, you centralize controls and move away from the actual physical process. If you don't have an understanding of what is happening and how all the pieces interact, it is more difficult. You need a new learning capability, because when you operate with the computer, you can't see what is happening. There is a difference in the mental and conceptual capabilities you need—you have to do things in your mind."

Zuboff pointed out that there is an important difference between the kind of "cause-and-effect" knowledge that experienced operators seemed to have and the kind of knowledge required when the computer system was installed:

> In plants like Piney Wood and Tiger Creek, where operators have relied upon action-centered skill, management must convince the operator to leave behind a world in which things were immediately known, comprehensively sensed, and able to be acted upon directly, in order to embrace a world that is dominated by objective data, is removed from the action context, and requires a qualitatively different kind of response. In this new world, personal interpretations of how to make things happen count for little. The worker who has relied upon an intimate knowledge of a piece of equipment—the operators talk about having "pet" knobs, or knowing just where to kick a machine to make it hum—feels adrift.

Zuboff noted that a new kind of learning was required when these wood-pulp factories yielded to computer control, a kind of learning where "Hammers and wrenches have been replaced by numbers and buttons." As a direct example of this kind of vocational disembodiment, Zuboff quoted from his interview with an operator with thirty years of service in the Piney Wood plant:

> "Anytime you mash a button you should have in mind exactly what is going to happen. You need to have in your mind where it is at, what it is doing, and why it is doing it. Out there in the plant, you can know things just by habit. You can know them without knowing that you know them. In here you have to watch the numbers, whereas out there you have to watch the actual process."

"You need to have in your mind where it is at"—it is a simple phrase, but deceptive. What it takes to have things "in your mind" is far different from the knowledge associated with action-centered skill.

This does not imply that action-centered skills exist independent of cognitive activity. Rather, it means that the processes of learning, remembering, and displaying action-centered skills do not necessarily require that the knowledge they can contain be made explicit. Physical cues do not require inference; learning in an action-centered context is more likely to be analogical than analytical. In contrast, the abstract cues available through the data interface do require explicit inferential reasoning, particularly in the early phases of the learning process. It is necessary to reason out the meaning of those cues—what is their relation to each other and to the world "out there"?

Returning mobility and the use of the body, the hands, the skin, the visual system, the ears, all the channels of information for our biological processors, might have a rehumanizing effect on the "informated" workers in chemical plants and steel mills from Jakarta to Chicago who have been moved out of the automated factory floor into a control room, dealing with a computer. It remains to be seen whether a useful model of a truly complex process like oil refining could be created and tested in cyberspace. And if such VR front ends to process-control systems could be built, it remains to be seen whether this is a step toward making industries more human-centered, or another step toward the roboticization of the human race. The stakes, however, are high enough to make it very likely that somebody is already looking into it.

When I was in London, en route to my encounter with Spitting Image, I visited Kees van der Heijden, a strategic planner at Shell Centre, an unremarkable, even somewhat drab building, across the Thames from the Houses of Parliament, that houses a $100 billion/year corporation. He was particularly interested in the notion of combining an architectural walk-through with some kind of rough process-control simulation. "We spend a lot of money building models of a refinery before we invest a half billion dollars in it. If we can build computer models that give us more insight into planning the facility, at lower costs than the mechanical models we make now, sure we're interested," van der Heijden said. He also mentioned something else I had heard on the net, that Bechtel, the people who built Hoover Dam, the first nuclear reactors, and the port of Jubail, are actively investigating VR. It makes sense, financially. Companies that big have already invested a great deal of money in computing power. If a relatively expensive front end can enable their people to dive into the extremely expensive computers they already have, and come back with better decisions, then VR might be driven quietly at the high end, by those companies that can afford to have their own hardware and software built.

Industry depends upon technology, but is driven by finance, and that brought me to consider another idea that sounds esoteric and futuristic today, but which might transform the world more quickly than anyone suspects—*financial visualization,* the use of VR systems in conjunction with economic models and real-time information about financial transactions, as a way of dealing with the increasingly complex global economy.

VR AND THE GLOBAL INFORMATION ECONOMY

A teletype operator must monitor a hundred minor details of method and machinery, but the process quickly becomes semi-automatic. Money moves in cycles, pulsing between trickle and torrent as shifts change and banks open around the world; at peak traffic hours it took two fast operators to keep the incoming paper from burying the receiving machines, and a dozen more to retransmit. The tape constantly piled up and snaked across the floor, to be processed and sent out again, creating rivers of paper tape for the operators at the other ends of the lines. Tapes and paper rolls were changed rapidly, like pit stops on an endless road race. After a while, I caught the feel of gigantic economic tempos, literally sensing the pulse of international trade through my fingertips.

HOWARD RHEINGOLD,
"The Ultimate Cashflow," 1976

Sex, drugs, weaponry, industry, science, medicine—what about money? Sure enough, another kind of cyberspace has been incubating in the world of global finance, far from the world of head-mounted displays and 3D visualization. I've already met the first brokers, traders, and futurists who hope to plunge into it directly, via VR interfaces to real-time financial data, as soon as such systems are developed. The relationship between VR and big bucks is connected with the way electronic communication technologies have changed the nature of finance at a fundamental level, almost without anyone noticing. The notion of the global information economy is no longer a forecast or a scenario; most economists would agree that the basis for wealth has changed subtly, profoundly, and irrevocably over the past three decades, due to the increasing linkages between financial transactions and global communication and computing systems. Transaction systems and representation systems are another convergence—or collision course—in the making.

I first got wind of this electronic virtualization of money when it was just beginning to gain momentum, in the mid-1970s. The vicissitudes of freelance writing brought me to the wire room of the Bank of America, then known as "Central Telegraph," as a swing-shift teletype operator. I glimpsed something back then that came back to haunt me when I started talking to some real experts on what has happened to wealth. This is what I wrote for the *San Francisco Chronicle* in 1976 about my job:

> I was a teletype operator at Central Telegraph, an obsolescent human bottleneck in the financial intercourse of nations. Teletype operators were replaced at the Bank of America by sleek consoles full of silicon and circuitry, so my coworkers and I were the last of the human money movers. To a teletype operator of the old school, money is nothing more nor less than a message, a message one banker sends to another. That part hasn't changed—only the communication technology is different.

What I missed at the time was the way the change in communication technology could change the nature of the game. The whole business of money, messages, teletypes, and automated trading systems came up when I talked with Peter Schwartz about VR and the world financial structure. He and I had been planning to have this conversation for quite some time; he was interested in what I knew about VR, and I was interested in what he knew about the future of the world economy. Schwartz was formerly the director of the Strategic Environment Center at what was then the Stanford Research Institute and is now known as SRI International, the first highly successful consulting futurists in the business world. He and his colleagues specialized in a way of helping high-level decision-makers reperceive the driving forces and critical uncertainties in their business environments by constructing "scenarios," structured thought experiments about the future of their business. After leaving SRI, Schwartz became a high-level strategic planner for Royal Dutch Shell, and from Shell, Schwartz went to the London Stock Exchange, which was in the process of converting its operations from an actual physical trading floor (on the site of a former restaurant—the brokers were still served by bewigged "waiters" when Schwartz came aboard). In 1986, an event that is still known as "the Big Bang" occurred: the trading floor disappeared overnight, and moved to the terminals of tens of thousands of traders worldwide. The London Stock Exchange became the International Stock Exchange. And its main business became the transportation of information.

Schwartz, a fellow who likes to grin and tell you something that is both true and astonishing the second you sit down to talk, has been described as "a cross between a sorcerer and a bartender." He likes to tell stories about the future and, through the telling, show how the world is changing.

"The movement of wealth through the world communication system has created a kind of cyberspace," Schwartz noted. In fact, Schwartz is a fan and a friend of William Gibson, whose vision of the future Schwartz finds "fascinating and ominous."

"One mind-blowing number tells its own story about the way the world works now," Schwartz said, warming up to the tale: "International foreign exchange transactions reached $87 trillion in 1986. That comes to twenty-three times the US gross national product. It's several times larger than the gross world product. The values of currencies are no longer determined by trade volumes or any of the physical activities normally associated with industrial economies. Trade is only about 10 percent of that $87 trillion. The rest of it is generated by electronic transactions." Peter Schwartz believes that money is now a kind of message, just as I had suspected in 1976. What I had not suspected then was that the message-money would soon dwarf the other kind and create a new kind of wealth that we have barely begun to understand. Schwartz continued, prompted by my inquiries about where this development started.

"The new kind of wealth—huge amounts, fluid, rapidly moving, and based on communications rather than precious metals, armaments, or industrial production—evolved over the past twenty years because of several interrelated factors. First, the exchange rates that determine the relative values of various currencies were cut loose when the Bretton Woods agreement fell apart in 1971. The agreement that had been established in the 1940s provided a relatively slow-moving mechanism for adjusting the exchange rates of currency. In 1971, the agreement disintegrated, the old rules for determining the values of currencies from day to day were no longer valid, and it became possible to make huge amounts of money absolutely risk-free by moving even huger amounts of money from one currency to another—a process that was facilitated by electronic communicaton."

The shift from teletypes to telecommunicating computers was a key step in that process, as it happened: "Money is an agreement. At a fundamental level, the agreement is in the minds of all the people who use currency. International currencies are more complex agreements built upon this fundamental social and psychological agreement, involving national and international regulations, market values, and the way in which transactions are consummated. And the way the transactions are consummated is via communication technology, which is, itself, evolving."

Over the past ten years, the wire rooms of all the major players on the world financial scene have traded their teletypes for computerized communication systems. Computers send coded electronic impulses through telephone wires, dedicated lines, fiberoptic cables, microwave links, satellites, and other channels of the world's increasingly interconnected communications infrastructure. With this new system, money-messages are delivered at the speed of light, which is considerably quicker than the teletype process that requires humans to tear, fold, staple, and walk across the room to deliver the messages as they came off the wire. "So when you replace teletype machines with computers, you boost enormously the rate at which financial messages can be transmitted and received," Schwartz pointed out, when we compared perceptions about money as messages.

This technological infrastructure for increasing the velocity of money transfer happened to be installed at the same time that large-scale international currency transactions became possible, Schwartz explained: "Changes in technology combined with changes in human agreements to create a new driving force. The amount of choices, and the complexity of keeping track of them, suddenly increased, which leads to more powerful information systems that boost the system to yet higher levels of complexity. As a result of these factors, the world economic system and all the local economic systems that plug into it have changed in a fundamental, qualitative way. We are only beginning to understand the dynamics of this system. It wasn't deliberately designed. It came about as institutions saw the new opportunities that technology and international agreements were making possible."

Schwartz also pointed out that Wall Street and Hollywood are the major customers for future information technologies: "The financial services industry worldwide is the biggest private customer of information technology and services. After finance comes the burgeoning global entertainment industry. Nearly half the planet is under twenty and if they don't all have personal cassette players yet, they soon will. The appetite for recorded entertainment seems boundless and not related to economic cycles. With the advent of new technologies such as CD, digital audio tape, satellite TV, high-definition TV, and video interactive personal computers, the revolution will pick up speed. Finance, entertainment, and education are the high leverage, high demand and therefore, high growth markets for information technology."

The complexity of the emerging global information economy, its global, dynamic, abstract, and technology-dependent characteristics, is where VR will enter the story, Schwartz believes. Schwartz knew about Engelbart and augmentation tools and the need for finding new ways of dealing with complexity. Having helped the London Stock

Exchange transform its trading floor into a computer network, Schwartz is particularly interested in new technologies that are still on the fringes but might cause sudden changes in the future. "VR is exactly the kind of major change that I look for on my clients' behalf and to satisfy my own curiosity," Schwartz noted. "Once I found out about it and contacted Jaron Lainer and tried it for myself, I realized that it is an extension of what we have already experienced in the evolution of the personal computer industry and this emergence of a global information economy. Commodity trading systems and Reuters have developed on-screen trading floors. I immediately thought of a scenario in which on-screen trading floors are just the first step toward VR trading." In other words, stockbrokers and bankers and even the ordinary citizens who buy a couple of shares here and there might eventually participate in some kind of global real-time video game with concrete financial consequences, just as military SIMNET crews play a different kind of video game with equally serious consequences.

I met a representative of American Express who was also interested in the idea of financial visualization, at the Austin conference on cyberspace. And in October, 1990, an article by Andrew Pollack in *The New York Times,* titled "Coming Soon: Data You Can Look Under and Walk Through," cited the work of Professor Steven Feiner of Columbia University, working on "similar tools for visualizing information," who is experimenting with the use of a VPL DataGlove in conjunction with electronic shuttered glasses to "reach into the computer world and manipulate 3D graphs." In the same *Times* article, Citicorp is mentioned as funding attempts to "portray portfolios of options as 3D structures to help options traders see how the value of a portfolio changes with variations in other factors."

Schwartz foresees cybernautic brokers of the future who zoom through landscapes that are the 3D depictions of marketplaces, as reported through global electronic transaction systems, in real time. The skill with which a company's programmers can portray key aspects of the changing economic landscape, and the skill with which the company's brokers use it to make decisions about the future of the market, will make the vital difference in large financial transactions. Magic lenses—proprietary 3D financial visualization modeling software that makes profitable or dangerous ventures visible early—will be jealously guarded secrets. Grabbing the gold ring in a market visualization might consummate a transaction in deutschmarks or petrodollars. Cruising through a forest and catching a lizard and eating it might be the way corporate raiders of the future will play out their games of merger and takeover.

It's too early to tell whether Disney or Penthouse, the Pentagon or

Wall Street, Hollywood or Sony, computers or telecommunication networks, or some unanticipated coalition, will occupy the dominant niches in the ecology of a maturing VR industry. But there is little doubt that an increasingly important survival element in just about every scenario is the ability to learn and adapt, to pay attention to the environment and oneself, to respond flexibly in complex and uncertain situations. It is possible, as Peter Schwartz suggested to me, that education will be a growth industry under such circumstances. "Individuals and nations that learn how to learn are more likely to prosper in the 'informated' world of the coming decades," he conjectured. Just as sex, work, war, entertainment, science, and commerce all seem to be affected in some way by VR research, education might be one of the more optimistic arenas of possible change.

THE IMPORTANCE OF PLAY

A child, as well as an adult, needs plenty of what in German is called Spielraum. Now, Spielraum is not primarily "a room to play in." While the word also means that, its primary meaning is "free scope, plenty of room"—to move not only one's elbows but also one's mind, to experiment with things and ideas at one's leisure, or, to put it colloquially, to toy with ideas.

BRUNO BETTELHEIM,
"The Importance of Play," 1987

If you want to change your life for less than ten dollars, stop at a toy store and buy a package of those small plastic building blocks that stack and snap together in various ways. Keep them in your desk drawer, and when you feel the need for creative inspiration, take out the blocks, sit on the floor with them, and . . . play. Be sure to lock your office door, though, because "play" is fraught with taboos in our culture. It still shocks me to discover how many parents never get down on the floor with their children and simply play. It might be that "play" answers one of the biggest questions raised by virtual reality technology: What are human beings meant to do?

Play has been seen as an antonym to "work," and that's probably where it got its bad reputation. Ever since John Calvin linked work and salvation, "just playing around" has been seen as something close to sinful. Creative work, however, requires a delicate balance of aimless play and focused, directed activity. If it isn't turned into toil by a relentless instrumental viewpoint, play is our most important thinking tool—particularly when we are learning to think in new ways.

Paleontologist John Pfeiffer was also interested in the origins of play and its relation to the maturation of human culture. His observations seem to evoke a resonance with some of the same visceral feelings I have about VR, a topic he most probably knew nothing about:

> Art seems somehow to have arisen from play, in a uniquely human spinoff process which has acquired a life of its own. Both involve imitation, pretending, a measure of fantasy, the freedom to improvise, to make and break rules and create surprise. But this insight, however plausible and frequently voiced over the years, raises more questions than it answers, a major reason being that play is a complex and poorly understood activity. The word itself is highly misleading, an intellectual boobytrap. It implies something useless, something not to be taken seriously; and yet everything we know about evolution argues that it is to be taken very seriously indeed.
>
> Play is something new under the sun. It appeared after 4.3 billion virtually play-free years, in times when the planet's surface consisted of a single great ocean surrounding a single land mass or super island just beginning to split into continents. It appeared some 200 million years ago with the appearance of warm-blooded animals. . . .
>
> Play must have been extremely important in recent evolution, all the more so because it entails a number of major disadvantages. It uses up energy which might be better devoted to feeding, resting, or less vigorous socializing, and often results in serious injuries from falls and collisions with rocks and trees. Furthermore, young animals may be so completely absorbed in their games that they become careless and extra-vulnerable to predators. To outweigh the risks play must offer especially high survival premiums, such as providing training or practice for real-life fighting and escape tactics, and promoting friendships and cooperation among individuals who will be spending many years together.

If you look very far into the literature on play, you encounter Johan Huizinga, the rector of Leyden University in Holland, who published a book in 1938 with the title *Homo Ludens* ("man, the player"). Huizinga followed the strands of subject matter that are woven into the topic of play—philological, mythological, anthropological, psychological— and ended up in roughly the same place as some of today's cognitive scientists. Play, particularly symbolic play, is where cognition and culture meet. It's a mental can-opener for liberating new ideas. It is also the first thing most people do when they find themselves immersed in a virtual world.

Huizinga had been trying to understand why medieval Christendom continued to maintain archaic cultural elements such as codes of honor, heraldry, chivalric orders, etc. Huizinga began to discover that these were vestiges of primitive intiation rites—sacred play. People did them because they were fun or deeply engaging, and because, in Huizinga's view, play was a vehicle for creating culture.

Only recently have psychologists become interested in studying the role of play in early development and adult life. After Piaget revolutionized developmental psychology by studying his own children and others as they engaged in various kinds of play, psychologists of various flavors contributed to a growing understanding of play as a vital part of every person's cognitive, social, and emotional development. To summarize and oversimplify their findings, these psychologists discovered that play is a way of organizing our models of the world and models of ourselves, of testing hypotheses about ourselves and the world, and of discerning new relationships or patterns in the jumble of our perceptions. Play, like the scenarios that Peter Schwartz and his colleagues use to prepare themselves for an uncertain future, is a way of thinking ahead, of running a mental simulation.

Consider the notion that play is a computer simulation that we run inside our head. "What if I pile these blocks up as high as I can?" "What does it feel like to smoosh together everything in the spice drawer?" "This isn't a cardboard box. It's a castle." I sensed that something important was happening when my daughter, at around eighteen months, lifted a rattle to the mouth of her teddy bear and started to feed him a "bottle." She was learning to use one object to symbolize another. She was learning to schematize information in a qualitatively new way—by using her memories of previous experiences to create a representation of a new kind of experience. She was playing with her mind.

A tendency to play around with ideas, to play "let's pretend," to imagine outcomes, isn't important only to children. Are you the kind of worker, communicator, decision-maker, designer, artist, business person, or engineer who builds a detailed plan for each creation, then follows it step-by-step? Or do you just start pottering around with ideas or materials until some kind of order begins to emerge? If you are the latter kind of worker, you've probably suffered the jibes and innuendos of your more analytically minded colleagues, and if you are any good at "thinking by the seat of your pants," you also know that you are capable of putting together concoctions on the fly that the careful preplanners will probably never achieve in years of deliberate effort. Anthropologist Claude Lévi-Strauss, writing about the way "primitive" cultures approach theory-building in their empirical disciplines, used the term *bricolage* as a model for the way all humans, "primitive" or "civilized," build scientific theories by pottering around with natural objects in various combinations. A *bricoleur*, in this sense, is a kind of intuitive technician, who plays with concepts and objects in order to learn about them.

Bricolage brings the development of first-person educational media full circle. The word was picked up by the computer scientist and educational expert Seymour Papert. In his book *Mindstorms*, Papert pointed out how the term can extend to an entire way of thinking:

> The process reminds one of tinkering; learning consists of building up a set of materials and tools that one can handle and manipulate. Perhaps most central of all, it is a process of working with what you've got. We're all familiar with this process on the conscious level, for example, when we attack a problem empirically, trying out all the things that we have ever known to have worked on similar problems before. But here I suggest that working with what you've got is a shorthand for deeper, even unconscious learning processes. . . . Here I am suggesting that in the most fundamental sense, we, as learners, are all bricoleurs.

The same statement could be made today about cyberspace.

Papert, one of the founders of artificial intelligence research at MIT, coined the term "microworlds," a concept that bridges theories of education and the capabilities of computer simulations, personal computers, and ultimately VR technologies. Papert spent five years in Switzerland, working with the developmental psychologist JeanPiaget. Piaget had triggered a revolution in learning theory by spending decades watching how children learn, starting with his own children. He concluded that learning is not simply something adults impose upon their offspring through teachers and classrooms, but is a deep part of the way children are innately equipped to react to the world, and that all children construct their notions of how the world works, from the material available to them, in definite stages. The mental processes they use in this vital intellectual construction project are their tools for thinking. These tools, which include experiment and systematic curiosity as well as multiplication tables and alphabets, are the essential elements of educaton

Piaget was especially interested in how different kinds of knowledge are acquired by children, and concluded that children are scientists— they perform experiments, formulate theories, and test their theories with more experiments. To the rest of us, this process is known as "playing," but to children it is a vital form of research.

Papert recognized that the responsiveness and representational capacities of computers might allow children to conduct their research on a scale never possible in a sandbox or on a blackboard. But the human interface had to be redesigned to be a children interface before preliterate and very young children would have an opportunity to make use of these capacities.

Advances in computer interfaces and new theories about human learning might converge in a whole new kind of educational tool, Papert conjectured. This was not simply the failed "computer-assisted instruction" concept from the 1960s. In *Mindstorms,* he asserted:

> Stated most simply, my conjecture is that the computer can concretize (and personalize) the formal. Seen in this light, it is not just another powerful educational tool. It is unique in providing us with the means for addressing what Piaget and many others see as the obstacle which is overcome in the passage from child to adult thinking. I believe that it can allow us to shift the boundary separating concrete and formal. Knowledge that was accessible only through formal processes can now be approached concretely. And the real magic comes from the fact that this knowledge includes those elements one needs to become a formal thinker.

Apparently, Nintendo Corporation, who already have computer-based simulators in the guise of video games in one out of every five American homes, think Papert's ideas still have some merit. In 1990, the president of Nintendo announced in Kyoto a $3 million grant for Papert's continuing work at Media Lab in the field of hands-on education amplifiers, specifically to support studies of how children learn while at play.

R. L. Gregory, a specialist in human visual perception, a field previously unrelated to educational psychology, had something similar to say about the place of play not just in human learning but in our most basic perceptions:

> The Exploratory aim is to amplify and extend first-hand experience to enrich perception and understanding for children and throughout adult life. The effectiveness of the hands-on approach for teaching has been questioned. But in any case, surely capturing interest is the first essential for more formal methods to be effective. It is hard to believe that learning has to be serious; it is far more likely that play is vitally important for primates to learn how to exist in the world in which they find themselves. It is fascinating to watch children and adults in this play-experiment situation of individual discovery. Although research is needed to be sure, they certainly give every indication of thinking and learning by doing.

As Gregory noted, much design and research remain to be done before anyone knows whether VR could furnish a magical sandbox for children and teachers to directly experience simulated worlds of history or biology, but Frederick Brooks pointed back to the American educational pioneer John Dewey when he summed up the educational potential of VR in the words "learning by doing." What if educational microworlds could be designed in the form of the cyberspace play-houses Jaron Lanier and Randal Walser suggest?

For those who wonder why kids love video games and hate school, and wonder if there's something to be learned there, the future of the commercial electronic education industry will be instructive. If the representational capabilities of VR simulations evolve, and a talented team like Ann McCormick and Warren Robinett, who created *Rocky's Boots* and other educational software in the early 1980s, or Brenda Laurel and Scott Fisher, who founded Telepresence Research in 1990, are given the tools and opportunity to create "learning worlds," we may see some positive results from what has generally been a dismal field. Whatever else might be improving in America and some other industrial societies, education isn't one of them. Perhaps technology and commerce can succeed where public school systems have failed. If video games do create a hidden infrastructure for a Nintendo VR educational technology of the future, or Fujitsu's plans to create educational VR simulations come to fruition, the key obstacle to getting an education in the future might simply be lack of access to direct-experience education technologies. If you have to attend deteriorated public schools and your parents can't afford Nintendo educational supplements, you might end up with the other informationally disen-franchised earthlings of the data age.

The effects of VR are so widespread, scattered over so many different scientific disciplines, potential commercial applications, and social roles, that the larger pattern that connects it all into something mean-ingful across all dimensions seems elusive. What is needed, Douglas Engelbart would point out, is a conceptual framework. One way to begin building such a framework is to look for a foundation in the past, to examine history in search of long-term patterns that might help make sense of tomorrow's complex mix of possibilities. Asking where it all began, and why, might prove helpful when it comes to discussing where it is all going, and why.

Chapter Seventeen

CYBERSPACE AND HUMAN NATURE

The great release, the breaking away which is our uniqueness, came during the Upper Paleolithic. Furthermore, the release is not something remote that happened in times out of mind, in an interesting but essentially irrelevant past. From the standpoint of a human lifespan any change dating back 30,000 or more years must indeed seem like ancient history. But it is the blink of an eyelid ago on the evolutionary time scale. We are very young, an infant species just beginning to wonder and observe and explore. The process which gathered momentum among the Cro-Magnons continues to accelerate in our times. The same forces which drive us today drove our recent ancestors underground with paints, engraving tools, lamps, and notions about their place in the scheme of things. . . .

The underlying theme is continuity. Tomorrow will be a working out, at increasing intensity and on a larger and larger scale, of forces unleashed yesterday during the Upper Paleolithic. Information is still piling up, and faster than ever. The task of processing and analyzing it is still crucial, still formidable, and so is the task of communication—creating new, more compact symbols, more sophisticated images on television-type screens, more sophisticated chunking methods. Survival still depends on using all our resources, art, and ceremony as well as technology, to build stable societies out of increasing numbers of rugged and unpredictable individuals.

JOHN PFEIFFER,
The Creative Explosion, 1982

Consider the feudal person, unaware that he lived on a planet loaded with natural resources like fossil fuels, which could power machines which would create more complex machines and produce chemical-electrical energy. . . . Today, at the end of the industrial age, at the dawning of the cybernetic age, most digital engineers and most managers of the computer industry are not aware that we live in a cyber-culture surrounded by limitless deposits of information which can be digitized and tapped by the individual equipped with cyber-gear. . . . There are no limits on virtual reality. It's all about access to information. The donning of computer clothing will be as significant in human history as the donning of outer clothing was in the Paleolithic.

TIMOTHY LEARY,
quoted by David Sheff in *Upside*, 1990

Thirty thousand years ago, outside a deceptively small hole in a limestone formation in the area now known as southern France, several adolescents shivered in the dark, awaiting initiation into the cult of toolmakers. The weeks of fasting and abstinence, the ordeals of silence and pain, the rituals of drumming, chanting, and dancing, were about to reach their climax. The first virtual reality awaited below. We will never know with absolute certainty what went on there, but at least one contemporary paleontologist has argued that the activities in those caves were closely related to a series of changes in human thought and action that continue to reverberate to this day.

At some point between ten and thirty thousand years ago, most of our species changed its way of life, not because of biological mutation or selection, but because people learned something new. The caves at Lascaux and other sites might be the places where that learning took place. If the theories of paleontologist John Pfeiffer are correct, primitive but effective cyberspaces may have been instrumental in setting us on the road to computerized world-building in the first place. Toolmaking was at the beginning of the road that led to the opening of cyberspace, and toolmaking, using human attention mechanisms and high-resolution video displays instead of ochre paintings on limestone walls, may be the ultimate future purpose of cyberspace, as well.

The novices had been selected carefully. Once a year, the candidates who had come of age were abducted by a shadowy group of toolmakers, shamans, and artists, whose activities were changing irrevocably the way the human race worked and lived. The mentors took the novices, one by one, down into the cave. Hours of crawling through deep, narrow, labyrinthine, utterly dark passageways led to the special chambers. After the chanting, the prostrations, the whispered myths and texts, the darkness was pierced by torches and lamps placed at strategic intervals. The novices, who were lying or standing in precisely predetermined positions, suddenly saw the supernatural figures floating in space in front of them—bisons, birds, symbols, human figures jumped out of the darkness to fill their fields of view.

In this moment of audiovisually induced fear and trembling, Pfeiffer proposes, the first technological secrets were imparted. The deliberately sensitized psyches of the novices were reframed by carefully crafted sequences of sights and sounds, imprinted with the secrets of fire and metal, the connections between seeds and stars. An early form of the most powerful idea at the basis of technological civilization—that it is possible to observe the world and learn from it and apply that knowledge to the requirements of daily life—was burned into our forebears' brains to the accompaniment of a three-dimensional sound and light show.

The earliest virtual realities on earth were constructed laboriously, by lamplight, deep underground. Their purpose is unknown, but John Pfeiffer presents evidence in his scientific publications and his book, *The Creative Explosion*, that these subterranean cyberspaces may have been created to imprint information on the minds of the first technologists. Although the origins of the human species are still a matter of controversy, current dating methods point to a common ancestor about 200,000 years ago. Until several tens of thousands of years ago, a relatively sparse human population led more or less the same kind of lives, millennium in, millennium out. Around the time the first cave paintings were created, however, a new way of living was born. The most successful revolutionaries of all time, the first paleolithic agriculturist/technologists, were transforming and ultimately dooming the hunting-gathering way of life that had sustained small bands of humans for the 200,000 years the species had existed.

Why did people suddenly start making paintings deep in caves, tens of thousands of years ago? The question is more than academic. As Pfeiffer pointed out in his book, people didn't go to this much trouble back then without having a reason:

> What brought about the use of art after hundreds of millenniums of unembellished and artless cave and open-air dwelling, and why in western Europe and apparently not as early in the rest of the world? Something must have happened to account for the difference, and it is a most significant difference. One wonders how society had changed, how people had changed. What new needs had to be satisfied, what new wants fulfilled? Were they more sensitive, more self-aware, more esthetic—and, above all, how could they have possibly benefited from such "impractical" behavior?

> The problem becomes several orders of magnitude more complex when one considers the deep art: the art located in utter darkness, far from daylight and twilight zones and living places, on wide expanses of wall or doubly hidden inside tiny chambers, caves within caves, secrets within secrets. The purpose of this sort of art differed enormously from the purposes of the domestic variety. It suggests such things as intense rituals, ordeals, journeys underground for mystical reasons. The burst of art marked a burst of ceremony and, again, we wonder about its evolutionary payoffs, about the needs and wants it involved, about its selective and adaptive value. There is every reason to believe that individual artists benefited directly from their rare talents, that art would have withered if this had not been the case. But at the same time art also favored the advancement of the group, and the band and tribe.

> The coming of art by itself would have been more than enough to distinguish the Upper Paleolithic, but a number of other major de-

velopments were under way during the same period. For one thing, a familiar newcomer appears for the first time in the fossil record, a new species. The Neanderthals vanish without a trace, and their place is taken by the remains of modern-type individuals like ourselves, Cro-Magnon people and related breeds referred to collectively and somewhat euphemistically as "doubly wise," *Homo sapiens sapiens.*

New lifestyles appeared along with the new species. Changes were taking place across the board, in practically all areas of human endeavor. The Cro-Magnons made more tools and more kinds of tools than the Neanderthals, using a wider variety of raw materials more efficiently, exploiting a wider range of plants and animals. They seem to have been coming together in larger groups, perhaps for somewhat longer periods, a hint of full-scale settlements to come, and communicating with one another over greater distances. There are signs of a crumbling of age-old traditions. Things were happening faster than ever before.

Pfeiffer noted that many of the paintings were "anamorphic," painted in a precisely distorted manner on natural protuberances and depressions in the limestone in order to give the rendering a three-dimensional appearance when viewed from the proper light and angle. Other images were incised in the walls in such a way that they reveal themselves only when a light is moved across them from the proper angle. From the placements of the paintings, the three-dimensional illusions, and external evidence of other human activities during the same period, Pfeiffer assembled evidence for his hypothesis that the purpose of these underground light shows was to cause a specific state of consciousness.

The reason for shocking people into an altered state of consciousness, Pfeiffer contends, might have been to aid in imparting a growing body of information needed for the expansion of a new way of life. Seed saving, herbal medicine, stone toolmaking, stargazing, animal husbandry, and all the other bodies of lore that were necessary to end the age-old cycles of migration and create permanent settlements were relatively new in human culture. The new techniques brought a need to transmit additional cultural information to future generations—the instruction manuals for the new labor-saving devices. An information explosion and communication crisis resulted, Pfeiffer asserts:

> The crisis, as severe as the threat of famine, threatened chaos. The collective knowledge of the band society was on the rise, demanding far more powerful systems of record-keeping—and there was no writing. Writing might have helped to some extent, and steps that would eventually lead to the earliest notations with complex signs

and symbols were already well under way. But writing as we know it lay some twenty millenniums in the future. The here-and-now emergency gave rise to measures of a different sort, measures created to transmit the expanding contents of "the tribal encyclopedia" intact and indelibly from generation to generation. . . .

One way of preparing people for imprinting has been known for a long time by tribes everywhere, modern as well as prehistoric: bring them into unfamiliar, alien, and unpleasant places, part of the procedure known in recent times as brainwashing. This is designed to erase or undermine his everyday world as completely as possible, apparently serves as an effective preliminary to making him remember. . . .

So many things are going on in such spots, so many impressions and the combined impact of all the impressions. The art itself is only part of the experience, often a rather small part at that. Except for a relatively small proportion of outstanding paintings and engravings, the figures may be more or less interesting but they are not exciting, especially viewed out of context and in two dimensions, reproduced in books or projected on screens. The setting is a major factor in the effect of the figures, and the process of getting there, a series of successive narrowings, from the outside world into the cave mouth, and on into galleries, side chambers and niches in or beyond the chambers. A kind of extended zooming or closing in enhances the figures. Then they are exciting. . . .

Imprinting enormous amounts of information in memory called for every device I have discussed so far—the use of confined spaces, obstacles and difficult routes, and hidden images to heighten the natural strangeness of underground settings—and a great deal more. We are only beginning to appreciate the subtlety of what was going on. Considering the technologies available at the time, the people of the Upper Paleolithic seem to have made use of every trick in the book, piling special effect upon special effect in an effort to ensure the preservation and transmission of the tribal encyclopedia.

The caves at Lascaux may have been among the earliest mental-imprinting ceremonies that made use of subterranean, three-dimensional art, myth and ritual, and possibly mind-altering substances. The use of underground initiation chambers filled with architectural and pictorial symbols of technological and spiritual principles, for special theatrical rituals and light-and-sound spectacles intended to alter the consciousness, was not confined to the Mediterranean. The oldest continuously inhabited human settlements in North America, the Hopi mesas at Oraibi in northern Arizona, still contain such rooms, known to the Pueblo tribes as *kivas:* A round, square, or

hexagonal, usually underground chamber, the kiva contains a hole in
the floor, a ledge around the perimeter, and a ladder extending
through a smokehole in the roof. In these kivas, as in the caves at
Lascaux or the initiation chambers at Eleusis, young people in altered
states of consciousness were educated in the use of core tools for
creating and maintaining a new way of life, both spiritually and technologically. As Frank Waters described sacred kiva architecture:

> Roughly corresponding to Buddhist cosmography, both the Na-
> vaho and Pueblo universes embrace four successive underworlds
> below the earth. . . .
> All this is abstractly symbolized in the Pueblo kiva—the secret,
> underground ceremonial chamber. For the kiva itself, with all its
> many variations, recapitulates in structural form this four-world uni-
> verse common to all. . . .
> In the kiva, man is ever reminded that he lives in the whole of the
> immense and naked universe. And he is constantly made aware of
> the psychic, universal harmony which he must help to perpetuate by
> his ceremonial life.
> For the kiva is not only an architectural symbol of the physical
> universe. The universe, with its great axis rock and its great sipapu
> canyon, is itself but a structural symbol of the mystical soul-form of
> all creation. And both are duplicated in man himself.

The most secret kiva ceremonials are still unknown outside the
Pueblo clans entrusted with their knowledge. But enough is known
about the public aspects of the ritual to conclude it involves an explicit
map of human origins and goals, theatrical symbolic rituals, and hard
information about a technology necessary to sustain a new way of life
for the culture. The Hopis, like the other Western Hemisphere descendants of bands of hunter-gatherers who migrated across the Ber-
ing land bridge, had wandered for centuries before settling down to
build urban societies. The technology that made these settlements
possible was the same technology that made most civilization in the
Western Hemisphere possible: the cultivation of corn.

It is now known that corn was not a wild plant but an artificial one,
deliberately crossbred from a wild corn and a native grass, a task that
highly expert agriculturists in central Mexico accomplished over the
course of generations, at about the same time the Eleusinian mysteries
were flourishing in the Mediterranean. The kiva rituals imprint a
ceremonial cycle that encodes a sophisticated agricultural system. First,
corn is planted at exactly the right time, as indicated by the stellar
patterns that guide the ceremonials. Then beans are planted next to
the corn; beans help the corn roots fix nitrogen, and the corn stalks

provide a trellis for the beans to climb. Then squash is planted, and in the heat of the summer the broad squash leaves conserve moisture in the soil and create a microclimate inhospitable to pests. This utterly technological, precisely timed, and culturally vital practice was one component of the external information that accompanied the psychological and spiritual components of the kiva ceremony. The uses of mimesis, masked characters, painted three-dimensional objects of symbolic significance, are some of the techniques known to be associated with kiva ceremonials. Sound familiar?

Something about caves and petroglyphs has always resonated with me. In Arizona, at the age of seventeen or eighteen, I encountered *Kokopelli* for the first time. I didn't know who he was, what he meant, or that he had a name, but when I saw the crude, almost stick-figure, humpbacked flute player incised in a rock in a shallow cave overlooking a broad river valley in northern Arizona, it was as if somebody had reached out and rung a chime in me I had never known about before. I found out about him later—associated with corn, a bearer of seeds and information, a horny devil like Shiva and Dionysius. An evolutionary technician, a dreamweaver. A civilization amplifier.

During my research for this book, I ended up at a scientific conference in Santa Barbara, California, one of those intense affairs where the information flow doesn't cease between sessions. So I took advantage of one of the two time slots allocated to personal time and found another cybernaut who was interested in seeing Painted Cave, about an hour out of town, in the Santa Ynez foothills. The Chumash Indians, who had lived in the Santa Barbara area and throughout a fairly wide swath of southern California, were known for their powerful shamans and their cave paintings—which were not unrelated phenomena. I saw a film of the spirals, stars, wavy lines, humanoid figures once. Very different from the startling realism of some of the older cave paintings in the south of France. A brochure I picked up in the vicinity noted that the shamans who did the paintings were known to use *datura*, a powerful plant hallucinogen, and other plant substances.

Painted Cave was right off the road, in a side canyon that faced the mountains, yet was only a short hike from a place where the Pacific was visible. A heavy iron gate barred the cave mouth. It was too dark to see anything, so we climbed around and waited for the sun to rise higher. When we peered in we could see the reason for the iron bars, atop spirals and an occluded sundisk that is believed to commemorate (or predict?) a solar eclipse known to have happened in the fourteenth century: "John loves Mary" and more prosaic contemporary graffiti marred the older, more serious layers. In many ways, a disappointing trip. After looking through the bars at those paintings that were visible from the entrance, we decided to follow one of the trails and climbed up onto the rocks atop the cave. We sat there for a while and talked about what it must have been like a thousand years ago. As we were leaving, and I was resting on a small ledge after negotiating a narrower section, I looked across the gap to the next rock, and caught my breath. The sun was in just the right position, and I was in just the right place, and my gaze was aimed in the proper direction—and the clear outline of a thunderbird design, two feet across, became visible on the surface of the boulder. Attention is the ultimate instrument.

"The metaphorical shape of existence permeates the culture," Brenda Laurel called to tell me, when she had returned from New Mexico, "and that shouldn't surprise us. That's what this stuff we do with our bodies and our emotions is for." She had visited Chaco Canyon, where foot drums are dug into the mesas, and the kivas were large enough to hold hundreds. The very shape of the Anasazi cities and the architecture of their ritual places, like the Eleusinian temples, the Paleolithic caves, the Dionysian amphitheaters, focuses the power of mimesis.

In response to Pfeiffer's hypothesis of VR as an imprinting medium, my speculations about corn and kivas, and the information-encoding aspects of drama, Laurel responded: "The transmission of values and cultural information is one face of VR. The other face is the creation of Dionysian experience. The piece I find important to both of those functions is this notion of being in the living presence of something. With the ceremony in the kiva, one is in living presence not just of other people but of an event that is happening in real time. No matter whether you look at the informational function or the Dionysian one, the idea of it happening in real time and in the present location, both activities require people to be in the same space at the same time. What VR has done that is so discontinuous is get rid of real space as a requirement. Cyberspace has the potential of being able to make space go away as a mediator of collective experiences." We're still going to use the new communication media that come with cyberspace, Laurel insists, as we have used every previous medium, to conjure up transformative powers, to propel us beyond the boundaries of our minds and push our cultural evolution into new territories.

The use of cyberspaces to influence, educate, and trigger ecstatic experience is an issue for the future as well as the past. The social codes by which we create values, the social mechanisms by which those values guide actions, and the ultimate uses we choose for the tools we create are questions about the future that VR technologies are likely to force us to confront. The question of what human beings are for (what our purpose is in the scheme of things) is not the same as where human beings seem to be aimed (which events in our past could have shaped our present capabilities). In order to exert influence on the direction of events, a sufficient number of people need to understand that difference between the trajectory that got us here and the pathways we would prefer to choose for our future. One way to zero-in questions about the many implications of VR technology and the future is to look at our use of these thinking tools we have been using for a few years now.

THE HUMAN USE OF THINKING TOOLS

If we could rid ourselves of all pride, if, to define our species, we kept strictly to what the historic and prehistoric periods show us to be the constant characteristic of man and of intelligence, we should not say Homo sapiens *but* Homo faber. In short, intelligence considered in what seems to be its original feature, is the faculty of manufacturing artificial objects, especially tools for making tools.

HENRI BERGSON,
Creative Evolution, 1911

We are evidently unique among species in our symbolic ability, and we are certainly unique in our modest ability to control the conditions of our existence by using these symbols. Our ability to represent and simulate reality implies that we can approximate the order of existence and bring it to serve human purposes. A good simulation, be it a religious myth or scientific theory, gives us a sense of mastery over our experience. To represent something symbolically, as we do when we speak or write, is somehow to capture it, thus making it one's own. But with this approximation comes the realization that we have denied the immediacy of reality and that in creating a substitute we have but spun another thread in the web of our grand illusion.

HEINZ PAGELS,
The Dreams of Reason, 1988

Virtual reality brings with it a set of questions about the industries and scientific capabilities it makes possible. It also brings with it a set of questions about human uses of technology, particularly the technologies that don't yet exist but are visible on the horizon. VR vividly demonstrates that our social contract with our own tools has brought us to a point where *we have to decide fairly soon what it is we as humans ought to become,* because we are on the brink of having the power of creating any experience we desire. The first cybernauts realized very early that the power to create experience is also the power to redefine such basic concepts as identity, community, and reality. VR represents a kind of new contract between humans and computers, an arrangement that could grant us great power, and perhaps change us irrevocably in the process.

The looming Faustian bargain involves certain changes in the partnership we have enjoyed with our machines. We might decide that we wouldn't mind becoming a little or a lot more machinelike in exchange for laborsaving devices, lifesaving tools, attractive conveniences and seductive entertainments. Such a decision would be a radical change, but not an abrupt one. Our minds, our senses, our consensual reality has been shaped for a century, to the point where billions of us are trained and ready to embrace our silicon partners more intimately than ever before. Trillions of human-hours have been logged so far in the virtual worlds of *I Love Lucy* and *Dallas*, FORTRAN and fax, computer networks, comsats, and mobile telephones. The transformations in our psyches triggered by the electronic media thus far may have been mere preparation for bigger things to come. The hinge of change seems to be connected with these machines we've created and the kind of partnership we are coevolving with our information tools.

VR is an important threshold in the evolution of human-computer symbiosis. But symbiosis is a two-way exchange; when one organism exists at the expense of another, without contributing something vital to the partnership, the relationship is parasitic. Two questions that emerge from an examination of VR are closely interrelated: How will cyberspace tools and environments affect the way we live, think and work? And how will cyberspace affect the way we apprehend the world, the way we define ourselves as sensing, thinking, communicating beings?

"Electronic media alter the ratios between the senses," was one of Marshall McLuhan's key epigrams: The ratios and amount of audio and visual input to the dominant reality recipe were altered by radio and telephones and then altered again by television; we see and hear and thus apprehend the world differently as a result. Those effects are taken for granted by now, three decades after *Understanding Media*. The cyberspace experience is destined to transform us in other ways because it is an undeniable reminder of a fact we are hypnotized since birth to ignore and deny—that our normal state of consciousness is itself a hyperrealistic simulation. We build models of the world in our mind, using the data from our sense organs and the information-processing capabilities of our brain. We habitually think of the world we see as "out there," but what we are seeing is really a mental model, a perceptual simulation that exists only in our brain. That simulation capability is where human minds and digital computers share a potential for synergy. Give the hyperrealistic simulator in our head a handle on computerized hyperrealistic simulators, and something very big might happen.

Cognitive simulation, mental model-making, is one of the things humans do best. We do it so well that we tend to become locked into our own models of the world by a seamless web of unconscious beliefs and subtly molded perceptions. And computers are model-making tools par excellence, although they are only beginning to approach the point where people might confuse simulations with reality. Computation and display technology are converging on hyperreal simulation capability. That point of convergence is important enough to contemplate in advance of its arrival. The day computer simulations become so realistic that people cannot distinguish them from nonsimulated reality we are in for major changes.

We are approaching a breakpoint where the quantitative improvement in that model-building interface will trigger a qualitative quantum leap. In coming years, we will be able to put on a headset, or walk into a media room, and surround ourselves in a responsive simulation of startling verisimilitude. Our most basic definitions of reality will be redefined in that act of perception; as Jean Baudrillard claims: "Abstraction today is no longer that of the map, the double, the mirror or the concept. Simulation is no longer that of a territory, a referential being or a substance. It is the generation by models of a real without origin or reality: a hyperreal. The territory no longer precedes the map, nor survives it. Henceforth, it is the map that precedes the territory."

The advent of technology-generated hyperreality could be the nightmarish "consensual hallucination" described by William Gibson in the novel *Neuromancer,* where the word *cyberspace* originated. Or the result might be an increase in human freedom and power, akin to the after-effects of printing and communication technologies. Which way it will go—dystopia or empowerment—depends in part upon how people react to the unmasking of reality as a cognitive-perceptual construct. People tend to react in different ways to the news that reality might be an illusion, depending on their personal emotional attachment to their brand of reality. Denial, cognitive dissonance, resistance, and satori are all possible psychological reactions to the truth we are forced to face in the illusory realm of cyberspace, in roughly descending order of popularity.

If humans and computers are poised on the verge of a symbiotic relationship, as computer pioneer J.C.R. Licklider prophesied more than three decades ago, shouldn't we take the time to discuss what we are getting into, before we get any farther into it? If we get a handle on what we are changing into, what should we *want* to change into, what should we *not* want to change into, and how do we gain the power to make that decision? After a year and a half wrestling with the issues raised by the emergence of virtual reality technology, I still find myself asking questions I thought I had left back in philosophy class in college.

In my travels, I found a researcher who intends to use VR as a probe for mapping the limits of human potential, for understanding what human beings are best suited to do in a world increasingly dominated by machines.

Psychologist Nathaniel Durlach is from neither the Atari nor the mainframe generation. In fact, he started down the path of inquiry that led him to VR by studying bats. His interest in tactile communication led him to the idea of mapping human senses onto mechanical transducers—his area of overlap with Margaret Minsky, whose laboratory is walking distance from Durlach's. Like others I've met who found themselves drawn into the world of virtual worlds research in unexpected ways, Durlach didn't realize he was going to meet roboticists coming down the same path he was pursuing, only from a different direction. As a psychologist, he is interested not only in the "human factors" element in VR systems, but in the prospect of using VR systems as a kind of psychological microscope for examining the deepest questions about human nature.

Human nature?

"Yes," Durlach told me, when I asked him to explain the connection between VR and human nature, "questions about how you can map human perceptions onto virtual worlds are really about the limits of human capability. What are the constraints on our ability to adapt to artificially extended senses? That's a question about human nature. And it can be tested by using the proper instrument in a well-designed experiment."

Durlach is an amiable fellow, informal, steel-gray-haired, perpetually smiling. Any conversation I've had with him has been a surprising ramble over a dozen territories utterly tangential to the main point, but all completely relevant. The day I walked into his office in Cambridge, he told me a little about his background and the tactile communication of speech. A pure mathematician at the beginning of his career, he got involved in radar search and development, which brought him to Lincoln Laboratory, the spawning ground of Ivan Sutherland's Sketchpad. If you want to understand how radar and particularly how sonar work, you inevitably end up interested in bats. Bat sonar studies led Durlach to the question of binaural hearing— how do two-eared creatures use their physiology to detect relevant signals among a soup of background noise?

"I started working with hearing," Durlach told me, "and through that I became interested in people who had trouble hearing, and eventually in people who were totally deaf. That led me to work with people who were totally deaf and blind. People who had lost all sight and all hearing at the age of eighteen months were people whose entire world came through the tactile sense. They learned everything through the tactile sense." The deaf/blind use a variety of methods for communicating, but the one that interested Durlach involved the "listener" putting a hand on the face of the talker, and by monitoring airflow, laryngeal vibration, lip position, and muscle tension in the speaker, actually decoding the speech. They "hear" speech, for all practical purposes, through their fingertips. And that led Durlach to an intriguing conclusion: "People are a lot more plastic than most people realize." Perhaps we can hear ultrasound or radar, see infrared or ultraviolet, by mapping the proper mechanical transducer to the appropriate sense, and letting human plasticity do the rest.

If we can see the invisible and hear with our fingers, what is "normal" human capability? That's where VR meets age-old questions about human nature, and that's why Durlach started thinking about VR, and particularly VR's sister subject, telerobotics.

"Suppose I have a slave robot that senses different wavelengths of light than I normally sense, and I can receive information from the robot's sensors through a head-mounted display. How would that information be displayed? As colors? Sound? Tactile feedback?" After spending several hours with him over a period of days, and learning about his career, it became clear that Durlach loves to ask odd questions that lead him new places. "Or suppose I have a slave robot that is very tiny and goes inside somebody's body as a microrobot. Or I have a slave robot that is enormously strong. There is some kind of transformation that takes place between my human senses and information processing and physical actions, on the one hand, and the software and computers and supersensory or superpowerful robots, on the other. What do you actually call that? It's a real world that you are sensing and acting upon, the solid physical world. But the way you are perceiving it is virtual. What kind of reality is that?"

Durlach sees VR systems of the near future as "ideal systems for experimental psychology. Every university that has an experimental psychology department is going to have a virtual world system," he forecast. "Experimental psychology involves controlling people's environments and observing how their response relates to the input. A virtual world could be an ideal environment, and there's no reason why you can't monitor much more than the motor inputs that gloves and joysticks provide. Heart rate, pupil size, changes in the skin resistance, could all be used. For my purposes, a good VR system could be an all-purpose research instrument."

"Film is truth at 24 frames per second," Jean-Luc Godard used to say. That projection rate is the threshold at which the separate photographic images projected on a movie screen fuse in the human perceptual system into the consensual hallucination we know as cinema. Cyberspace is where the human interface to digital computers is approaching 24 frames per second. The advancement of many of the key qualities we think of as human is linked to the evolution of worldviews—to the emergence and invention of new ways to see the world. As Jacob Bronowski put it, "We cannot separate the special importance of the visual apparatus of man from his unique ability to imagine, to make plans, and to do all the other things which are generally included in the catchall phrase 'free will.' What we really mean by free will, of course, is the visualizing of alternatives and making a choice between them. In my view . . . the central problem of human consciousness depends on the ability to imagine."

Given a tool for visualizing and modeling, how might we use it to help us make plans, imagine, and otherwise exert conscious influence on an increasingly complex environment? Can we imagine ways to apply it to the very real problems of the world?

As Brenda Laurel put it (1986):

Reality has always been too small for human imagination. The impulse to create an "interactive fantasy machine" is only the most recent manifestation of the age-old desire to make our fantasies palpable—our insatiable need to exercise our imagination, judgment, and spirit in worlds, situations, and personae that are different from those of our everyday lives. Perhaps the most important feature of human intelligence is the ability to internalize the process of trial and error. When a man considers how to climb a tree, imagination serves as a laboratory for "virtual" experiments in physics, biomechanics, and physiology. In matters of justice, art, or philosophy, imagination is the laboratory of the spirit.

As with other technologies, cyberspace is not an either-or case. It will be both-and. People will use it as a hybrid of entertainment, escape, and addiction. And other people will use it to navigate through the dangerous complexities of the twenty-first century. It might be the gateway to the Matrix. Let us hope it will be a new laboratory of the spirit—and let's see what we can do to steer it that way.

REFERENCES

Chapter 1

James J. Batter and Frederick P. Brooks. Jr., "GROPE-1," *IFIP Proceedings 71*, 1972, p. 759.

F.P. Brooks, Jr., "The Computer Scientist as Toolsmith: Studies in Interactive Computer Graphics," *Information Processing 77*, B. Gilchrist, ed., Amsterdam: North-Holland, 1977, pp. 625–634.

F.P. Brooks, Jr., *The Mythical Man-Moth; Essays in Software Engineering*, Reading, MA: Addison-Wesley, 1975.

F.P. Brooks, Jr., "Grasping Reality Through Illusion: Interactive Graphics Serving Science," *CHI'88 Proceedings*, Reading, MA: Addison-Wesley, 1988, pp. 1–11.

F.P. Brooks, Jr., M. Ouh-Young, J.J. Batter, and P. Jerome Kilpatrick, "Project GROPE—Haptic Displays for Scientific Visualization," *ACM Computer Graphics*, vol. 24, no. 4, August, 1990, pp. 177–185.

H. Fuchs, J. Poulton, J. Eyles, and T. Greer, "Coarse-Grain and Fine-Grain Parallelism in the Next Generation Pixel Planes Graphics System," *Proceedings of the International Conference and Exhibition on Parallel Processing for Computer Vision and Display*, New York: Springer-Verlag, 1988.

H. Fuchs, S.M. Pizer, J.L. Creasy, J.B. Renner, and J.G. Rosenman, "Interactive, Richly Cued Shaded Display of Multiple 3D Objects in Medical Images," *Proceedings SPIE's Medical Imaging II Conference*, 914(2), 1988.

William Gibson, *Neuromancer*, New York: Berkley Publications Group, 1984.

C. Levinthal, "Molecular Model-Building By Computer," *Scientific American*, vol. 214, no. 6, June, 1966, pp. 42–52.

Heniz Pagels, *The Dreams of Reason*, New York: Simon & Schuster, 1988.

Ivan Sutherland, "The Ultimate Display," *Proceedings of the IFIP Congress*, 1965, pp. 506–508.

Chapter 2

Morton Heilig, "The Cinema of the Future," *Espacios*, Mexico City, January, 1955.

Morton Heilig, *US Patent #2,955,156*, "Stereoscopic Television Apparatus for Individual Use," October 4, 1960.

Morton Heilig, *US Patent #3,050,870*, "Sensorama Simulator," August 28, 1962.

L. Lipton, "Sensorama," *Popular Photography*, July, 1964.

D.N. Perkins, "Pictures and the Real Thing," *Project Zero*, Cambridge, MA: Harvard University, 1979.

Alan Rifkin, "Mort Heilig's Feelie Machine," *L.A.Weekly*, March 12–18, 1982, p. 10.

Chapter 3

Susan Brennan, "Conversation as Direct Manipulation," *The Art of Human-Computer Interface Design*, Brenda Laurel, ed., Menlo Park, CA: Addison-Wesley, 1990, pp. 394–395.

Vannevar Bush, "As We May Think," *Atlantic*, August, 1945.

Douglas Engelbart, "A Conceptual Framework for Augmenting Man's Intellect," *Vistas in Information-Handling*, vol. 1, Paul W. Howerton and David C. Weeks, eds., Washington, D.C.: Spartan Books, 1963, pp. 1–29.

Alan Kay, "Microelectronics and the Personal Computer," *Scientific American*, vol. 237, no. 3, September, 1977, p. 230.

Alan Kay, "User Interface: A Personal View," *The Art of Human-Computer Interface Design*, Brenda Laurel, ed., Menlo Park, CA: Addison-Wesley, 1990, p. 192.

J.C.R. Licklider, "Man-Computer Symbiosis," *IRE Transactions on Human Factors in Electronics*, vol. HFE-1, March, 1960, pp. 4–11.

Marvin Minsky, "Toward a Remotely-Manned Energy and Production Economy," Massachusetts Institute of Technology Artificial Intelligence Laboratory, A.I. Memo No. 544, September, 1979.

Ted Nelson, *The Home Computer Revolution*, self-published, 1977, pp. 120–123.

David Canfield Smith, Charles Irby, Ralph Kimball, and Eric Harslem, "The Star User Interface: An Overview," in *Office Systems Technology*, El Segundo, CA: Xerox Corporation, 1982.

Chapter 4

R.A. Bolt, "Gaze-Orchestrated Dynamic Windows," *Computer Graphics* 15 (3), August, 1981, pp. 109–119.

Richard Bolt, " 'Put-That-There': Voice and Gesture at the Graphics Interface," *Computer Graphics* 14 (3), 1980, pp. 262–270.

Stewart Brand, *The Media Lab: Inventing the Future at MIT*, New York: Viking, 1987.

Scott Fisher, "Viewpoint Dependent Imaging: An Interactive Stereoscopic Display," *Proceedings SPIE* 367, 1982.

C. Herot, "Spatial Management of Data," *ACM Transactions on Database Systems*, vol. 5, no. 4, 1980.

Alan Kay, "User Interface: A Personal View," *The Art of Human-Computer Interface Design*, Brenda Laurel, ed., Menlo Park, CA: Addison-Wesley, 1990, p. 192.

K.C. Knowlton, "Computer Displays Optically Superimposed on Input Devices," *The Bell System Technical Journal*, vol. 56, no. 3, March, 1977, pp. 367–383.

Andrew Lippman, "Movie-Maps: An Application of the Optical Videodisc to Computer Graphics," *Computer Graphics*, vol. 14, no. 3, 1980.

Nicholas Negroponte, *The Architecture Machine*, Cambridge: MIT Press, 1970.

Nicholas Negroponte, "Media Room," *Proceedings of the Society for Information Display*, vol. 22, no. 2, 1981, pp. 109–113.

F.H. Raab, et al., "Magnetic Position and Orientation Tracking Systems," *IEEE Transactions in Aerospace and Electronic Systems*, vol. AES 15 (5), 1979.

Chapter 5

H. Greenfield, D. Vickers, I. Sutherland, W. Kolff, and K. Reemtsma, "Moving Computer Graphic Images Seen From Inside the Vascular System," *Transactions of the American Society of Artificial Internal Organs,* vol. 17, 1971, pp. 381–385.

Myron Krueger, "Responsive Environments," *Proceedings of the National Computer Conference,* 1977, pp. 423–433.

Myron Krueger, *Artificial Reality,* Reading, MA: Addison-Wesley, 1983.

Myron Krueger, "VIDEOPLACE: A Report from the ARTIFICIAL REALITY Laboratory," *Leonardo,* vol. 18, issue 3, October, 1985.

Andrew Pollack, "What Is Artificial Reality? Wear a Computer and See," *The New York Times,* April 10, 1989, p. 1

Ivan Sutherland, "The Ultimate Display," *Proceedings of the IFIP Congress,* 1965, pp. 506–508.

Ivan Sutherland, "A Head-Mounted Three-Dimensional Display," *Proceedings of the Fall Joint Computer Conference,* 1968, pp. 757–764.

Daniel L. Vickers, "Sorcerer's Apprentice: Head-Mounted Display and Wand," unpublished Ph.D Dissertation, University of Utah, 1971.

Frank Yeaple, "Live Video and Animated Graphics Are Interfaced Effortlessly," *Design News,* August 18, 1986, pp. 98–102.

Chapter 6

S. Fisher, M. McGreevy, J. Humphries, and W. Robinett, "Virtual Environment Display System," *ACM Workshop on Interactive 3D Graphics,* Chapel Hill, NC, Oct. 23–24, 1986.

Scott Fisher, "Telepresence Master Glove Controller for Dextrous Robotic End-Effectors, *Advances in Intelligent Robotics Systems,* D.P. Casasent, ed., *Proceedings of the SPIE,* 726, 1986.

Scott Fisher, "Virtual Interface Environments," *The Art of Human-Computer Interface Design,* B. Laurel, ed., Menlo Park, CA: Addison-Wesley, 1990.

Scott Fisher, "Virtual Environments, Personal Simulation & Telepresence," *Multimedia Review: The Journal of Multimedia Computing,* vol. 1, no. 2, Summer, 1990.

J.D. Foley, *Interfaces for Advanced Computing, Scientific American,* 257(4), October, 1987, pp. 126–135.

E.M. Wenzel, F.L. Wightman, and S.H. Foster, "A Virtual Acoustic Display for Conveying Three-Dimensional Information," *Proceedings of the Human Factors Society,* 1988.

E.M. Wenzel, Scott H. Foster, Frederic L. Wightman, and Doris J. Kistler, "Real-time Synthesis of Localized Auditory Cues," presented at meeting of the Association for Computing Machinery, Special Interest Group, Computer Human Interface (SIGCHI), 1989.

Chapter 7

John Barlow, "Life in the DataCloud: Scratching Your Eyes Back In," *Mondo 2000,* Summer, 1990, p. 36.

Priscilla Burgess, *MacWEEK,* August 2, 1988, p. 38.

Cover, *Scientific American,* September, 1984.

Steve Ditlea, "Grand Illusion," *New York,* August 6, 1990, p. 32.

James D. Foley, "Interfaces for Advanced Computing," *Scientific American,* October, 1987, pp. 127–135.

Howard Levine and Howard Rheingold, *The Cognitive Connection,* New York: Prentice-Hall, 1987, pp. 232–233.

Sherry Posnick-Goodwin, "Dreaming for a Living," *Peninsula,* July, 1988, p. 58.

VPL Research, Inc., *Virtual Reality at Texpo '89,* Redwood City, CA: VPL Research, Inc., 1989.

Chapter 8

John Perry Barlow, "Being in Nothingness," *Mondo 2000,* Summer, 1990, p. 44.

Eric Gullichsen and Randal Walser, "Cyberspace: Experiential Computing," *Nexus '89 Science Fiction and Science Fact,* 1989.

Theodor Nelson, "Interactive Systems and the Design of Virtuality," *Creative Computing,* November-December, 1980, pp. 56–62.

John Walker, *Through the Looking Glass,* Sausalito, CA: Autodesk, Inc., 1988.

John Walker, "Through the Looking Glass," *The Art of Human-Computer Interface Design,* Brenda Laurel, ed. Menlo Park, CA: Addison-Wesley, 1990.

Randal Walser, "Elements of a Cyberspace Playhouse," *Proceedings of National Computer Graphics Association '90,* 1990.

Chapter 9

Thomas A. Furness, III, "Fantastic Voyage," *Popular Mechanics,* December 1986, pp. 63–65.

Thomas A. Furness, III, "Harnessing Virtual Space," *Society for Information Display Digest,* 1988, pp. 4–7.

Ralph Norman Haber, "Flight Simulation," *Scientific American,* 255(1), July, 1986, pp. 96, 103.

Stephen L. Thompson, "The Big Picture," *Air & Space,* April/May, 1987, pp. 75–83.

Chapter 10

Richard Bolt, *The Human Interface: Where People and Computers Meet,* Belmont, CA: Lifetime Learning Publications, 1984.

E.M. Forster, "The Machine Stops," *The Eternal Moment and Other Stories,* Harcourt Brace Jovanovich, 1929.

T. Hatada, H. Sakata, and H. Kusaka: "Psychological Analysis of the 'Sensation of Reality' Induced by a Visual Wide-field Display," *SMPTE* (Society of Motion Picture and Television Engineers) *Journal,* vol. 89, 1980, pp. 560–569.

Hermann Hesse, *Magister Ludi (The Glass Bead Game),* New York: Bantam, 1982.

Alan Kay, "Computer Software," *Scientific American,* September, 1984.

Yukio Kobayashi, Introduction, "Artificial Intelligence Department 1988 Special Report," Advanced Telecommunications Research Institute International, Communication Systems Research Laboratories, Kyoto, Japan, 1989.

Myron Krueger, *Artificial Reality,* Reading, MA: Addison-Wesley, 1983, p. 3.

Howard Rheingold, *Tools for Thought,* New York: Simon and Schuster, 1985, pp. 260–261.

Ivan Sutherland, "The Ultimate Display," *Proceedings of the IFIP Congress,* 1965, pp. 506–508.

Terry Winograd, "Computer Software for Working with Languages," *Scientific American,* September, 1984, p. 59.

Hiroyuki Yamaguchi, Akiro Tomono, and Yukio Kobayashi, "Proposal for a Large Visual Field Display Employing Eye Movement Tracking," Kyoto, Japan: ATR Communication Systems Research Laboratories, *Proceedings SPIE,* 1989.

Chapter 11

Rudolf Arnheim, *Visual Thinking,* Berkeley, CA: University of California Press, 1969.

Jacob Bronowski, *The Origins of Knowledge and Imagination,* New Haven: Yale University Press, 1978, p. 18.

Takaya Endo and Hiroshi Ishii, "NTT Human Interface Laboratories" NTT publication, Kanagawa, Japan, 1989.

Myron Krueger, *Artificial Reality,* Reading, MA: Addison-Wesley, 1983.

Robert H. McKim, *Thinking Visually,* Belmont, CA: Lifelong Learning Publications, 1980, p. 7.

NTT Visual Media Laboratory, *Annual Report,* 1988.

NTT Visual Media Laboratory, *Annual Report,* 1989.

Yoshinobu Tonomura: private communication with author.

Chapter 12

Robert Heinlein, *Three By Heinlein: The Puppet Masters; Waldo; Magic, Inc.,* Garden City, NY: Doubleday, 1965.

J.C.R. Licklider, "Man-Computer Symbiosis," *IRE Transactions on Human Factors in Electronics,* vol. HFE-1, March, 1960, pp. 4–11.

Larry Leifer, Machiel Van der Loos, and Stefan Michalowski, "Telerobotics in Rehabilitation: Barriers to a Virtual Existence," paper presented at Conference on Human-Machine Interfaces for Teleoperators and Virtual Environments, March 4–9, 1990, Santa Barbara, CA, sponsored by the Engineering Foundation.

Marvin Minsky, "Toward a Remotely-Manned Energy and Production Economy," Massachusetts Institute of Technology Artificial Intelligence Laboratory, A.I. Memo No. 544, September, 1979.

Robert Stone, "Human Factors Research at the U.K. National Advanced Robotics Research Centre," *The Ergonomist,* no. 239, May, 1990.

Susumu Tachi, Hirohiko Arai, and Taro Maeda, "Development of an Anthropomorphic Tele-existence Slave Robot," *Proceedings of the International Conference on Advanced Mechatronics,* Tsukuba Science City, Mechanical Engineering Laboratory, MITI, May, 1989, p. 385.

Susumu Tachi, K. Tanie, K. Komoriya, and M. Kanego, "Tele-existence (I): Design and Evaluation of a Visual Display with Sensation of Presence," *Proceedings of RoManSy '84,* The Fifth CISM-IFToMM Symposium, Udine, Italy, June, 1984, Kogan Page, London: Hermes Publishing, p. 245.

Richard Waite, "Thatcher Lends a Hand," *London Observer Sunday Supplement Magazine*, March 4, 1990, pp. 44–46.

Chapter 13

Joseph Campbell, "Day of the Dead Lecture," *Magical Blend*, 16, 1988, pp. 58–62.

Alain Danielou, *Shiva and Dionysus: The Religion of Nature and Eros*, translated by K.F. Hurry, New York: Inner Traditions International, 1984.

Fujitsu Laboratories, *Research and Development*, Kawasaki, Japan: Fujitsu Laboratories, Ltd., 1989.

William James, *The Varieties of Religious Experience*, Garden City, NY: Image Books, 1978.

Brenda Laurel, *Computers as Theatre*, Menlo Park, CA: Addison-Wesley, 1991.

Brenda Laurel, "Interface as Mimesis," *User-Centered System Design: New Perspectives on Human-Computer Interaction*, D.A. Norman and S. Draper, eds., Hillsdale, NJ: Lawrence Erlbaum Associates, 1986.

Koichi Murakami, "Activities on Artificial Reality in Fujitsu," personal presentation to author, March, 1990.

Masako Nishijima and Yuji Kijima, "Learning a Sense of Rhythm with a Neural Network: The Neuro-Drummer," *Proceedings of The First International Conference on Music Perception and Cognition*, Kyoto, Japan, October, 1989, pp. 77–80.

Karl Schoenberger, "Nintendo Investing in Research on Children," *Los Angeles Times*, May 16, 1990.

David Sheff, "The Virtual Realities of Timothy Leary," *Upside*, April, 1990, p. 70.

Randal Walser, "Elements of a Cyberspace Playhouse," *Proceedings of National Computer Graphics Association '90*, Anaheim, CA, March, 1990.

Chapter 14

Diane Ackerman, *A Natural History of the Senses*, New York: Random House, 1990, pp. 80–81.

James J. Batter and Frederick P. Brooks, Jr., "GROPE-1," *IFIP Proceedings 71*, 1972, p. 759.

Claude Cadoz, Jean-Loup Florens, and Annie Luciani, "Responsive Input Devices and Sound Synthesis by Simulation of Instrumental Mechanisms: the CORDIS System," *Computer Music Journal*, 8, no. 3, 1984, pp. 60–73.

Claude Cadoz, Leszek Lisowski, and Jean-Loup Florens, "Modular Feedback Keyboard," *Computer Music Journal*, 14, no. 2, 1990, pp. 47–51.

Claude Cadoz and Christophe Ramstein, "Capture, Representation, and Composition of the Instrumental Gesture," *Proceedings of ICMC 90*, Glasgow, 1990.

Frank Geldard, "Adventures in Tactile Literacy," *American Psychologist*, vol. 12, no. 3, March, 1957, p. 117.

B. Hannaford, "A Design Framework for Teleoperators with Kinesthetic Feedback," IEEE Transactions on Robotics and Automation, vol. 5, no. 4, August, 1989.

J. Huizinga, *Homo Ludens: A Study of the Play Element in Culture*, New York: Roy Publishers, 1950.

Laboratoire d'Informatique Fondamentale et d'Intelligence Artificielle, 1990, brochure published by l'Institut d' Informatique et de Mathématiques Appliquées de Grenoble.

Margaret Minsky, Ming Ouh-Young, Oliver Steele, Frederick P. Brooks, Jr., and Max Behensky, "Feeling and Seeing: Issues in Force Display," *ACM Computer Graphics*, vol. 24, no. 2, March, 1990, pp. 235–243.

Michael Noll, "Man-Machine Tactile Communication," *Journal of the Society for Information Display*, July, 1972.

M.A. Srinivasan, "Tactile Sensing in Humans and Robots: Computational Theory and Algorithms," *Neuman Laboratory for Biomechanics and Human Rehabilitation, Department of Mechanical Engineering, MIT Technical Report*, October, 1988.

M.A. Srinivasan, J.M. Whitehouse, and R.H. LaMotte, "Tactile Detection of Slip: Surface Microgeometry and Peripheral Neural Codes," *Journal of Neurophysiology*, February, 1990.

Chapter 15

Steven Levy, *Hackers: Heroes of the Computer Revolution*, New York: Doubleday, 1984, pp. 194, 206, 250.

Chapter 16

Bruno Bettelheim, "The Importance of Play," *The Atlantic Monthly*, March, 1987.

Jacob Bronowski, *The Origins of Knowledge and Imagination*, New Haven: Yale University Press, 1978, p. 18.

Richard L. Gregory, "Seeing by Exploring," *Spatial Displays and Spatial Instruments*, Stephen R. Ellis, Mary K. Kaiser, and Arthur Grunwald, eds., 1989 NASA Conference publication 10032, pp. 5–11.

Yotaro Hatamura and Hiroshi Miroshita, "Direct Coupling System Between Nanometer World and Human World," Department of Mechanical Engineering for Production, The University of Tokyo, Tokyo, Japan, 1990.

R.L. Hollis, S. Salcudean, and D.W. Abraham, "Toward a Tele-Nanorobotic Manipulation System with Atomic Scale Force Feedback and Motion Resolution," IBM Thomas J. Watson Research Center, Yorktown Heights, NY, 1990.

Brenda Laurel, *Computers as Theatre*, Menlo Park, CA: Addison-Wesley, 1991.

Brenda Laurel, "Toward the Design of a Computer-Based Interactive Fantasy System," Ph.D. dissertation, Ohio State University, 1986, p. 1.

Marshall McLuhan, "The Playboy Interview," *Playboy*, March, 1969. 1964.

Heinz Pagels, *The Dreams of Reason*, New York: Simon & Schuster, 1988.

Seymour Papert, *Mindstorms: Children, Computers, and Powerful Ideas*, New York: Basic Books, 1980, pp. 21, 76.

Andrew Pollack, "Coming Soon: Data You Can Look Under and Walk Through," *New York Times*, October 14, 1990.

John E. Pfeiffer, *The Creative Explosion: An Inquiry into the Origins of Science and Religion*, Ithaca, NY: Cornell University Press, 1982, p. 250.

Howard Rheingold, "The Ultimate Cashflow," *California Living*, September 26, 1976.

E.H. Spain and D. Coppock, "Toward Performance Standards for Remote Manipulation," *Proceedings of the 11th Annual Meetings of the IEEE Engineering in Biology and Medicine Society*, Seattle, WA, Nov., 1989, vol. IV, pp. 923–924.

Allucquére Rosanne Stone, private electronic correspondence with author.

Allucquére Rosanne Stone, "Sex and Death Among the Disembodied," paper presented at the Conference of the International Association for Philosophy and Literature, 1990.

J.A. Thorpe, "The New Technology of Large Scale Simulator Networking: Implications for Mastering the Art of Warfighting," Ninth Interservice Industry Training Systems Conference, 1987.

William R. Uttal, "Teleoperators," *Scientific American*, December, 1989, pp. 124–129.

Shawna Vogel, "Smart Skin," *Discover*, April, 1990.

Joseph Weigenbaum, *Computer Power and Human Reason*, San Francisco: Freeman, 1976.

Shoshana Zuboff, *In the Age of the Smart Machine: The Future of Work and Power*, New York: Basic Books, 1988, pp. 71–73.

Chapter 17

Baudrillard, Jean, *SIMULATIONS*, New York: Semiotext(e)/Columbia University Press, 1983, p. 2.

Henri Bergson, *Creative Evolution*, translated by Arthur Mitchell, New York: Henry Holt, 1911, p. 139.

Jacob Bronowski, *The Origins of Knowledge and Imagination*, New Haven: Yale University Press, 1978, p. 18.

Brenda Laurel, *Computers as Theater*, Menlo Park, CA: Addison-Wesley, 1991.

Brenda Laurel, "Toward the Design of a Computer-Based Interactive Fantasy System," Ph.D. dissertation, Ohio State University, 1986, p. 1.

Paul C. Mangelsdorf, "The Origin of Corn," *Scientific American*, August, 1986.

Marshall McLuhan, *Understanding Media: The Extensions of Man*, New York: McGraw-Hill, 1964.

Heinz Pagels, *The Dreams of Reason*, New York: Simon & Schuster, 1988.

John E. Pfeiffer, *The Creative Explosion: An Inquiry into the Origins of Art and Religion*, Ithaca, NY: Cornell University Press, 1982, p. 205.

David Sheff, "The Virtual Realities of Timothy Leary," *Upside*, April, 1990, p. 70.

Frank Waters, *Masked Gods*, New York: Ballantine, 1970, pp. 170–171.

INDEX

ABOUT THE AUTHOR

Howard Rheingold is author of *Tools for Thought, Excursions to the Far Side of the Mind, They Have a Word for It: A Lighthearted Lexicon of Untranslatable Words and Phrases;* coauthor of *Higher Creativity,* and *The Cognitive Connection;* and editor of *Whole Earth Review.* He is multimedia columnist for *Publish* magazine and has been consultant to the US Congress Office of Technology Assessment. He has written for *The New York Times, Esquire, Psychology Today, Playboy, The San Francisco Chronicle,* and *Omni.*